T0179372

Particle Swarm Optimisation

Classical and Quantum Perspectives

CHAPMAN & HALL/CRC
Numerical Analysis and Scientific Computing

Aims and scope:
Scientific computing and numerical analysis provide invaluable tools for the sciences and engineering. This series aims to capture new developments and summarize state-of-the-art methods over the whole spectrum of these fields. It will include a broad range of textbooks, monographs, and handbooks. Volumes in theory, including discretisation techniques, numerical algorithms, multiscale techniques, parallel and distributed algorithms, as well as applications of these methods in multi-disciplinary fields, are welcome. The inclusion of concrete real-world examples is highly encouraged. This series is meant to appeal to students and researchers in mathematics, engineering, and computational science.

Proposals for the series should be submitted to one of the series editors above or directly to:
CRC Press, Taylor & Francis Group
4th, Floor, Albert House
1-4 Singer Street
London EC2A 4BQ
UK

Published Titles

Classical and Modern Numerical Analysis: Theory, Methods and Practice
Azmy S. Ackleh, Edward James Allen, Ralph Baker Kearfott, and Padmanabhan Seshaiyer

Computational Fluid Dynamics
Frédéric Magoulès

A Concise Introduction to Image Processing using C++
Meiqing Wang and Choi-Hong Lai

Decomposition Methods for Differential Equations: Theory and Applications
Juergen Geiser

Discrete Variational Derivative Method: A Structure-Preserving Numerical Method for Partial Differential Equations
Daisuke Furihata and Takayasu Matsuo

Grid Resource Management: Toward Virtual and Services Compliant Grid Computing
Frédéric Magoulès, Thi-Mai-Huong Nguyen, and Lei Yu

Fundamentals of Grid Computing: Theory, Algorithms and Technologies
Frédéric Magoulès

Handbook of Sinc Numerical Methods
Frank Stenger

Introduction to Grid Computing
Frédéric Magoulès, Jie Pan, Kiat-An Tan, and Abhinit Kumar

Iterative Splitting Methods for Differential Equations
Juergen Geiser

Mathematical Objects in C++: Computational Tools in a Unified Object-Oriented Approach
Yair Shapira

Numerical Linear Approximation in C
Nabih N. Abdelmalek and William A. Malek

Numerical Techniques for Direct and Large-Eddy Simulations
Xi Jiang and Choi-Hong Lai

Parallel Algorithms
Henri Casanova, Arnaud Legrand, and Yves Robert

Parallel Iterative Algorithms: From Sequential to Grid Computing
Jacques M. Bahi, Sylvain Contassot-Vivier, and Raphael Couturier

Particle Swarm Optimisation: Classical and Quantum Perspectives
Jun Sun, Choi-Hong Lai, and Xiao-Jun Wu

Particle Swarm Optimisation

Classical and Quantum Perspectives

Jun Sun
Choi-Hong Lai
Xiao-Jun Wu

CRC Press
Taylor & Francis Group
Boca Raton London New York

CRC Press is an imprint of the
Taylor & Francis Group, an **informa** business

A CHAPMAN & HALL BOOK

CRC Press
Taylor & Francis Group
6000 Broken Sound Parkway NW, Suite 300
Boca Raton, FL 33487-2742

First issued in paperback 2019

© 2012 by Taylor & Francis Group, LLC
CRC Press is an imprint of Taylor & Francis Group, an Informa business

No claim to original U.S. Government works

ISBN-13: 978-1-4398-3576-0 (hbk)
ISBN-13: 978-0-367-38193-6 (pbk)

Library of Congress Cataloging-in-Publication Data

Sun, Jun, 1971-
    Particle swarm optimisation : classical and quantum perspectives / Jun Sun, Choi-Hong Lai, Xiao-Jun Wu.
        p. cm. -- (Chapman and Hall/CRC numerical analysis and scientific computation series)
    Includes bibliographical references and index.
    ISBN 978-1-4398-3576-0 (hardback)
    1. Mathematical optimization. 2. Particles (Nuclear physics) 3. Swarm intelligence. I. Lai, Choi-Hong. II. Wu, Xiao-Jun. III. Title.

QC20.7.M27S86 2011
531'.16--dc23   2011033831

Visit the Taylor & Francis Web site at
http://www.taylorandfrancis.com

and the CRC Press Web site at
http://www.crcpress.com

Contents

Preface

The particle swarm optimisation (PSO) algorithm, inspired by the social behaviour of bird flocking or fish schooling, is a typical population-based random search algorithm developed by Eberhart and Kennedy in 1995. The algorithm has been widely used by many researchers in various fields due to its low computational complexity, easy implementation and relatively few parameters. However, PSO also has several major shortcomings—notably it is not a global convergent algorithm. In order to overcome these shortcomings, we have developed the concept of quantum-behaved particles, inspired by quantum mechanics, which leads to the quantum-behaved particle swarm optimisation (QPSO) algorithm. This novel algorithm has fewer parameters, faster convergence rate as well as stronger searchability for complex problems, and, most of all, it is a global convergent algorithm.

This book intends to provide a detailed description and the state of the art of PSO and QPSO algorithms to a level suitable to be used as a reference by research scientists and engineers interested in its theory and applications.

The main themes of this book are on the fundamental principles and applications of PSO and QPSO algorithms. Some advanced topics on QPSO such as the behaviour of individual particles, global convergence, time complexity and rate of convergence are also discussed. The main contents are outlined as follows.

Chapter 1 introduces the concepts of optimisation problems and random search methods for optimisation. A background of metaheuristics, including evolutionary computation and particle swarm optimisation, is also included. In particular, the main procedures of most of the metaheuristics are described.

Chapter 2 discusses the principles behind the PSO algorithm, including the motivation, basic concepts and implementation procedures. A simple example is used to illustrate the process of the PSO algorithm for the solution of an optimisation problem. Further examples with increasing complexity are solved by means of three variants of the PSO algorithm, with a performance comparison between the three algorithms included.

Chapter 3 discusses some shortcomings of the PSO algorithm and analyses the reasons behind these shortcomings. In order to tackle these shortcomings, several variations of the algorithm are examined. These variations, including inertia-weighed version, constriction version, local best model, probabilistic algorithm and other revised versions, are briefly explained with the aim of enhancing or improving PSO in some respects.

Chapter 4 gives a thorough overview of the literature on QPSO with the formulation based on the motivations of the QPSO algorithm. It also discusses the fundamental model—a quantum δ potential well model—for the QPSO algorithm. The applications of the algorithm to solve typical optimisation problems have been detailed. Several variations of the QPSO method are also included toward the end of this chapter.

Chapter 5 presents some advanced topics on QPSO and is targeted at readers interested in the theory of QPSO. First, the behaviour of an individual particle is analysed using probability theory, and stochastic simulation is used to validate these theoretical results. Second, theoretical results showing global convergence of the QPSO algorithm using three different mathematical tools are provided. Third, the computational complexity and convergence rate of a stochastic search algorithm are defined, followed by empirical analysis of the computational complexity and convergence rate of QPSO on a specific function. Finally, parameter selection of the algorithm is discussed and tested on several benchmark functions. Readers who wish only to understand how to use the algorithm may skip this chapter, apart from the last section.

Chapter 6 covers several real-world applications using the QPSO algorithm, including inverse problems, optimal design of digital filters, economic dispatch problems, biological multiple sequence alignment and image processing, showing details of implementation aspects so that readers may be better equipped to apply the algorithm.

The source codes of the main algorithms using MATLAB®, Fortran and C are included in a CD-ROM provided for the benefit of readers.

We would not have succeeded in producing this book without the help of various people. In particular, we would like to thank Professor Wenbo Xu, Dr. Wei Fang, Dr. Ji Zhao, Dr. Xiaoqing Luo, Dr. Na Tian,

Dr. Di Zhou, Dr. Haixia Long and Dr. Hao Gao. We are extremely grateful to various senior academics, including Professor Jingyu Yang, Professor Zhihua Zhou and Professor Josef Kittler, who have participated in the discussion of several important concepts.

Jun Sun
Jiangnan University
Wuxi, Jiangsu Province, China

Choi-Hong Lai
University of Greenwich
London, United Kingdom

Xiao-Jun Wu
Jiangnan University
Wuxi, Jiangsu Province, China

For MATLAB® product information, please contact:

The MathWorks, Inc.
3 Apple Hill Drive
Natick, MA, 01760-2098 USA
Tel: 508-647-7000
Fax: 508-647-7001
E-mail: info@mathworks.com
Web: www.mathworks.com

Authors

Jun Sun is an associate professor in the Department of Computer Science and Technology, Jiangnan University, Wuxi, Jiangsu Province, China. He holds a PhD in control theory and control engineering, an MSc in computer science and technology, and a BSc in mathematics and applied mathematics. He is also a researcher at Key Laboratory of Advanced Process Control for Light Industry (Ministry of Education), China. His major research areas include work related to computational intelligence, numerical optimisation and machine learning. From 2005 to 2006, he was a visiting scholar at the University of Greenwich, working with Professor Choi-Hong Lai.

Choi-Hong Lai is a professor of numerical mathematics in the Department of Mathematical Sciences, University of Greenwich, London, United Kingdom. He has a BSc first class degree in mathematics and engineering and a PhD in computational aerodynamics and partial differential equations (PDEs). His major research interests include numerical partial differential equations, numerical algorithms, and parallel algorithms for industrial applications such as aeroacoustics, inverse problems, computational finance, and image processing. He teaches two senior undergraduate courses, advanced algorithms and industrial applied mathematics, and is the programme manager for the mathematics masters' degree programmes at the University of Greenwich.

Xiao-Jun Wu received his BS in mathematics from Nanjing Normal University, Nanjing, China, in 1991, and his MS and PhD in pattern recognition and intelligent systems from Nanjing University of Science and Technology, Nanjing, China, in 1996 and 2002, respectively. He was a fellow of the United Nations University, International Institute for Software Technology (UNU/IIST) from 1999 to 2000. From 1996 to 2006, he taught at the School of Electronics and Information, Jiangsu University of Science and Technology, where he was an exceptionally promoted professor. He joined Jiangnan University in 2006, where he currently works as a professor. Dr. Wu won the most outstanding postgraduate award from Nanjing University of Science and Technology in 1996. He has published more than 150 papers in his fields of research. He was a visiting researcher at the Centre for Vision, Speech, and Signal Processing (CVSSP), University of Surrey, United Kingdom, from 2003 to 2004. His research interests include pattern recognition, computer vision, fuzzy systems, neural networks, and intelligent systems.

CHAPTER 1

Introduction

1.1 OPTIMISATION PROBLEMS AND OPTIMISATION METHODS

Optimisation methods are widely used in various fields, including engineering, economics, management, physical sciences, social sciences, etc. The task is to choose the best or a satisfactory one from amongst the feasible solutions to an optimisation problem, providing the scientific basis of decision-making for decision-makers. The process of using optimisation methods to solve a practical problem mainly involves these two steps. First, formulate the optimisation problem which involves determining the decision variables, objective function and constraints, and possibly an analysis of the optimisation problem. Second, select an appropriate numerical method, solve the optimisation problem, test the optimal solution and make a decision accordingly. This book concentrates on methods of solving optimisation problems numerically and does not address the construction of the optimisation problem for a practical problem throughout the text.

Mathematically, an optimisation problem may be summarised as follows [1]:

$$\max f(x) \quad \text{or} \quad \min f(x)$$

$$\text{subject to } x \in S \subset R^N,$$

where
  $f(x)$ is the objective function
  x is an N-dimensional vector consisting of the decision variables
  $S \subset R^N$ is the feasible region or feasible solution space of the problem

Let $\| \cdot \|$ represent the Euclidean norm. The point (solution) $x^* \in S$ is said to be the local minimum solution of the problem, if there exists an $\varepsilon > 0$, such that for all $x \in S, f(x^*) \leq f(x)$ whenever $\|x - x^*\| \leq \varepsilon$. x^* is accepted as the global minimum solution if for all $x \in S, f(x^*) \leq f(x)$. Similarly, x^* is called the local maximum solution (or the global maximum solution) if for all $x \in S \cap \{x \big| \| x - x^* \| \leq \varepsilon \}$ (or for all $x \in S$), $f(x^*) \geq f(x)$. The global minimum value of $f(x)$ is denoted as $\min\{f(x) : x \in S\}$ or $\min_{x \in S} f(x)$, while the global maximum value of $f(x)$ is denoted as $\max\{f(x) : x \in S\}$ or $\min_{x \in S} f(x)$.

Now consider the following minimisation problem:

$$\min_{x \in S} f(x), \tag{1.1}$$

where $S = \{x \in R^N : g_i(x) \leq 0, i = 1, 2, \ldots, q\}$ such that $g_i : A \to R$ is a function defined on a set A (usually $A = R^N$). A maximisation problem can always be included in the minimisation model (1.1) because $\max\{f(x) : x \in S\} = -\min\{-f(x) : x \in S\}$. Furthermore, one can easily see that the minimisation problem in (1.1) can include many different types of constraints since $g_i(x) = 0$ is equivalent to $-g_i(x) \leq 0$ and $g_i(x) \leq 0$, and $g_i(x) \geq 0$ is equivalent to $-g_i(x) \leq 0$. If the goal of solving the aforementioned problem is to find its global optimal solution, the problem in Equation 1.1 is known as a global optimisation problem. When $S = R^N$, the global optimisation problem mentioned earlier turns into an unconstrained optimisation problem. In particular Equation 1.1 is known as a linear constrained optimisation if S is a polyhedron. Furthermore, if the objective function is also linear, the problem is referred to as a linear programming problem or a linear optimisation problem. Equation 1.1 is known as a non-linear programming problem or non-linear optimisation problem when one of the objective functions and/or constraint functions is non-linear. When $S = Z^N$, where Z^N is the N-dimensional integer space, the aforementioned problem is known as an integer programming problem. If S is a set of combinatorial codes, the problem is referred to as a combinatorial optimisation problem. There are many other kinds of optimisation problems that may be included in the scope of the problem defined in Equation 1.1, including multi-objective optimisation, min–max problem, stochastic programming problem and so forth.

Different types of optimisation problems can be solved by using different methods. For linear optimisation problems, there are many mature methods of solving them, including the simplex method proposed by

Dantzig in 1947 and the interior point or Newton barrier method developed by Karmarkar in 1984 [2,3]. On the other hand, non-linear programming problems turn out to be more difficult to solve than the linear ones, especially when a problem has many local optimal solutions in the feasible region. Commonly used methods for solving non-linear programming problems usually only guarantee a local optimal or sub-optimal solution of the problem and the final results are generally influenced by the initial trial solutions [4].

In the past several decades, both processing power and storage capacity of computers have increased by roughly a factor of a million, making it possible to solve those considered to be 'hard' optimisation problems in the past, such as non-linear optimisation, combinatorial optimisation and integer programming problems. The main advantage of computer-based methods is that the entire feasible solution space of the problem can be searched quickly and exhaustively. These computer-based search techniques can be divided into the following four types in accordance with the search strategy: enumeration methods, gradient methods, direct search methods and random search methods. The former three are deterministic methods and the latter are stochastic methods.

Enumeration methods, also known as exhaustive search techniques, are one of the oldest computer approaches of solving optimisation problems. They are implemented by generating and inspecting all points in the feasible solution space of the problem. Exhaustive search techniques, though conceptually simple and may be effective for some problems, require high computational costs and the knowledge of a finite feasible solution region of the problem. These methods are generally employed to solve integer programming and combinatorial problems. Typical enumeration methods include dynamic programming, branch-bound method, etc. [5].

Gradient methods of finding solutions to non-linear optimisation problems are also suitable. They take advantage of the geometric features of the problems such as first-order and second-order derivatives in the solution process. Generally, the gradient direction is used to guide the search of the algorithm using a step length proportional to the negative or positive of the gradient of the function at the current iterative approximate solution. Such approaches are also known as 'gradient descent' or 'gradient ascent' methods. However, these methods belong to the local search techniques and usually can only guarantee the local optima rather than the global optima. Newton's method and conjugate gradient method are typical gradient methods [4].

Gradient methods may become powerless when the objective function appears to be very complex or cannot be expressed in explicit functions. In this case, direct search methods can be used to find the optimal and suboptimal solutions of the problem after executing a few iterations of the search in the absence of geometric feature or gradient of the problem. At each step, a set of trial points is generated with their function values being compared with the best solution in the previous step. This information is then used to determine the next set of trial points. Hill-climbing method and Powell method are two members of this type of approaches [4,5].

Random search methods, including metaheuristc methods, are direct search techniques that have incorporated stochastic strategies into the search processes, enabling the algorithms to jump out the local optima with high probability [6,7]. More detailed information about this kind of approach is described in the following sections.

1.2 RANDOM SEARCH TECHNIQUES

A random search method refers to an algorithm that uses the randomness or probabilistic features, typically in the form of pseudo-random number generator, and is often known as a Monte Carlo method or a stochastic method in the literature. The random search method can be used in many cases, including those objective functions that are not continuous or differentiable, because the gradient of the problem is not required. At each search step, the random search method measures the objective function values at more than one point sampled from a probability distribution over the entire feasible solution space and takes the point with the smallest value of the object function as an approximation to the minimum. There are mainly two versions of random search techniques, namely, blind random search and localised random search. Refs. [8,9] are among the many literatures that discuss these random search methods.

The blind random search method is the simplest random research method in which the current sampling for the solution does not take into account the previous samples. In other words, it does not adapt the current sampling strategy to the information that has been garnered during the previous search process. This approach can be implemented either in a batch (non-recursive) form by laying down a number of points in the feasible solution space S, yielding the best objective function value as the estimate of the optimum, or in the form of a series of recursive steps [10].

Step 0: (Initialization) Choose an initial value of x, that is, $x_0 \in S$ either randomly (usually a uniform distribution on S is used) or deterministically. Calculate $f(x_0)$ and set $n = 0$.

Step 1: Generate a new independent point $x_{new} \in S$, according to the chosen probability distribution. If $f(x_{new}) < f(x_n)$, set $x_{n+1} = x_{new}$, else take $x_{n+1} = x_n$.

Step 2: Stop if the maximum number of objective function evaluations has been reached, or $f(x_n)$ has reached a user-specified satisfactory value, or other stopping criteria; else, return to Step 1 with new n set as $n + 1$.

FIGURE 1.1   The procedure of the blind random search method.

Note that the blind search can be used for either continuous or discrete problems. The algorithm is illustrated in Figure 1.1.

The localised random search approach is different from the blind search in that the random sampling strategy is a function of the position of the current best estimate for x. The term 'localised' is related to the sampling strategy and does not mean that the algorithm can only find the local optima of the problem. The procedure of the localised algorithm outlined in Figure 1.2 was first presented by Matyas [11].

In contrast to the deterministic search techniques, such as those belonging to enumeration, gradient-based and direct search algorithms, random search methods generally ensure convergence in probability to the global optima of the problem. With the trade-off being in terms of computational effort, random search algorithms can easily provide a relatively good solution.

The previous two versions of random search approaches have led to many efficient random algorithms, the most important one being the so-called metaheuristics [12]. These methods have become increasingly popular

Step 0: (Initialization) Pick an initial guess $x_0 \in S$ either randomly or with prior information. Set $n = 0$.

Step 1: Generate an independent vector $d_n \in R^N$, add it to the current x value x_n. Check if $x_n + d_n \in S$. If $x_n + d_n \in S$, generate a new d_n and repeat, or alternatively, move $x_n + d_n$ to the nearest valid point within S. Let x_{new} equal $x_n + d_n \in S$ or the aforementioned valid point within S.

Step 2: If $f(x_{new}) < f(x_n)$, set $x_{n+1} = x_{new}$, else take $x_{n+1} = x_n$.

Step 3: Stop if the maximum number of objective function evaluations has been reached, or $f(x_n)$ has reached a user-specified satisfactory value, or other stopping criteria; else, return to Step 1 with new n set as $n + 1$.

FIGURE 1.2   The procedure of the localised random search method.

because of their potential in solving large-scale problems efficiently in a way that is impossible by using deterministic algorithms. On the other hand, they are easy to implement for complex problems without gradient information.

1.3 METAHEURISTIC METHODS

A metaheuristic or a metaheuristic method is formally defined as an iterative generation process which guides a subordinate heuristic by combining intelligently different concepts for exploring and exploiting the search space. Learning strategies are used to structure information in order to find efficiently near-optimal solutions [12–14]. Here, exploration and exploitation are referred to as global search and local search of the algorithm, respectively. In other words, a metaheuristic can be seen as a general algorithmic framework which can be applied to different optimisation problems with relatively few modifications to make them adapt to a specific problem. Most metaheuristic methods are stochastic in nature and commonly associated with the localised random search algorithms. They are subject to the *no free lunch* theorem which states that any two optimisation algorithms are equivalent when their performance is averaged across all possible problems [15]. Therefore, it is of significance to design particular algorithms aiming at a particular class of problems. Examples of metaheuristic methods include simulated annealing, evolutionary algorithms, tabu search, differential evolution and swarm intelligence. A brief description of these metaheuristic methods is given in the remaining part of this chapter.

1.3.1 Simulated Annealing

Simulated annealing (SA) techniques are based upon the physical analogy of cooling crystal structures (a thermodynamic system) which spontaneously attempt to arrive at certain stable (globally or locally minimal potential energy) thermodynamic equilibrium [16,17]. In SA algorithms, the current state of the thermodynamic system is analogous to the current solution to the optimisation problem, the energy equation for the thermodynamic is analogous to the objective function, and the ground state is analogous to the global minimum. The search process of SA simulates the cooling process controlled by the cooling schedule. The algorithm presented in Figure 1.3 outlines the procedure of a typical SA originally proposed by Kirkpatrick [17].

In the previous procedure, the cooling schedule is defined as $T_{n+1} = \alpha T_n$ ($0 < \alpha < 1$) and rand(0, 1) denote random numbers generated uniformly in the range (0, 1). It should be noted that the smaller the value of α the slower the cooling schedule or the rate of decreasing in the temperature. This means

Step 0: Generate an initial solution x_0 randomly or deterministically; set the initial temperature $T_0 = T_{max}$ and $n = 0$;

Step 1: Pick randomly a new vector x_{new} in the neighbourhood of x_n and calculate $\Delta f = f(x_{new}) - f(x_n)$. If $\Delta f \leq 0$, $x_n = x_{new}$, else if $\exp(-\Delta f/T_n) > \text{rand}(0,1)$, $x_n = x_{new}$.

Step 2: Update the temperature T_n according to the given cooling schedule and set $n = n + 1$. If the termination condition is not met, return to Step 1.

FIGURE 1.3   The procedure of simulated annealing.

that the algorithm is more likely to find an optimal or near-optimal solution. The procedure is generally applicable to both discrete and continuous global optimisation problems under mild structural requirements [18–20].

1.3.2 Evolutionary Algorithms

Evolutionary algorithms (EAs) are random search methods inspired by the natural selection and survival of the fittest in the biological world [21]. The algorithm involves searching a population of solutions, not from a single point as in other random search techniques. Each iterative step of an EA requires a competitive selection that eliminates poor solutions. The solutions with high fitness are recombined with other solutions by swapping parts of the approximate solution with those of the others. Mutations are also exerted on solutions by making a small change to a single element of the solution vector containing the decision variables. Recombination and mutation are employed to yield new solutions that are biased towards regions of the search space where good solutions have already been seen. A brief description of several different types of evolutionary algorithms, including genetic algorithm, genetic programming, evolution strategy and evolution programming, is given as follows.

1.3.2.1 Genetic Algorithm

Genetic algorithms (GAs) are the most important sub-class of EAs. A GA generates solutions to an optimisation problem using operations such as selection (reproduction), crossover (recombination) and mutation, with each individual or a candidate solution in the population represented by a binary string of 0s and 1s or by other forms of encodings [21]. The evolution process (or the search process) starts from a population of individuals generated randomly within the search space and continues for generations. In each generation, the fitness of every individual is evaluated, and multiple individuals are randomly selected from the current population based on their fitness and

---

Step 0: Encode the M individuals of the populations for the problem, generate the initial population $P(0)$ and set $n = 0$.

Step 1: Evaluate the fitness of each individual $x_i(n)$ in $P(n)$ by

$$f_i = \text{fitness}(x_i(n)).$$

Step 2: If the termination condition is met, the algorithm terminates; Otherwise, calculate the selection probability for each individual by

$$p_i = \frac{f_i}{\sum_{j=1}^{M} f_j}, \quad (i = 1, 2, \dots, M)$$

And using roulette wheel selection method, select multiple individuals from $P(n)$ with the above probability distribution to form a population

$$P'(n + 1) = \{x_j(n) \mid j = 1, 2, \dots, M\}$$

Step 3: Perform crossover operation on population $P'(n + 1)$ to form a population $P''(n + 1)$.

Step 4: Mutate a single element (called a gene) of an individual with probability p_m to form a population $P'''(n + 1)$.

Step 5: Set $n = n + 1$, $P(n + 1) = P'''(n + 1)$ and return to Step 1.

FIGURE 1.4   The procedure of genetic algorithm.

modified by recombination and mutation operation to form a new population, which is then used in the next generation of the evolution. In general, the search process terminates when either a maximum number of generations have been produced or a satisfactory fitness level has been reached for the population. The framework of GAs is described as in Figure 1.4.

In the previous procedure, a fitness function must be defined in order to evaluate the fitness of each individual. The fitness function is associated with the objective function of the problem and the fitness value of an individual should generally be positive. For the problem in Equation 1.1, the smaller the objective function value of the individual, the larger its fitness value. The fitness value of the individual is used to determine the probability with which the individual is selected into the new population. Although the selection used in the previous GA procedure is known as roulette wheel selection [21], other alternatives of selection operators, such as tournament selection, are possible.

GAs are the most widely used evolutionary algorithms. The solid theoretical foundation has been established and many variants have been proposed to improve the performance of the algorithms. More detailed information about GAs can be found from the literature [22–26].

1.3.2.2 Genetic Programming

The prototype genetic programming (GP) methodology was originally proposed by Fogel when he applied evolutionary algorithms to the problem of discovering finite-state automata. Crammer presented the first modern tree-based GP [27] in 1985 and the method was later extended by Koza [28]. Various applications of GP for several complex optimisation and search problems [29–32] were examined. GP is well known to be computationally intensive and was originally used to solve relatively simple problems. The method has many novel and outstanding results recently in areas such as quantum computing, electronic design, game playing and sorting due to improvements in GP technology and the exponential growth in CPU power [33–36]. The theory of GP has also received a formidable and rapid development in early 2000s. In particular, exact probabilistic models, including schema theories and Markov chain models, have been built for GP.

GP is in essence a specification of genetic algorithms where each individual is a computer program represented as a tree structure in the memory. It optimises a population of computer programs according to a fitness landscape that is determined by the ability of a computer program to perform a given computational task [29]. Although GP evolves the population of computer programs through a similar procedure as that for GA, it uses the crossover and mutation specified for the tree-based representation. Crossover is applied on an individual by simply switching one of its nodes with another node from another individual selected from the population. With an individual represented by a tree, replacing a node means replacing the whole branch under this node, which makes the crossover operator more effective. Mutation works on an individual by replacing a whole node in the selected individual or just replacing the information of a node. To maintain integrity, all the operations must be fail-safe or the type of information the node holds must be taken into account. For instance, the mutation operator must be aware of binary operation nodes or must be able to handle missing values.

1.3.2.3 Evolution Strategy

Evolution strategy (ES) was first introduced in the 1960s by Rechenberg [37] and further developed by Schwefel and his co-worker [38–42]. It is a random search technique based on the ideas of adaptation and evolution and belongs to the general class of evolutionary algorithms. Evolution strategies use natural problem-dependent representations and recombination, mutation and selection as search operators. In real-valued search

---

Step 0: Determine the representation for the problem to be solved.

Step 1: Generate randomly μ individuals to form the initial population $P(0)$, evaluate the fitness of each individual and set $n = 0$.

Step 2: If the termination condition is met, terminates the algorithm; otherwise, perform recombination on $P(n)$ to generate λ individuals to form an offspring population $P'(n + 1)$.

Step 3: Mutate the λ individuals in $P'(n + 1)$ to generate a new offspring population $P''(n + 1)$.

Step 4: Evaluate the fitness of individuals in $P''(n + 1)$.

Step 5: Perform offspring selection (i.e., select μ individuals from $P''(n + 1)$), or perform parents and offspring selection, that is, plus selection (select μ individuals from $P''(n + 1)$ and $P(n)$, to generate population $P'''(n + 1)$.

Step 6: Set $n = n + 1$, $P(n + 1) = P'''(n + 1)$ and return to Step 2.

FIGURE 1.5 The procedure of evolution strategy.

spaces, mutation is normally performed by adding a normally distributed random value to each vector element. The step size or mutation strength is often governed by self-adaptation. Individual step sizes for each coordinate or correlations between coordinates are either governed by self-adaptation or by co-variance matrix adaptation.

There are two canonical versions of ES, namely the offspring selection and the parents and offspring selection (or plus selection) [42]. The parents are deterministically selected from one of the aforementioned selection methods. The procedure of ES algorithm is outlined in Figure 1.5. Some of the details of generating offspring selection and plus selection are skipped for simplicity.

1.3.2.4 Evolutionary Programming

Evolutionary programming (EP) was also first used by Fogel to simulate evolution as a learning process in order to evolve finite state machines as predictors [43]. It is an evolutionary computing dialect without fixed structure or representation, in contrast with some of the other dialects [44,45]. In essence it is very difficult to distinguish EP from evolution strategies nowadays.

The main variation operator in EP is mutation and there is no recombination operator. Individuals of the population are viewed as part of a specific species rather than members of the same species. Therefore, μ parent generates μ offspring, using a $(\mu + \mu)$ survivor selection, that is, selecting μ individuals from 2μ candidates to form a new population. The EP algorithm can be described as in Figure 1.6.

```
Step 0: Determine the representation for the problem to be solved.
Step 1: Generate randomly μ individuals to form the initial popula-
tion P(0), evaluate the fitness of each individual and set n = 0.
Step 2: If the termination condition is met, terminates the algo-
rithm; otherwise, mutate each individual in P(n) to form an off-
spring population P'(n + 1).
Step 3: Evaluate the fitness of individuals in P'(n + 1).
Step 4: Select μ individuals from 2μ ones in P'(n + 1) and P(n), to
generate population P"(n + 1).
Step 5: Set n = n + 1, P(n + 1) = P"(n + 1) and return to Step 2.
```

FIGURE 1.6 The procedure of evolution programming.

1.3.3 Tabu Search

Tabu search (TS) is a metaheuristic algorithm originally proposed by Glover and his co-worker [46–49]. In TS, a local or neighbourhood search procedure is used iteratively moving from one approximate solution (x) to another (x') in the neighbourhood, denoted as $N(x)$, of the approximate solution until certain stopping criterion is satisfied. In this algorithm, the neighbourhood structure of each approximate solution in the search process is to be modified according to certain rules in order to best explore regions of the search space that might have been left unexplored by the local search procedure. Suppose $N^*(x)$ denotes the new neighbourhood of the approximate solution x. The approximate solutions admitted to the new neighbourhood are determined through the use of memory structures which contain previous approximate solutions that have been visited in the previous iterations. The search then progresses iteratively again.

One important type of memory structure is the tabu list which may be used to determine an updated approximate solution admitted to $N^*(x)$. A tabu list usually keeps those iterative approximate solutions of the previous n iterations. TS excludes iterative approximate solutions in the tabu list from $N^*(x)$. Figure 1.7 provides the basic procedure of TS.

TS was traditionally applied to combinatorial optimisation (e.g. scheduling, routing, travelling salesman) problems [48–51]. The technique can be made, at least in principle, directly applicable to continuous global optimisation problems by a discrete approximation (encoding) of the problem. Readers interested in advanced topics of the algorithm with applications to different types of problems should consult Refs. [19,52,53].

```
Step 0: Generate the initial solution x₀, evaluate its objective
function value, set the tabu list H = Φ and n = 0.

Step 1: If the termination condition is met, terminates the algo-
rithm; otherwise, execute the following steps.

Step 2: Select the solutions in N(xₙ) that are not 'tabu' and put
them into the allowed set N*(xₙ).

Step 3: Evaluate the objective function values of each solution
in N*(xₙ) and find the solution with the best objective function
value, xₙₑw.

Step 4: Set n = n + 1, xₙ = xₙₑw, update the tabu list H and return
to Step 2.
```

FIGURE 1.7 The procedure of tabu search.

1.3.4 Differential Evolution

Differential evolution (DE) is a metaheuristic method developed by Storn and Price and it solves real-valued problems based on the principles of natural evolution [54–57]. Here an easy approach using simple updating formula of the decision variables involved is introduced. Let P denote a population of size M. The population P consists of floating point encoded individuals that evolve over n generations to reach an optimal solution. Each individual X_i in P is a vector containing as many components as the number of decision variables or the dimensionality of the problem. The population size M is an algorithmic control parameter provided by the user which remains constant throughout the optimisation process. It governs the number of individuals in the population and must be large enough to provide sufficient diversity to search the solution space.

Let $P_n = [X_{1,n}, X_{2,n}, ..., X_{M,n}]$ be the population at the nth generation (iteration) and $X_{i,n} = (X_{i,n}^1, X_{i,n}^2, ..., X_{i,n}^N)$ be the ith individual in P_n, where N is dimensionality of the problem. The optimisation process in DE involves three basic operations: mutation, crossover and selection. The algorithm starts with an initial population of M vectors. Each component of an individual in the initial population is assigned with random values generated as follows:

$$X_{i,0}^j = X_{min}^j + \eta^j(X_{max}^j - X_{min}^j), \tag{1.2}$$

where

$i = 1, 2, ..., M$
$j = 1, 2, ..., N$
X_{min}^j and X_{max}^j are the lower and upper bound of the jth decision variable
$\eta^j (j = 1, 2, ..., N)$ are random numbers uniformly distributed on (0,1)
$X_{i,0}^j$ is the jth component of the ith individual in the initial population

The mutation operator creates mutant vectors $(X'_{i,n})$ by perturbing a randomly selected vector $(X_{a,n})$ with the difference of two other randomly selected vectors (say $X_{b,n}$ and $X_{c,n}$).

$$X'_{i,n} = X_{a,n} + F(X_{b,n} - X_{c,n}), \quad i = 1, 2, \ldots, M, \tag{1.3}$$

where $X_{a,n}$, $X_{b,n}$ and $X_{c,n}$ are randomly selected vectors from $P_n = [X_{1,n}, X_{2,n}, \ldots, X_{M,n}]$. $X_{a,n}$, $X_{b,n}$ and $X_{c,n}$ are selected anew for each parent vector. Here F is an algorithmic control parameter used to control the size of the perturbation in the mutation operator and to improve the convergence of the algorithm. The value of F is a value chosen in the range $(0,2]$.

The crossover operation generates the trial vector, $X''_{i,n}$, by mixing the components of the mutant vectors with the target vectors, $X_{i,n}$, according to a selected probability distribution.

$$X''^{j}_{i,n} = \begin{cases} X'^{j}_{i,n} & \text{if } \eta'^{j} \le C_R \quad \text{or} \quad j = q, \\ X^{j}_{i,n} & \text{otherwise,} \end{cases} \tag{1.4}$$

where
$i = 1, 2, \ldots, M$
$j = 1, 2, \ldots, N$
q is a randomly chosen index from $\{1, 2, \ldots, M\}$ that guarantees that the trial vector receives at least one component from the mutant vector
η'^{j} is a uniformly distributed random number within $(0,1)$

The crossover constant C_R is an algorithmic parameter which may be used to control the diversity of the population. It can also guide the optimisation process to escape from local optima. $X^{j}_{i,n}$, $X'^{j}_{i,n}$ and $X''^{j}_{i,n}$ are the jth components of the ith target vector, mutant vector and trial vector at generation n, respectively.

The selection operator determines the population by choosing between the trial vectors and their predecessors, also known as target vectors, those individuals that either present a better fitness or are more optimal according to the following formula:

$$X_{i,n+1} = \begin{cases} X''_{i,n} & \text{if } f(X''_{i,n}) \le f(X_{i,n}), \\ X_{i,n} & \text{otherwise,} \end{cases} \quad i = 1, 2, \ldots, M. \tag{1.5}$$

This process repeats, allowing individuals to improve their fitness as they explore the solution space in the search for optimal solutions. The search process completes until the maximum number of objective function evaluations has been reached, the value of $f(X_{best,n})$ has reached a user-specified satisfactory value or other stopping criterion is met. Here $X_{best,n}$ is the best solution found so far in the optimisation process.

Readers who are interested to read various variants of the DE method should consult the Refs. [58–67]. An example of extending the DE method for global optimisation is to take the best solution found so far to be perturbed using two difference vectors based on a binomial distribution crossover scheme such as the one given in the following:

$$X'_{i,n} = X_{best,n} + F(X_{a,n} - X_{b,n} + X_{c,n} - X_{d,n}), \quad i = 1,2,\ldots,M, \qquad (1.6)$$

where $X_{a,n}$, $X_{b,n}$, $X_{c,n}$ and $X_{d,n}$ are randomly chosen vectors from $P_n = [X_{1,n}, X_{2,n}, \ldots, X_{M,n}]$ and $a \neq b \neq c \neq d$.

1.3.5 Swarm Intelligence Algorithms

The collective intelligence (i.e. swarm intelligence) emerging from the behaviour of a group of social insects seems to offer insight into meta-heuristics. For example, in the community of insects there are very limited individual capabilities but many complex tasks can be performed for their survival. Problems like finding and storing foods, selecting and picking up materials for future usage, require a detailed planning and could be treated by insect colonies without any kind of supervisor or controller. Since the introduction of swarm intelligence algorithm in the 1980s, it has been studied and used widely in fields such as economic analysis and decision-making, biology and industry. An example of particularly successful research direction in swarm intelligence is the ant colony optimisation (ACO) [68–71], which focuses on discrete optimisation problems, and has been applied successfully to a large number of NP hard discrete optimisation problems, including the travelling salesman, the quadratic assignment, scheduling, vehicle routing, routing in telecommunication networks, etc. Particle swarm optimisation (PSO) is another very popular swarm intelligence algorithm for global optimisation over continuous search spaces [72]. Since its introduction in 1995, PSO has attracted the attention of many researchers resulting in many variants of the original algorithm and many parameter automation strategies for the algorithm.

1.4 SWARM INTELLIGENCE

1.4.1 Properties of Swarm Intelligence

Swarm intelligence refers to a class of algorithms that simulates natural and artificial systems composed of many individuals that coordinate using decentralised control and self-organisation. The algorithm focuses on the collective behaviours that result from the local interactions of the individuals with each other and with the environment where these individuals stay. Some common examples of systems involved in swarm intelligence are colonies of ants and termites, fish schools, bird flock, animal herds.

A typical swarm intelligence system has the following properties:

1. It consists of many individuals.

2. The individuals are relatively homogeneous (i.e. they are either all identical or they belong to a few typologies).

3. The interactions between the individuals are based on simple behavioural rules that exploit only local information that the individuals exchange directly or via the environment.

4. The global behaviour of the system results from the local interactions of individuals with each other and with their environment.

The main property of a swarm intelligence system is its ability to act in a coordinated way without the presence of a coordinator or of an external controller. Despite the lack of individuals in charge of the group, the swarm as a whole does show intelligent behaviour. This is the result of interactions of spatially located neighbouring individuals using simple rules.

The swarm intelligence has already been used in many different fields since it was first introduced. The studies and applications of swarm intelligence mainly focus on the clustering behaviour of ants, nest-building behaviour of wasps and termites, flocking and schooling in birds and fish, ant colony optimisation, particle swarm optimisation, swarm-based network management, cooperative behaviour in swarms of robots, etc.

1.4.2 Ant Colony Optimisation

Ant colony optimisation (ACO) simulates the search process of a group of ants for food source [69–71]. When ACO is used, the optimisation problem should be transformed into a problem of finding the best path on a weighted graph. The artificial ants in ACO incrementally build solutions by randomly moving on the graph, biased by a pheromone model, which is known as a set of parameters associated with graph components (either nodes or edges) whose values are modified at runtime by the ants.

The procedure of the ACO algorithm can be outlined as follows. A colony of ants, that is, a set of computational concurrent and asynchronous agents, moves through states of the problem corresponding to those partial solutions of the problem. They move as according to a stochastic local decision policy based on two parameters, known as *trails* and *attractiveness*. These movements allow each ant to construct a solution to the problem incrementally. When an ant completes a solution or is in the construction phase, it evaluates the solution and modifies the trail value of the components used in its solution, and this pheromone information will guide the search of the future ants.

In addition, the ACO algorithm involves two more mechanisms [73,74]: *trail evaporation* and, optionally, *daemon actions*. Trail evaporation decreases all trail values over time so as to avoid unlimited accumulation of trails over certain components. Daemon actions can be employed to perform centralised actions that cannot be implemented by single ants. The centralised actions can be the invocation of a local optimisation procedure, or the update of global information to be used to decide whether to bias the search process from a non-local perspective. Trails are usually updated when all ants have completed their solutions. Increasing or decreasing the level of trails corresponds to moves that were part of good or bad solutions respectively.

Early applications of ACO were in the domain of NP-hard combinatorial optimisation problems and the routing programming in telecommunication networks [73,74]. Current research in ACO algorithms covers both the development of theoretical foundations and the application of this metaheuristic to new challenging problems. Gutjahr first proved the convergence in probability of an ACO algorithm [75]. In terms of applications, the use of ACO for the solution of scheduling [76–80], data mining [81,82], robotics [83–86], colouring problems [87], biological science [88] and other combinatorial optimisation problems [89,90] are current hot topics. On the other hand, the creation of parallel implementations capable of taking advantage of the new available parallel hardware [91] is also an important area of research.

1.4.3 Particle Swarm Optimisation

PSO algorithm is another important member of swarm intelligence algorithms originally developed by Kennedy and Eberhart in 1995 [72]. It was motivated by social behaviour of bird flock or fish schooling and the technique shares many similarities with evolutionary computation techniques such as genetic algorithms (GA). As in other population-based intelligence systems, PSO requires an initial population of random solutions. The search for optima is obtained by updating generations without evolution operators such as crossover and mutation. The potential solutions are usually called particles in PSO. These particles fly through the solution space by following their own experiences and the current optimum particles. It was shown that the PSO algorithm is comparable in performance with and may be considered as an alternative method to GA [92].

Various improvements and modification have been done on the PSO algorithm since it was first introduced. Quantum-behaved particle swarm optimisation (QPSO) is one important PSO variant motivated from quantum mechanics and trajectory analysis of PSO [93,94]. This book focuses on the principles and applications of PSO and QPSO. More detailed survey on this type of swarm intelligence algorithm is given in subsequent chapters.

REFERENCES

1. R. Horst, P.M. Pardalos, N.V. Thoal. *Introduction to Global Optimization*, 2nd edn. Kluwer Academic Publisher, Dordrecht, the Netherlands, 2000.
2. J.E. Beasley (Ed.). *Advances in Linear and Integer Programming*. Oxford Science, Oxford, U.K., 1996.
3. A. Dmitris, M.W. Padberg. *Linear Optimization and Extensions: Problems and Extensions*. Universitext, Springer-Verlag, Berlin, Germany, 2001.
4. M. Avriel. *Nonlinear Programming: Analysis and Methods*. Dover Publishing, New York, 2003.
5. R. Horst, H. Tuy. *Global Optimization: Deterministic Approaches*, 3rd edn. Springer-Verlag, Berlin, Germany, 1996.
6. D.C. Karnopp. Random search techniques for optimization problems. *Automatica*, 1963, 1: 111–121.
7. T.G. Kolda, R.M. Lewis, V. Torczon. Optimization by direct search: New perspectives on some classical and modern methods. *SIAM Review*, 2003, 45: 385–482.
8. F.J. Solis, R.J.-B. Wets. Minimization by random search techniques. *Mathematics of Operations Research*, 1981, 6(1): 19–30.
9. A.A. Zhigljavsky. *Theory of Global Random Search*. Kluwer Academic, Boston, MA, 1991.

10. J.C. Spall. *Introduction to Stochastic Search and Optimization. Estimation, Simulation, and Control.* Wiley, Hoboken, NJ, 2003.

11. J. Matyas. Random optimization. *Automation and Remote Control*, 1965, 26: 244–251.

12. Z. Michalewicz, D.B. Fogel. *How to Solve It: Modern Heuristics.* Springer-Verlag, New York, 2000.

13. F. Glover, G.A. Kochenberger. *Handbook of Metaheuristics. International Series in Operations Research & Management Science.* Springer, New York, 2003.

14. E.-G. Talbi. *Metaheuristics: From Design to Implementation.* Wiley, Hoboken, NJ, 2009.

15. D.H. Wolpert, W.G. Macready. No free lunch theorems for optimization. *IEEE Transactions on Evolutionary Computation*, 1997, 1: 67–82.

16. N. Metropolis, A. Rosenbluth, M. Rosenbluth, A. Teller, E. Teller. Equation of state calculations by fast computing machines. *Journal of Chemical Physics*, 1953, 21(6): 1087–1092.

17. S. Kirkpatrick, C.D. Gelatt, Jr., M.P. Vecchi. Optimization by simulated annealing. *Science*, 1983, 220(4598): 671–680.

18. P.J. van Laarhoven, E.H. Aarts. *Simulated Annealing: Theory and Applications.* Kluwer, Dordrecht, the Netherlands, 1987.

19. I.H. Osman, J.P. Kelly (Eds.). *Meta-Heuristics: Theory and Applications.* Kluwer, Dordrecht, the Netherlands, 1996.

20. E. Aarts, J.K. Lenstra (Eds.). *Local Search in Combinatorial Optimization.* Wiley, Chichester, U.K., 1997.

21. D.B. Fogel. *Evolutionary Computation: Towards a New Philosophy of Machine Intelligence.* IEEE Press, New York, 2000.

22. J.H. Holland. *Adaptation in Natural and Artificial Systems.* University of Michigan Press, Ann Arbor, MI, 1975.

23. D.E. Goldberg. *Genetic Algorithms in Search, Optimization, and Machine Learning.* Addison-Wesley, Reading, MA, 1989.

24. Z. Michalewicz. *Genetic Algorithms + Data Structures = Evolution Programs*, 3rd edn. Springer-Verlag, New York, 1996.

25. M. Mitchell. *An Introduction to Genetic Algorithms.* MIT Press, Cambridge, MA, 1996.

26. C.R. Reeves, J.E. Rowe. *Genetic Algorithms—Principles and Perspectives: A Guide to GA Theory.* Kluwer Academic, Boston, MA, 2003.

27. N.L. Cramer. A representation for the adaptive generation of simple sequential programs. In *Proceedings of an International Conference on Genetic Algorithms and the Applications*, Grefenstette, J.J. (Ed.), Carnegie Mellon University, Pittsburgh, PA, 1985.

28. J.R. Koza. Genetic programming: A paradigm for genetically breeding populations of computer programs to solve problems, Computer Science Department, Stanford University, Technical Report STAN-CS-90-1314, 1990.

29. J.R. Koza. *Genetic Programming: On the Programming of Computers by Means of Natural Selection.* MIT Press, Cambridge, MA, 1992.

30. J.R. Koza. *Genetic Programming II: Automatic Discovery of Reusable Programs.* MIT Press, Cambridge, MA, 1994.

31. J.R. Koza, F.H. Bennett, D. Andre, M.A. Keane. *Genetic Programming III: Darwinian Invention and Problem Solving*. Morgan Kaufmann, San Francisco, CA, 1999.

32. J.R. Koza, M.A. Keane, M.J. Streeter, W. Mydlowec, J. Yu, G. Lanza. *Genetic Programming IV: Routine Human-Competitive Machine Intelligence*. Kluwer Academic Publishers, Boston, MA, 2003.

33. B. Mckay, S.-H. Chen, X.H. Nguyen. Genetic programming: An emerging engineering tool. *International Journal of Knowledge-Based Intelligent Engineering System*, 2008, 12(1): 1–2.

34. M. Korns. Large-scale, time-constrained, symbolic regression-classification. In *Genetic Programming Theory and Practice V*, R. Riolo, T. Soule, B. Worzel (Eds.), Springer, New York, 2007.

35. M. Korns. Symbolic regression of conditional target expressions. In *Genetic Programming Theory and Practice VII*, R. Riolo, U. O'Reilly, T. McConaghy (Eds.), Springer, New York, 2009.

36. M. Korns. Abstract expression grammar symbolic regression. In *Genetic Programming Theory and Practice VIII*, R. Riolo, T. McConaghy, E. Vladislavleva (Eds.), Springer, New York, 2010.

37. I. Rechenberg. *Evolutionsstrategie: Optimierung technischer Systeme nach Prinzipien der biologischen Evolution*. Frommann-Holzboog, Stuttgart, Germany, 1973.

38. H.-P. Schwefel. Evolutionsstrategie und numerische optimierung. Dissertation, Technical University of Berlin, Berlin, Germany, 1975.

39. H.-P. Schwefel. Binäre optimierung durch somatische mutation. Technical University of Berlin and Medical University of Hannover, Berlin, Germany, Technical Report, 1975.

40. H.-P. Schwefel. Collective phenomena in evolutionary systems. In *Preprints of the 31st Annual Meeting of the International Society for General System Research*, Budapest, Hungary, Vol. 2, 1987, pp. 1025–1033.

41. H.-P. Schwefel. *Evolution and Optimum Searching*. Wiley Interscience, John Wiley & Sons, New York, 1995.

42. H.-G. Beyer, H.-P. Schwefel. Evolution strategies—A comprehensive introduction. *Natural Computing: An International Journal*, 2002, 1(1): 3–52.

43. L.J. Fogel, A.J. Owens, M.J. Walsh. *Artificial Intelligence through Simulated Evolution*. John Wiley, New York, 1966.

44. L.J. Fogel. *Intelligence through Simulated Evolution: Forty Years of Evolutionary Programming*. John Wiley, New York, 1999.

45. A.E. Eiben, J.E. Smith. *Introduction to Evolutionary Computing*. Springer, Berlin, Germany, 2003.

46. F. Glover. Future paths for integer programming and links to artificial intelligence. *Computers and Operations Research*, 1986, 5: 533–549.

47. F. Glover. Tabu search: A tutorial. *Interfaces*, 1990, 20(4): 74–94.

48. F. Glover, M. Laguna. Tabu search. In *Modern Heuristic Techniques for Combinatorial Problems*, C.R. Reeves (Ed.). John Wiley & Sons, Inc., New York, 1993.

49. F. Glover, M. Laguna. *Tabu Search*. Kluwer Academic Publishers, Boston, MA, 1997.

50. M. Dell'Amico, M. Trubian. Applying tabu search to the job-shop scheduling problem. *Annals of Operations Research*, 1993, 41(3): 231–252.

51. C. Rego. A subpath ejection method for the vehicle routing problem. *Management Science*, 1998, 44(10): 1447–1459.

52. F. Glover, M. Laguna. *Tabu Search*. Kluwer, Dordrecht, the Netherlands, 1996.

53. S. Voss, S. Martello, I.H. Osman, C. Roucairol (Eds.). *Meta-Heuristics: Advances and Trends in Local Search Paradigms for Optimization*. Kluwer, Dordrecht, the Netherlands, 1999.

54. K. Price. Differential evolution: A fast and simple numerical optimizer. In *Proceedings of the Biennial Conference of the North American Fuzzy Information Processing Society*, NAFIPS, Berkeley, CA, 1996, pp. 524–527.

55. R. Storn. On the usage of differential evolution for function optimization. In *Proceedings of the Biennial Conference of the North American Fuzzy Information Processing Society*, NAFIPS, Berkeley, CA, 1996, pp. 519–523.

56. R. Storn, K. Price. Differential evolution—A simple and efficient adaptive scheme for global optimization over continuous spaces. *Journal of Global Optimization*, 1997, 11: 341–359.

57. K. Price, R.M. Storn, J.A. Lampinen. *Differential Evolution: A Practical Approach to Global Optimization*. Springer, New York, 2005.

58. R. Gamperle, S. Muller, P. Koumoutsakos. A parameter study for differential evolution. In *Proceedings of the WSEAS International Conference on Advances in Intelligent Systems, Fuzzy Systems, Evolutionary Computation*, WSEAS Press, Interlaken, Switzerland, 2002, pp. 293–298.

59. V. Feoktistov. *Differential Evolution: In Search of Solutions*. Springer, New York, 2006.

60. U.K. Chakraborty (Ed.). *Advances in Differential Evolution*. Springer, Heidelberg, Germany, 2008.

61. J. Liu, J. Lampinen. On setting the control parameter of the differential evolution method. In *Proceedings of the Eighth International Conference on Soft Computing*, Brno, Czech Republic, 2002, pp. 11–18.

62. D. Zaharie. Critical values for the control parameters of differential evolution algorithms. In *Proceedings of the Eighth International Conference on Soft Computing*, Brno University of Technology, Faculty of Mechanical Engineering, Brno, Czech Republic, 2002, pp. 62–67.

63. M.E.H. Pedersen. Tuning & simplifying heuristical optimization, PhD thesis, School of Engineering Sciences, Computational Engineering and Design Group, University of Southampton, 2010.

64. M.E.H. Pedersen. Good parameters for differential evolution, Hvass Laboratories, Technical Report HL1002, 2010.

65. J. Liu, J. Lampinen. A fuzzy adaptive differential evolution algorithm. *Soft Computing*, 2005, 9(6): 448–462.

66. A.K. Qin, P.N. Suganthan. Self-adaptive differential evolution algorithm for numerical optimization. In *Proceedings of the IEEE Congress on Evolutionary Computation*, Edinburgh, U.K., 2005, pp. 1785–1791.

67. J. Brest, S. Greiner, B. Boskovic, M. Mernik, V. Zumer. Self-adapting control parameters in differential evolution: A comparative study on numerical benchmark functions. *IEEE Transactions on Evolutionary Computation*, 2006, 10(6): 646–657.

68. A. Colorni, M. Dorigo, V. Maniezzo. Distributed optimization by ant colonies. In *Proceedings of the First European Conference Artificial Life*, MIT Press, Cambridge, MA, 1991, pp. 134–142.

69. A. Colorni, M. Dorigo, V. Maniezzo. An investigation of some properties of an ant algorithm. In *Proceedings of the Parallel Problem Solving from Nature*, Brussels, Belgium, 1992, pp. 509–520.

70. M. Dorigo, M. Luca. A study of some properties of ant-Q. In *Proceedings of the Fourth International Conference on Parallel Problem Solving Problem from Nature*, Springer Verlag, Berlin, Germany, 1996, pp. 656–665.

71. M. Dorigo, V. Maniezzo, A. Colorni. Ant system: Optimization by a colony of cooperating agents. *IEEE Transactions on Systems, Man, Cybernetics, Part B*, 1996, 26(1): 28–41.

72. J. Kennedy, R.C. Eberhart. Particle swarm optimization. In *Proceedings of the IEEE International Conference on Neural Networks*, IEEE Service Center, Piscataway, NJ, Vol. IV, 1995, pp. 1942–1948.

73. L.M. Gambardella, M. Dorigo. Ant-Q: A reinforcement learning approach to the traveling sales man problem. In *Proceedings of the 12th Machine Learning Conference*, Morgan Kaufmann, San Francisco, CA, 1995, pp. 252–260.

74. C.-H. Chu, J. Gu, X. Hou. A heuristic ant algorithm for solving QoS multicast routing problem. In *Proceedings of the 2002 Congress on Evolutionary Computation*, Honolulu, HI, 2002, pp. 1630–1635.

75. W.J. Gutjahr. A graph-based ant system and its convergence. *Future Generation Computer Systems*, 2000, 16(9): 873–888.

76. N. Shervin. Agent-based approach to dynamic task allocation. In *Proceedings of the Third International Workshop on Ant Algorithm*, Brussels, Belgium, 2002, pp. 28–39.

77. Y.-J. Li, T.-J. Wu. A nested ant colony algorithm for hybrid production scheduling. In *Proceedings of the American Control Conference*, Anchorage, U.K., 2002, pp. 1123–1128.

78. Y.-J. Li, T.-J. Wu. A nested hybrid ant colony algorithm for hybrid production scheduling problems. *Acta Automatica Sinica*, 2003, 29(1): 95–101.

79. A. Colorni, M. Dorigo, V. Maniezzo. Ant system for job-shops scheduling. *Belgian Journal of Operations Research, Statistics and Computer Science*, 1994, 34(1): 39–53.

80. T. Stutzle. An ant approach to the flow shop problem. In *Proceedings of the Sixth European Congress on Intelligent Techniques & Soft Computing*, Germany, 1997, pp. 1560–1564.

81. C.-F. Tsai, H.-C. Wu, C.-W. Tsai. A new data clustering approach for data mining in large databases. In *Proceedings of the International Symposium on Parallel Architectures, Algorithms and Networks*, Manila, Philippines, 2002, pp. 315–321.

82. R.S. Parpinelli, H.S. Lopes, A.A. Freitas. Data mining with an ant colony optimization algorithm. *IEEE Transactions on Evolutionary Computation*, 2002, 6(4): 321–332.
83. A.W. Israel, L. Michael, M.B. Alfred. Distributed covering by ant-robot using evaporating traces. *IEEE Transactions on Robotics and Automation*, 1999, 15(5): 918–933.
84. R. Hoar, J. Penner, C. Jacob. Ant trails—An example for robots to follow. In *Proceedings of the 2002 Congress on Evolutionary Computation*, Honolulu, HI, 2002, pp. 1910–1915.
85. R.A. Russell. Ant trails—An example for robots to follow. In *Proceedings of the 1999 IEEE International Conference on Robotics and Automation*, Detroit, MI, 1999, pp. 2698–2703.
86. M.J.B. Krieger, J.B. Billeter, L. Keller. Ant-like task allocation and recruitment in cooperative robots. *Nature*, 2000, 406(31): 992–995.
87. D. Costa, A. Hertz. Ant can colour graphs. *Journal of the Operational Research Society*, 1997, 48(3): 295–305.
88. S. Ando, H. Iba. Ant algorithm for construction of evolutionary tree. In *Proceedings of the Genetic and Evolutionary Computation Conference*, New York, 2002, pp. 1552–1557.
89. T. Stutzle, H. Hhoos. Max-min ant system and local search for combinational optimization problems. In *Proceedings of the Second International Conference on Metaheuristics*, Springer-Verlag, Wien, Austria, 1997.
90. T. Stutzle, H. Hoos. The max-min ant system and local search for the traveling salesman problem. In *Proceedings of the IEEE International Conference on Evolutionary Computation*, Piscataway, NJ, 1997, pp. 309–314.
91. E.G. Talbi, O. Roux, C. Fonlupt. Parallel ant colonies for the quadratic assignment problem. *Future Generation Computer Systems*, 2001, 17(4): 441–449.
92. P.J. Angeline. Evolutionary optimization versus particle swarm optimization: Philosophy and performance differences. *Evolutionary Programming VII*. Lecture Notes in Computer Science, 1998, 1447: 601–610.
93. J. Sun, B. Feng, W.B. Xu. Particle swarm optimization with particles having quantum behavior. In *Proceedings of the Congress on Evolutionary Computation*, Portland, OR, 2004, pp. 326–331.
94. J. Sun, W.B. Xu, B. Feng. A global search strategy of quantum-behaved particle swarm optimization. In *Proceedings of the IEEE Conference on Cybernetics and Intelligent Systems*, Singapore, 2004, pp. 111–116.

CHAPTER 2

Particle Swarm Optimisation

2.1 OVERVIEW

As a terminology first introduced in the context of cellular robotic systems [1], swarm intelligence (SI) was originally used to describe the collective behaviour of decentralised and self-organised systems, which are either natural or artificial. Now, it generally refers to a class of meta-heuristics. A typical SI system consists of a population of simple agents interacting locally with one another and the environment where they live. The individual agent shows no intelligence and follows very simple rules without centralised control structure dictating how it should behave. Interactions between such agents are local and even show a certain degree of randomness. However, coherent functional global patterns or 'intelligent' global behaviour which is unknown to individual agents emerges due to such local interactions. There are natural examples of SI systems, such as ant colonies, bird flocking, bee swarm, animal herding, bacterial growth, and fish schooling and human, to mention a few.

The particle swarm optimisation (PSO) algorithm falls into the category of SI algorithms and is a population-based optimisation technique originally developed by Kennedy and Eberhart in 1995 [2]. It was motivated by social behaviour (i.e. collective behaviour) of bird flocking or fish schooling and shares many similarities with evolutionary computation techniques such as genetic algorithms (GA). The optimisation process of a PSO system begins with an initial population of random solutions and searches for optima by updating various properties of the individuals in each generation. However, unlike GA, PSO does not have evolution operators such as crossover and mutation. The potential solutions are known as particles which fly through the solution space by following their own experiences and the current best particles. It was shown in [3] that the PSO algorithm is comparable in performance with and may be considered as an alternative method to GA.

In the last decade, PSO has been frequently used as an optimisation algorithm because of its effectiveness in performing difficult optimisation tasks. In addition, the scheme obtains better results in a faster and cheaper way compared to several other methods with fewer parameters to adjust. Application areas can be found in multi-objective optimisation problems [4–9], min–max problems [10,11], integer programming problems [12], combinatorial optimisation problems [13,14], clustering and classification problems [15–17] and numerous engineering applications [18–44].

The algorithm has been analysed by several people. Kennedy analysed the simplified particle behaviour [45] and demonstrated different particle trajectories for a range of design choices. Ozcan and Mohan showed that a particle in a simple one-dimensional (1D) PSO system follows a path defined by a sinusoidal wave, randomly deciding on both its amplitude and frequency [46]. They also generalised the results to obtain closed form for the trajectory equations of particles in a multidimensional search space [46]. The first formal analysis of the particle trajectory and the stability properties of the algorithm was probably by Clerc and Kennedy [47]. A deterministic coefficient instead of a random coefficient is used in the analysis. The discrete iterative update becomes a second-order differential equation when it is converted into a continuous process. The stability analysis relies on the eigenvalues of the auxiliary equation of the differential equation. A similar analysis based on the deterministic version of PSO was also carried out in identifying regions in the parameter space that guarantees stability [48]. The issue of convergence and parameter selection was also addressed in [49,50]. However, all researchers mentioned earlier acknowledged the limitations of their results, which did not take into account the stochastic nature of the algorithm. An analysis of a continuous-time version of PSO was presented in [51]. A Lyapunov analysis approach was adopted in [52] for the social foraging swarms, different to the standard PSO, in a continuous-time setting. Using Lyapunov stability analysis and the concept of passive systems, Kadirkamanathan et al. [53] analysed the stability of the particle dynamics and derived sufficient conditions for stability, without the restrictive assumption that all parameters are deterministic. A formal stochastic convergence analysis was carried out for the canonical PSO algorithm as described in [54], which involved randomness. By considering the position of each particle in each evolutionary step as a stochastic vector, Jiang et al. analysed the canonical PSO, derived the stochastic convergent condition of the particle system and provided guidelines on the parameter selection in the stochastic process theory. Unfortunately, van den Bergh proved that the canonical PSO is not a global search algorithm [48], even not a local one, using the convergence criterion provided by Solis and Wets [55].

In addition to the analyses mentioned earlier, there has been a considerable amount of work performed in developing empirical simulations of the original version of PSO. Shi and Eberhart introduced the concept of inertia weight into the original PSO, in order to balance the local and global search during the optimisation process [56]. Clerc proposed an alternative version of PSO incorporating a parameter known as the constriction factor which replaces the restriction on velocities [57]. Angeline introduced a tournament selection into PSO based on the particle's current fitness so that the properties that make some solutions superior were transferred directly to some of the less effective particles [58]. This technique improves the performance of the PSO algorithm in some benchmark functions. Suganthan proposed a neighbourhood operator for PSO, which divided the swarm into multiple 'neighbourhoods', where each neighbourhood maintained its own local best solution [59]. This approach belongs to another general form of PSO referred to as the local best model, and is less prone to becoming trapped in local minima, but typically has slower convergence. Inspired by the social–psychological metaphor of social stereotyping, Kennedy applied cluster analysis to the previous best positions of the particles in PSO and improved the algorithm by substituting the previous best position of individual particle with the cluster centres [60]. Kennedy also investigated other neighbourhood topologies and concluded that the von Neumann topology results in superior performance [61]. Several researchers experimented with adaptive topologies, randomising, adding, deleting and moving links in response to aspects of the iterative search, and it seems very likely that the ultimate particle swarm adapts its social network depending on the situation [62–64]. van den Bergh and Engelbrecht proposed a variation on the canonical PSO, called the cooperative PSO, employing multiple swarms to optimise different components of the solution vector cooperatively [65].

It was demonstrated that their method significantly improved the performance of the PSO algorithm. Mendes developed a type of particle swarm where each particle used information from all its neighbours, but not from its own history [66]. This version is known as the fully informed particle swarm (FIPS) and was found to perform better than the canonical version on an aggregated suite of various test functions, when an appropriate topology was used. Janson and Middendorf designed a hierarchical particle swarm optimiser (H-PSO), in which the particles were arranged in a dynamic hierarchy that was used to define a neighbourhood structure and the particles moved up or down the hierarchy. It was shown that H-PSO and its variant exhibit comparable performances with PSO using different standard neighbourhood schemes [67]. In [68], Parrott and Li invented a particle swarm using the notion of species to determine its neighbourhood best values for solving multi-modal optimisation problems and for tracking multiple optima in a dynamic environment. Their method was demonstrated to be very effective in dealing with multi-modal optimisation functions in both static and dynamic environments.

Some researchers have attempted to simulate the particle trajectories using various ways such as direct sampling, using a random number generator with a certain probability distribution. For example, Krohling updated the velocity using the absolute value of a Gaussian distribution which leads PSO to generate good results [69]. Secrest and Lamont proposed a Gaussian PSO, in which the mean value of the local and global velocities are equal to the distances between the particle and the local best and global best positions, respectively; the standard deviation of the local and global velocities are equal to half the distance between the particle and the local and global bests, respectively [70]. Richer and Blackwell showed good results using a Lévy distribution [71]. Kennedy sampled the position of each particle with a Gaussian distribution [72] and, later, with various probability distributions [73,74]. Sun et al., inspired by quantum mechanics and trajectory analysis of PSO [47], used a strategy based on a quantum δ potential well model to sample around the previous best points [75]. They later introduced the mean best position into the algorithm and proposed a new version of PSO, known as the quantum-behaved particle swarm optimisation (QPSO) [76,77]. The iterative equation of QPSO is very different from that of PSO and leads to global convergent. It should be noted that QPSO does not require velocity vectors for particles and has fewer parameters to adjust, making it easier to implement, in contrast to PSO.

This review on PSO only covers a small part of this field. It is hoped that foundation knowledge could be established for PSO after the earlier brief overview. In Chapter 3, some important variants of PSO, including those mentioned earlier, are described in detail. To gain further details on various improvements and applications of PSO readers are referred to other review articles such as [78,79].

As the main work of this book concerns the convergence of the positions of particles, the meaning for 'particles' and 'positions' and their usage throughout this book may be interchangeable at many instances. Similarly, the meaning for 'swarm' and 'population' and their usage may be interchangeable as well.

2.2 MOTIVATIONS

Particle swarm optimisation is known to be rooted from two methodologies [2]. One obvious root is its ties with artificial life in general, and to bird flocking, fish schooling and swarming theory in particular. The other root is associated with evolutionary algorithms, such as GAs and evolutionary programming. Evolutionary algorithms are briefly described in Chapter 1. This section concentrates on the collective behaviour in bird flocks or fish schools, follows with a description of the relationship between PSO and the simulation of collective behaviour, including fishing schooling, bird flocking, and that of humans.

2.2.1 Collective Behaviour

In nature, collective (or social) behaviour of biological systems is a widespread phenomenon and has been observed in many social organisms, at very different scales and levels of complexity [80]. Bird flocks gathering over the roost at dusk [81], fish schools milling under water [82,83], swarms of insects [84], trails of foraging ants [85] and herds of mammals [86] are among the examples of familiar collective behaviour. These social organisms are genetically related and can cooperate. There are of course aggregations of unrelated 'selfish' individuals, like fish or birds or even human beings. A cohesive group is formed by individuals in many cases, acting as a whole with remarkable coordination and adaptability and being sustained spontaneously by the mutual attraction among the members.

Collective behaviour emerges without centralised control. Individual members act on the basis of some limited local information, which is due to their interaction with neighbours or chemicals deposition. This information flows through the system and produces collective patterns, that is, global patterns. One typical example is flocking birds. Each individual bird flies in the same direction as its neighbours. However, this local tendency gives rise to a coherent moving flock. This mechanism produces global patterns from local rules and is known as self-organisation [87–90].

Due to the complexity of biological systems, investigating collective behaviour is, therefore, a difficult task. When difference among species is significant, an apparently satisfactory qualitative explanation is not always consistent with a quantitative analysis of empirical data. As a result, a continuous feedback between empirical observations and mathematical modelling is essential to understand the origin of collective phenomena and appropriately characterise them.

Until recently, empirical data have been scarce and limited to small systems for three-dimensional (3D) animal groups such as bird flocks or fish schools. This has certainly restrained the possibility of a reliable statistical analysis of empirical data and a comparison with the predictions of the models.

2.2.2 Modelling and Simulation of Self-Organised Collective Behaviour

There are different types of self-organised collective behaviour models according to (1) the scale, space and time, at which the collective phenomenon is analysed; (2) the kind of local information individuals use to aggregate (direct response to other individuals or indirect cues); and (3) the mathematical complexity. The most widely used model for collective behaviour in biological systems is known as the agent-based model. This type of model is closely related to PSO and is addressed in the rest of this section. Other kinds of models based on probabilities and cellular automata can be referred to in literature such as [88].

Agent-based models were originally developed for fish schools [91–93] and bird flocks [94,95], and were eventually applied to mammal herds [96] and other vertebrate groups [97]. Very often cohesive and polarised groups are formed within such systems and these groups exhibit remarkable coordination and adaptability. In general, agent-based models use certain behavioural rules at the level of the individual by adjusting its position according to the rules. There are three simple behavioural rules leading to collective behaviour: move in the same direction as their neighbours, remain close to them and avoid collisions. These behavioural rules are modelled by using three distinct inter-individual interactions: (1) alignment of velocities, which makes neighbouring birds fly in the same direction; (2) attraction, which ensures no bird remains isolated; (3) short-range repulsion, which prevents dangerous proximity. The inter-individual interactions between the agents are projected to the entire biological system leading to system activities in directional polarity (alignment), aggregation cohesiveness (attraction) and individual integrity preservation (short-range repulsion).

These rules may be implemented in different ways depending on the target biological system and on the experience and points of view of the modeller. Heuristic reasons and behavioural assumptions are required to be established based on various knowledge and properties of the system. In general, the model results in an equation which is used in the update of the velocities of each individual. If $\vec{d}_i$ is the direction of motion of an agent i then an update equation typically looks like the following:

$$\vec{d}_i(t+1) = \frac{1}{n_{in}} \sum_{j=1}^{n_{in}} w_j \vec{d}_j(t) + \frac{1}{n_{in}} \sum_{j=1}^{n_{in}} f_{ij} \frac{\vec{r}_{ij}(t)}{|\vec{r}_{ij}(t)|} + \vec{\eta}_i(t), \qquad (2.1)$$

where
  $\vec{r}_{ij}(t)$ indicates the distance vector from agent i to agent j
  $\vec{\eta}_i$ is a stochastic noise modelling the uncertainty of the decision-making
    process

The first term on the right-hand side of (2.1) is due to the alignment property and is usually the weighted average taken over the directions of motion of the n_{in} interacting neighbours. The second term represents the positional response to neighbours with the 'force' f_{ij} specifying the way agent i is attracted or repelled by agent j. At small distances, the second term gives a negative increment in heading away from close neighbours. At larger distances, instead, the increment is positive and towards the average position of its neighbours. The current practice in the modelling of such behaviour is to assume a given region in space in a chosen neighbourhood of the focal individual. There are more sophisticated models which allow behavioural zones of different spatial extent where alignment, attraction and repulsion occur [91,93,94,98]. Some models assume a functional dependence on distance of the forces f_{ij} due to neighbours [99–102]. The typical behavioural results for the two kinds of models as seen from [80] are shown in Figure 2.1 as an example [80]. The distance between an individual and other members of the group is the essential metric to be considered in these models.

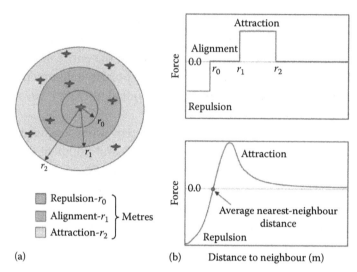

FIGURE 2.1 Behavioural rules in numerical models: (a) distance-dependent behavioural zones and (b) distance-dependent attraction/repulsion function. (From Giardina, I., *HFSP J.*, 2(4), 205, 2008.)

There are usually a large number of individuals in most agent-based models and the resulting set of coupled dynamical equations cannot be treated analytically. Computer simulations of various interpretations of movement of organism in a bird flock or fish school is one way out. The first artificial life program which simulates the flocking behaviour of birds is the BOID model developed by Reynolds [94]. The model is intrigued by the aesthetics of bird-flocking choreography and is an example of emergent behaviour—the complexity of BOID arises from the interaction of individual agents adhering to a set of simple rules. The rules in the BOID world are as follows: (1) separation: steer to avoid crowding local flockmates; (2) alignment: steer towards the average heading of local flockmates; (3) cohesion: steer to move towards the average position (centre of mass) of local flockmates. Based on the BOID model, Heppner and Grenander independently developed a stochastic model [95] taking the assumption that the birds are attracted by the habitat. The common features of both models are based on the consideration of the birds' effort to maintain an optimal distance between themselves and their neighbours.

2.2.3 From Simulation of Collective Behaviour to the PSO Algorithm

Wilson, a sociobiologist, gave an interesting statement [103]: 'In theory at least, individual members of the school can profit from the discoveries and previous experience of all other members of the school during the search for food. This advantage can become decisive, outweighing the disadvantages of competition for food items, whether the resource is unpredictably distributed in patches'. This statement suggests a common rule that underlies collective behaviour of herds, birds, schools and flocks and humans. It also indicates that social sharing of information among consecrates offers an evolutionary advantage, which is the fundamental hypothesis to the development of PSO [2].

PSO was originated from the simulation and modelling of human and social animal behaviour, which is certainly not identical to fish schooling or bird flocking. Birds and fish would adjust their physical movement in order to avoid collision, steer towards the average heading, and move towards the average position as shown by the agent-based model and simulation. They would also like to avoid predators, seek food and mates, optimise environmental parameters such as temperature, etc. Humans adjust cognitive and experiential variables as well as physical movements. In other words humans do not usually walk in step and turn in unison as birds and fish do, but tend to conform to the beliefs and attitudes of their peers [2].

Two individuals in a human society can hold identical attitudes and beliefs without banging together, but two birds or two fish cannot occupy the same position in space without colliding. This is the major distinction between collective behaviour of humans and bird flocking or fish schooling. However, 'it seems reasonable, in discussing human social behaviour, to map the concept of *change* into bird/fish analog of *movement*'. According to Kennedy and Eberhart—'This is consistent with the classic Aristotelian view of qualitative and quantitative changes as types of movement'. Therefore, modelling human collective behaviour requires not only movements in 3D physical space to avoid collisions, but also changes in abstract multidimensional space, collision-free. Note that the physical space may affect informational inputs, but it is an insignificant component of psychological experience. The PSO algorithm developed by Kennedy and Eberhart was based on such a modelling concept.

2.3 PSO ALGORITHM: BASIC CONCEPTS AND THE PROCEDURE

2.3.1 Simulation on Two-Dimensional Plane

A classical simulation example provided by Heppner and Grenander [95] was to graphically simulate the graceful but unpredictable choreography of a bird flock. In this classical simulation, a population of birds (also called agents) was randomly initialised with their positions on a torus pixel grid and two velocities, along X and Y coordinates, around a 'roost'; a position on the pixel screen was used to attract these birds until they all finally landed there. In each iterative step of the PSO algorithm a loop is used to determine those nearest neighbouring agents and their velocities along X and Y coordinates, that is, VX and VY, are then assigned to the bird in focus. In the simulation, those birds knew where their 'roost' was, but in a realistic environment, birds would land on any tree or any wire post that meets their immediate needs. In particular, the bird flock may land on places where food is found even though they have no previous knowledge about its location, appearance and so forth. Therefore, a dynamic force should be introduced into the simulation, as suggested by Wilson [103], to enable members of the flock to capitalise knowledge of one another. One possible modification to the simulation is to include the knowledge of the location of food, say a cornfield, defined at the position (X^*, Y^*). This modification involves the evaluation of the food distance of each agent according to the Euclidean distance between the agent and the location of food:

$$Eval = \sqrt{(presentX - X^*)^2 + (presentY - Y^*)^2} , \qquad (2.2)$$

where $(presentX, presentY)$ is the present position of each agent. The smaller the value of the $Eval$ is, the closer is the agent to the location where food is defined. Note that $Eval$ is zero at the point (X^*, Y^*).

Kennedy and Eberhart [2] added the aforementioned variation into the simulation and called the agents of the flock a *swarm*, which presents behaviour that is very different from those obtained by Heppner and Grenander [95]. Firstly, in the simulation performed by Kennedy and Eberhart [2], each agent 'remembered' its best value of $Eval$, and the position resulting in that value. Using pseudo-code variables, the best value and the position are denoted as *pbest*[] and (*pbestX*[], *pbestY*[]), respectively, where the square brackets indicate that these are arrays with number of components equal to the number of agents. The pseudo-code variable *pbest* resembles personal memory as each agent remembers its own experience. This phenomenon is called *personal cognition*. The velocities along X and Y directions (i.e. VX[] and VY[]) of an agent are adjusted in the following way:

If *presentX*[] $>$ *pbestX*[],   then VX[] $=$ VX[] $-$ rand()* *p_increment*

If *presentX*[] $<$ *pbestX*[],   then VX[] $=$ VX[] $+$ rand()* *p_increment*

If $presentY[] > pbestY[]$, then $VY[] = VY[] - rand()*p\_increment$

If $presentY[] < pbestY[]$, then $VY[] = VY[] + rand()*p\_increment$

where
    $p\_increment$ is a positive system parameter
    $rand()$ is a generator of random number uniformly distributed on (0,1)

Secondly, Kennedy and Eberhart assumed that each agent 'knows' the global best position and its *Eval* value that a member has found. They accomplished this by assigning an array index of the agent with the best value to a variable called *gbest*, so that the flock's best X position is denoted as $pbestX[gbest]$ and its best Y position is denoted as $pbestY[gbest]$. The variable *gbest* is conceptually similar to some publicised knowledge, or a group norm or standard, which individuals seek to obtain. This information was then informed to all the flock members. The member with this 'social cognition' adjusted its $VX[]$ and $VY[]$ as follows:

If $presentX[] > pbestX[gbest]$,

   then $VX[] = VX[] - rand()*g\_increment$

If $presentX[] < pbestX[gbest]$,

   then $VX[] = VX[] + rand()*g\_increment$

If $presentY[] > pbestY[gbest]$,

   then $VY[] = VY[] - rand()*g\_increment$

If $presentY[] < pbestY[gbest]$,

   then $VY[] = VY[] + rand()*g\_increment$

In the simulation, when $p\_increment$ and $g\_increment$ are set relatively high, the flock seemed to be sucked violently into the cornfield. On the other hand, with $p\_increment$ and $g\_increment$ being set relatively low, the flock swirled around the goal during the approach, swinging out rhythmically with subgroups synchronised and finally alighting on the target [2]. Furthermore, a relatively high $p\_increment$ compared to $g\_increment$ resulted in excessive wandering of isolated individuals through the solution space, while the verse (high *g_increment* relative to *p_increment*) led to the flock rushing towards local optima in premature convergence. They found that setting both of these variables to be approximately equal could result in most effective search of the solution domain.

2.3.2 Simulation of Multidimensional Search

The strategy illustrated earlier as adopted by Kennedy and Eberhart for simulation of bird flocking in a two-dimensional (2D) plane is very close to that of the PSO algorithm. They began their experiments on modelling multidimensional and collision-free social behaviour. To achieve this goal, a simple step to change *presentX[]* and *presentY[]* (and of course *VX[]* and *VY[]*) from 1D arrays to $M \times N$ matrices, where M is the number of agents and N is the number of dimensions, was adopted in the work. They performed multidimensional experiments by training a feed-forward multi-layer perceptron neural network (NN). They found that the algorithm performed well on the problem. However, other experiments performed by them revealed that further improvement could be achieved if velocities were adjusted according to their difference from the best locations in each dimension as shown in (2.3), rather than by simply testing the sign of the inequality as in the previous simulation.

$$VX[][] = VX[][] + \text{rand}() * p\_increment * (pbestX[][] - presentX[][]),$$

$$(2.3)$$

where *VX[][]* and *presentX[][]* have two sets of brackets because they are now matrices of agents by dimensions. Note that *increment* and *bestX* may be prefixed with *g* or *p* for different purposes.

However, Kennedy and Eberhart realised that it was impossible to assign the best choice for *p_* or *g_increment*. These terms were removed from the algorithm and the stochastic factors were multiplied by 2. On the other hand, each agent was assumed to possess 'personal cognition' and 'social cognition' simultaneously. A new version of the algorithm in which velocities adjusted according to the following formula was then used:

$$VX[][] = VX[][] + 2 * \text{rand}() * (pbestX[][] - presentX[][])$$

$$+ 2 * \text{rand}() * (pbestX[gbest][] - presentX[][]). \quad (2.4)$$

This version of the algorithm is known as the original PSO and it outperforms the previous version as described by (2.3).

Another version adjusted velocities according to the following formula which results from removing the momentum of $VX[][]$ on the right side of Equation 2.4.

$$VX[][] = 2 * \text{rand}() * (pbestX[][] - presentX[][])$$

$$+ 2 * \text{rand}() * (pbestX[gbest][] - presentX[][]). \quad (2.5)$$

which implies that velocity initialisation is not required and the velocities at each iteration are adjusted only according to the difference from $presentX[][]$ to $pbestX[][]$ and $pbestX[gbest][]$. This version, though simplified, turned out to be ineffective in finding the global optima since it is more prone to encounter premature convergence.

2.3.3 Formal Description of PSO

This section provides a full description of the PSO algorithm in a formal way. Suppose there are M agents, called particles, used in the PSO algorithm. Each agent is treated as a particle with infinitesimal volume with its properties being described by the current position vector, its velocity vector and the personal best position vector. These vectors are N-dimensional vectors containing properties based on the N decision variables. At the nth iteration, the three vectors describing the properties of particle i $(1 \leq i \leq M)$ are

1. The current position vector: $X_{i,n} = (X_{i,n}^1, X_{i,n}^2, \ldots, X_{i,n}^j, \ldots, X_{i,n}^N)$, where each component of the vector represents a decision variable of the problem and $1 \leq j \leq N$

2. The velocity vector: $V_{i,n} = (V_{i,n}^1, V_{i,n}^2, \ldots, V_{i,n}^j, \ldots, V_{i,n}^N)$, denoting the increment of the current position where $1 \leq j \leq N$

3. The personal best position vector: $P_{i,n} = (P_{i,n}^1, P_{i,n}^2, \ldots, P_{i,n}^j, \ldots, P_{i,n}^N)$ $(1 \leq j \leq N)$

Note that the personal best position vector, which is denoted as *pbest* in the pseudo-code notation used by Kennedy and Eberhart, returns the best objective function value or fitness value in the subsequent iterative process. The initial approximation of $X_{i,0}$ may be randomly generated within the search domain $[X_{min}^j, X_{max}^j]$ $(1 \leq j \leq N)$, where X_{min}^j and X_{max}^j are upper limit and lower limit of particle positions in the jth dimension. Generally, uniform distribution on $[X_{min}^j, X_{max}^j]$ in the jth dimension is used to generate the initial current position vector. In a similar way, the initial velocity vector $V_{i,0}$ may be initialised by choosing its jth component randomly on $[-V_{max}^j, V_{max}^j]$ $(1 \leq j \leq N)$, where V_{max}^j is the upper limit of velocities in the jth dimension. The initial approximation of $P_{i,n}$ can be set as the initial current position vector.

At the nth iteration, the particle with its personal best position which returns the best objective function value among all the particles is called the global best particle. This personal best position is denoted as $P_{g,n}$, which is recorded in a position vector $G_n = (G_n^1, G_n^2, \ldots, G_n^N)$ known as the global best position. Here g is the index of the global best particle. Without loss of generality, one can restate the minimisation problem in Equation 1.1 as follows:

$$\min f(x) \quad \text{s.t.} \quad x \in S \subseteq R^N, \tag{2.6}$$

where

$f(x)$ is an objective function continuous almost everywhere
S is the feasible space

The update of $P_{i,n}$ is given by

$$P_{i,n} = \begin{cases} X_{i,n} & \text{if } f(X_{i,n}) < f(P_{i,n-1}), \\ P_{i,n-1} & \text{if } f(X_{i,n}) \geq f(P_{i,n-1}), \end{cases} \tag{2.7}$$

and thus G_n can be found by

$$G_n = P_{g,n}, \quad \text{where } g = \arg \min_{1 \leq i \leq M} [f(P_{i,n})]. \tag{2.8}$$

At the $(n + 1)$th iteration, the component-wise properties of each particle are given by (2.4) which can be restated formally as

$$V_{i,n+1}^j = V_{i,n}^j + c_1 r_{i,n}^j (P_{i,n}^j - X_{i,n}^j) + c_2 R_{i,n}^j (G_n^j - X_{i,n}^j), \tag{2.9}$$

$$X_{i,n+1}^j = X_{i,n}^j + V_{i,n+1}^j \tag{2.10}$$

for $i = 1, 2, \ldots, M; j = 1, 2, \ldots, N$, where c_1 and c_2 are acceleration coefficients. In the simulation performed by Kennedy and Eberhart [2], c_1 and c_2 are chosen as 2. The parameters $r_{i,n}^j$ and $R_{i,n}^j$ form two different sequences of random numbers distributed uniformly over $(0,1)$, that is, $\{r_{i,n}^j : j = 1, 2, \ldots, N\} \sim U(0,1)$ and $\{R_{i,n}^j : j = 1, 2, \ldots, N\} \sim U(0,1)$. Generally, the vector $V_{i,n}$ has each of its components being restricted in the interval $[-V_{max}^j, V_{max}^j]$ $(1 \leq j \leq N)$. The PSO in which the velocities of the particles are updated according to (2.4) or (2.9) is called the original PSO.

Although the simulation performed by Kennedy and Eberhart [2] demonstrates that the original PSO performed well on some testing problems, its performance on most other widely used benchmark problems are far from satisfactory. A few examples are used to demonstrate the performance of the original PSO in this chapter. Many revised versions of PSO algorithm have been proposed in order to improve the performance as described at the beginning of this chapter. Here, two most influential improvements are discussed along with the original PSO.

The first revised version is the PSO with inertia weight (PSO-In), proposed by Shi and Eberhart [56], in which velocities of the particles are adjusted according to the formula

$$V_{i,n+1}^j = wV_{i,n}^j + c_1 r_{i,n}^j (P_{i,n}^j - X_{i,n}^j) + c_2 R_{i,n}^j (G_n^j - X_{i,n}^j), \qquad (2.11)$$

where $w < 1$ is called the inertia weight. It is usually set to be smaller than 1 in order to accelerate the convergence of the particle.

Another revised version developed by Clerc [57] uses the constriction factor χ to adjust the velocities. The velocity update in this version is given by

$$V_{i,n+1}^j = \chi[V_{i,n}^j + c_1 r_{i,n}^j (P_{i,n}^j - X_{i,n}^j) + c_2 R_{i,n}^j (G_n^j - X_{i,n}^j)], \qquad (2.12)$$

where χ is determined by the formula

$$\chi = \frac{2}{\left| 2 - \varphi - \sqrt{\varphi^2 - 4\varphi} \right|}, \qquad (2.13)$$

such that $\varphi = c_1 + c_2$ and c_1 and c_2 should be set according to $\varphi > 4$. This version is usually known as the particle swarm optimisation with constriction (PSO-Co).

These two versions of PSO, generally referred to as the canonical PSO collectively, accelerate the convergence speed of the swarm effectively with better performance in general. There are many other revisions for the PSO algorithms, but all of them are based on these two versions. A detailed description of the background for these modified PSO algorithms is provided in Chapter 3.

2.3.4 Procedure of the PSO Algorithm

The PSO algorithm begins with initialising each component of current positions of the particles and their velocities within $[X_{min}^j, X_{max}^j]$, and $[-V_{max}^j, V_{max}^j]$ $(1 \leq j \leq N)$, respectively, and setting their initial personal best positions as $P_{i,0} = X_{i,0}$. The value $f(X_{i,0})$ $(1 \leq i \leq M)$ is then computed; each $f(P_{i,0})$ is set to be $f(X_{i,0})$, and G_0 is finally computed. Then the search

```
Begin
Initialize the current position X^j_{i,0}, velocity V^j_{i,0} and personal best
position P^j_{i,0} (setting P^j_{i,0} = X^j_{i,0}) of each particle, evaluate their
fitness values, and find the global best position G_0; Set n = 0.
   While (termination condition = false)
   Do
      Set n = n + 1;
      Choose a suitable value of w; (for PSO-In)
      for (i =1 to M)
        for j=1 to N
          V^j_{i,n+1} = V^j_{i,n} + c_1 r^j_{i,n}(P^j_{i,n} - X^j_{i,n}) + c_2 R^j_{i,n}(G^j_n - X^j_{i,n}) (for the original PSO);
          (or V^j_{i,n+1} = wV^j_{i,n} + c_1 r^j_{i,n}(P^j_{i,n} - X^j_{i,n}) + c_2 R^j_{i,n}(G^j_n - X^j_{i,n}) (for PSO-In));
          (or V^j_{i,n+1} = χ[V^j_{i,n} + c_1 r^j_{i,n}(P^j_{i,n} - X^j_{i,n}) + c_2 R^j_{i,n}(G^j_n - X^j_{i,n})] (for PSO-Co));
          if V^j_{i,n+1} > V_{max}
            V^j_{i,n+1} = V_{max};
          end if
          if V^j_{i,n+1} < -V_{max}
            V^j_{i,n+1} = -V_{max};
          end if
          X^j_{i,n+1} = X^j_{i,n} + V^j_{i,n+1};
        end for
        Evaluate the fitness value of X_{i,n+1}, that is, the objective
        function value f(X_{i,n+1});
        Update P_{i,n} and G_n
      end for
   end do
end
```

FIGURE 2.2 The procedure of the QPSO algorithm.

process carries on until the termination condition is met. At each iteration of the search process, the current position of each particle is updated as according to Equations 2.9, 2.11, or 2.12. Each component of the new velocity vector should be confined within $[-V_{max}^j, V_{max}^j]$ $(1 \leq j \leq N)$. Each component of the current position vector is updated according to Equation 2.10. After that, the fitness value of the new present position is evaluated. Then its personal best position and the global best position are updated according to Equations 2.7 and 2.8, respectively, before the next iteration. The whole procedure of the PSO algorithm is described in the algorithm as shown in Figure 2.2.

2.4 PARADIGM: HOW TO USE PSO TO SOLVE OPTIMISATION PROBLEMS

2.4.1 Simple Benchmark Problem

A simple example is used to demonstrate the implementation of the PSO algorithm for applications in real-world problems. This example concerns the minimisation of the sphere function as examined by the original developers of the original PSO, PSO-In and PSO-Co algorithms. The definition of the sphere function is given by

$$f_1(x) = \sum_{i=1}^{N} x_i^2 \quad (-100 \leq x_i \leq 100, 1 \leq i \leq N) \tag{2.14}$$

The domain of the search process for the sphere function in each dimension is given by the interval $[-100, 100]$. The global minimum of the function is $x^* = (0, 0, \ldots, 0)$ and $f_1(x^*) = 0$. This function is a simple and strongly convex function and has been widely used in the development of the theory of evolutionary strategies [104] and in the evaluation of GAs as part of the test set proposed by De Jong [105]. Figure 2.3 shows the picture of the sphere function in two dimensions. Note that MATLAB® script codes are used in this book for the implementation of various algorithms.

The source code of the sphere function is shown in Figure 2.4 and is kept in the file sphere.m. Line 1 of the source code provides the definition of the function. The name of the function must be the same as the file name, so that it can be called by citing the function name in other calling routines. Line 2 is the main body of the function which calculates the function value of Equation 2.14. If matrix operations were not used, line 3 to line 8 should replace line 2.

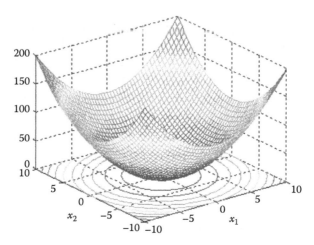

FIGURE 2.3   The sphere function.

```
%%%%%%%%%%%%%%%%%%%%%%%%%%%%%%%%%%%%%%%%%%%%%%%%%%%%%%%%%%%%%%%%%%%%%%%%%%%%
%                                          %
% The source codes for sphere function, where the variable x is
a vector %
%                                          %
% %%%%%%%%%%%%%%%%%%%%%%%%%%%%%%%%%%%%%%%%%%%%%%%%%%%%%%%%%%%%%%%%%%%%%%%
1. function y=sphere(x)

2. y=x*x'; % This line is a matrix operation.

%%%%%%%The above line can be replaced by the following lines
%%%%%%%%
3.% d=length(x);
4.% sum=0;
5.% for j=1:d
6.%    sum=sum+x(j)*x(j);
7.% end
8.% y=sum;
```

FIGURE 2.4   Source codes in MATLAB for the sphere function.

2.4.2 Implementation of the PSO Algorithm

Figure 2.5 presents the script code of the PSO algorithms and is saved in the file PSO.m. The script code can be roughly divided into two parts. The first part, including lines from line 1 to line 14, involves the definitions of parameters for the problem and the algorithm. The first line defines the population size M of the particle swarm, which is denoted as 'popsize'. For the sphere function the population size is set as 20 in this case. Line 2 specifies the maximum number of iterations, MAXITER, as 3000. The search domain of the particle in each dimension is defined

```
%%%%%%%PSO.m (Source code of PSO for the Sphere Function) %%%%%%%%%%
1. popsize=20; % population size
2. MAXITER=3000; % Maximum number of iterations
3. dimension=30; % Dimensionality of the problem
4. xmax=100;    % Upper bound of the search scope
5. xmin=-100;   % Lower bound of the search scope
6. irange_l=-100; % Lower bound of initialization scope
7. irange_r=100; % Upper bound of initialization scope
8. M=(xmax-xmin)/2; % The middle point of the search cope on
each dimension
9. vmax=xmax;   % The limit of the absolute particle velocity
10. c1=2; c2=2; % Set the values of the acceleration coefficients.
For PSO-Co, the both values are 2.05
11. k=0.729; % Set the value of constriction factor for PSO-Co
12. runno=50; % The runno is the times that the algorithm runs
13. data=zeros(runno, MAXITER); % The matrix record the fitness
value of gbest position at each iteration
%%%%%%%%%%%%%The following is that the algorithm runs for runno
times%%%%%%%%%%%%
14. for run=1:runno
%%%%%%%%%%%%%%%% Initialization of the particle swarm %%%%%%%%%%%%
%%%%%
15. x = (irange_r-irange_l)*rand(popsize,dimension,1) +
irange_l; % Initialize the particle's position
16. pbest = x;          %Set the pbest position as the current
position of the particle
17. gbest=zeros(1,dimension); % Initialize the gbest poistion
vector

18. for i=1:popsize
19.   f_x(i)=sphere(x(i,:)); %Calculate the fitness value of the
current position of the particle
20.   f_pbest(i)=f_x(i);    % Set the fitness value of the pbest
position to be that of the current position
21. end

22. g=min(find(f_pbest==min(f_pbest(1:popsize)))); % Find index
of the particle with gbest position
23. gbest=pbest(g,:);    % Determine the gbest position
24. f_gbest=f_pbest(g); % Determine the fitness value of the
gbest position
%%%%%%%%%%%%%% The following is the loop of the PSO's search
process %%%%%%%%%%%
25. for n=1:MAXITER
26.   w=(0.9-0.4)*(MAXITER-n)/MAXITER+0.4; % Determine the value
of alpha
```

FIGURE 2.5   The MATLAB source code for the PSO (including the original PSO, PSO-In and PSO-Co) algorithm.

(continued)

```
27.  for i=1:popsize %The following is the update of the
particle's position
28.    v(i,:)=v(i,:)+c1*rand(1,dimension).*(pbest(i,:)-
x(i,:))+c2*rand(1,dimension).*(gbest-x(i,:)));
29. %  v(i,:)=w*v(i,:)+c1*rand(1,dimension).*(pbest(i,:)-
x(i,:))+c2*rand(1,dimension).*(gbest-x(i,:)));
30. %  v(i,:)=k*(v(i,:)+c1*rand(1,dimension).*(pbest(i,:)-
x(i,:))+c2*rand(1,dimension).*(gbest-x(i,:))));
31.    v(i,:)=sign(v(i,:)).*min(abs(v(i,:)),vmax);
32.    x(i,:)=x(i,:)+v(i,:);
33. % x(i,:)=x(i,:)-(xmax+xmin)/2;% These tree lines are to
restrict the position in search scopes
34. % x(i,:)=sign(x(i,:)).*min(abs(x(i,:)),M);
35. % x(i,:)=x(i,:)+(xmax+xmin)/2;

36. f_x(i)=sphere(x(i,:)); % Calculate the fitness value of the
particle's current position
37. if (f_x(i)<f_pbest(i))
38.  pbest(i,:)=x(i,:); % Update the pbest position of the
particle
39.  f_pbest(i)=f_x(i); % Update the fitness value of the
particle's pbest position
40. end
41. if f_pbest(i)<f_gbest
42.    gbest=pbest(i,:); % Update the gbest position
43.    f_gbest=f_pbest(i); % Update the fitness value of the
gbes position
44.    end
45.  end
46.  data(run, n)=f_gbest; % Record the fitness value of the gbest
at each iteration at this run
47.  end
48. end
```

FIGURE 2.5 (continued)

in line 4 and line 5 where the upper bound 'x_{max}' and lower bound 'x_{min}' are set to be 100 and −100, respectively for the sphere function. Lines 7 and 8 are the initialisation range of the particle positions. Here, the upper bound and lower bound of the initialisation range are set as those of the search domain. In line 8, the middle point of the search domain is obtained for restricting the particle's position within the search domain in line 34. The variable 'x_{max}' denotes the limit of the absolute velocity, 'v_{max}', and is set to be equal to the upper bound of the search domain as shown in line 9. In line 10, the acceleration coefficients 'c_1' and 'c_2' are both fixed at 2.0 for the original PSO and PSO-In. For PSO-Co, these two values should be chosen as 2.05. Line 11 suggests that the value of constriction factor should be fixed at 0.729 for PSO-Co when the algorithm is running. In line 12, the variable 'runno' is the number of runs of each algorithm. In this example, each version of PSO runs for 50 times, with each run executing 3000 iterations. Line 13 involves initialisation of a matrix which records the fitness value of the global best position at each iteration for each single run.

The second part of the source code, from line 14 to line 48, is the main body of the PSO algorithm. Line 14 dictates the algorithm to run for 'runno' (=50) times. From line 15 to line 24 the particle swarm is initialised. In this MATLAB implementation of PSO, the current position of each particle is recorded in the matrix 'x' with the number of rows being 'popsize' and the number of columns being 'dimension'. Thus, '$x(i,:)$' means the current position of particle i and '$x(i,j)$' indicates the jth component of the current position of particle i. The personal best position of each particle is recorded in the matrix '$pbest$' of the same size as 'x'. Line 15 initialises the components of the current position vector using a random number uniformly distributed within the search domain. Note that other probability distributions can also be used depending on the actual problem. Line 16 concerns the initialisation of the personal best position of each particle. Generally, the initial personal best position of the particle is set to its initial current position. Line 17 initialises the global best position vector. From line 18 to line 21, the fitness value (objective function value) of the current position of each particle is calculated by calling the file sphere.m and the fitness value of the personal best position is set to be that of its current position. The index of the particle with the best fitness value of personal best positions, 'g', is obtained as in line 23, which follows with the global best position being set as the personal best position of the gth particle, that is, '$pbest(g,:)$', according to line 24. The next line sets the fitness value of global best position, that is, '$pbest(g,:)$'.

Line 25 and the subsequent lines form the loop of the PSO search process which lasts MAXITER iterations in a single run. If PSO-In is used in each iteration, the inertia weight 'w' should be computed as according to line 26, that is, decreasing linearly from 0.9 to 0.4 during the course of the simulation. From line 27 to line 45, for each particle, the velocity and the current position are adjusted, the fitness value of each particle is evaluated and the global best position is then updated. Lines 28, 29 and 30 represent the update equation of the velocity vector of the particle in the original PSO, PSO-In and PSO-Co, respectively. When executing the source code, the reader should select one of these lines for a particular version of PSO and comment on the

```
%%%%%The following source codes are the for update of particle's
position in each dimension%%%%%
1. for d=1: dimension
2.    v(i,d)=v(i,d)+c1*rand*(pbest(i,d)-x(i,d))+c2*rand*(gbest-
x(i,d)); % For Original PSO
3. % v(i,d)=w*v(i,d)+c1*rand*(pbest(i,d)-x(i,d))+c2*rand*(gbest-
x(i,d)); % For PSO-In
4. % v(i,d)=k*(v(i,d)+c1*rand*(pbest(i,d)-x(i,d))+c2*rand*(gbest-
x(i,d))); %For PSO-Co
5. % if v(i,d)<-vmax
          v(i,d)=-vmax;
       if v(i,d)>vmax
          v(i,d)=vmax;
       end
       x(i,d)=x(i,d)+v(i,d)
10. if x(i,d)>xmax;
11.    x(i,d)=xmax; % If the component of the position larger
than xmax, set it to be xmax
12. end
13. if x(i,d)<-xmin;
14.    x(i,d)=-xmin; % If the component of the position smaller
than xmin, set it to be xmin
15.  end
16. end
```

FIGURE 2.6 The source code in MATLAB for adjusting the velocity and position of a particle in each dimension.

other two lines with mark '%' in the front of each line. The three lines can be replaced by the codes in Figure 2.6, which update the particle's velocity and position in a loop for dimensions instead of in matrixes. Each component of the velocity vector should be restricted within $[-v_{max}, v_{max}]$ as shown in line 31. The update of the current position of the particle is done by adding its updated velocity to its current position at previous iteration. Lines 33, 34 and 35 are executed to restrict the current positions of the particles within the search domain. However, it has been indicated by many researchers that if the velocity of each particle is confined within $[-v_{max}, v_{max}]$, the restriction for the position is not necessary. Indeed, the algorithm may generate very poor results in some cases if the position of a particle is restricted. Therefore, it is suggested that when the PSO algorithm is used to optimise the sphere function and other benchmark functions presented in the next section, the particle's current position need not be restricted—let the particle fly freely!

Line 36 evaluates the current position of a particle. Again the file sphere.m is called. According to the procedure of PSO, the next step is to update the personal best position of the particle as implemented in lines 37–40. For minimisation problems when the fitness value of the current position of the particle is smaller than that of its previous personal best position, the personal best position and its fitness value should be set as the present position and the corresponding fitness value, respectively. If the condition in line 37 is not met, the personal best position is not updated. Lines 41–44 are used to update the global best position and its fitness value. It should be noted that they should be renewed after the update of each particle. In other words, even within the same iteration, different particles may use different global best positions to adjust their velocities in lines 28–30. For example, particle i has used the current global best position to determine its velocity by using one of the lines 28–30 and to update its current position. If the fitness value of its new current position is better than those of the personal best position and the global best position, then the global best position is updated. For the next particle, that is, particle $i + 1$, it uses the new global best position to calculate its velocity according to line 28, 29 or 30. This implies that all the particles interchange information among themselves during—instead of after—each iteration. The task of line 46 is to record the fitness value of the global best position after each iteration of a single run for the purpose of data processing and result analysis.

The readers may wish to create and type their own .m files as those shown in Figures 2.3 and 2.4. These files are also available in the CD accompanying this book.

2.4.3 Results and Analysis

2.4.3.1 Results

The program was executed for the versions of the original PSO, PSO-In and PSO-Co by commenting out suitable lines 28, 29 or 30, and 50 cases of the final fitness values for each version of PSO were obtained. The final fitness value for each run of a PSO algorithm was obtained after 3000 iterations of the search process. The average of the 50 best fitness values is known as the mean best fitness value. Note that in the code, the fitness value of the global best position after each iteration during a single run is recorded in the matrix 'data' and the best fitness value obtained in the ith ($1 \leq i \leq$ runno = 50) run of the algorithm is recorded in 'data(i, MAXITER)'. Thus the mean best fitness value can be obtained by executing the command 'mean(data(:,MAXITER))' in the command window of MATLAB. The reader may also run the command 'std(data(:, MAXITER))' to calculate the standard deviation of the 50 best fitness values for further comparative analysis.

Table 2.1 lists the results yielded after 50 runs of each of the three PSO algorithms for the sphere function. It is observed that the original PSO

TABLE 2.1   Mean and Standard Deviation of the Best Fitness Values
Obtained after 50 Runs of Each Algorithm for the Sphere Function

Algorithms	Mean Best Fitness Value	Standard Deviation of the Best Fitness Values
Original PSO	3.289×10^4	3.746×10^3
PSO-In	3.5019×10^{-14}	4.8113×10^{-14}
PSO-Co	8.6167×10^{-38}	3.1475×10^{-37}

generated very poor results even though the sphere function is a uni-modal function, while PSO-In and PSO-Co generated comparatively better results than the original PSO. In particular, the mean best fitness value of using PSO-Co was obtained as 8.6167×10^{-38} implying that the solution is extremely close to the optimum.

2.4.3.2 Statistical Comparison

It should be noted that a direct comparison between two mean best fitness values may be misguiding in most cases if the algorithms are all of stochastic nature. In other words, it would be sensible to perform a set of experiments in order to obtain a good sample of the best fitness values. Statistical tests such as estimating the confidence intervals and testing statistical hypothesis for the difference of means should be performed. An example of comparing PSO-Co and PSO-In is given as follows.

Samples of the difference of means from the tests obtained by using PSO-Co and PSO-In are used in the statistical test. The aim here is to determine the 95% confidence interval for the difference of the means. Using the mean values of the best fitness values generated by PSO-Co and PSO-In as listed in Table 2.1 one obtains their difference as -3.5019×10^{-14}. The standard error is obtained by using the formula

$$s = \sqrt{\frac{s_1^2}{n_1} + \frac{s_2^2}{n_2}}, \tag{2.15}$$

where
   s_1 and s_2 are standard deviations of the two samples
   n_1 and n_2 are sample sizes of the two samples

In this example, s_1 and s_2 are 3.1475×10^{-37} and 4.8113×10^{-14} as listed in Table 2.1, and $n_1 = n_2 = 50$. Hence, the computed standard error is 6.8042×10^{-15}. Finally, the 95% confidence interval for the difference of two means can be calculated according to

$$\overline{X}_1 - \overline{X}_2 \pm t_{0.025}(n_1 + n_2 - 1)s, \qquad (2.16)$$

where

$\overline{X}_1 - \overline{X}_2$ is the difference of two means

$t_{0.025}(n_1 + n_2 - 1)$ is the t value for two-tailed t-test of 95% confidence
level with the number of degrees of freedom $n_1 + n_2 - 1$

Since $n_1 + n_2 - 1$ is equal to 99, which is larger than 30, the two samples
are considered as large samples and t distribution can be approximated
by normal distribution. Thus, $t_{0.025}(n_1 + n_2 - 1)$ can be replaced by $z_{0.025}$
which is found to be 1.96 according to statistical tables. Equation 2.16 can
be written as

$$\overline{X}_1 - \overline{X}_2 \pm 1.96s. \qquad (2.17)$$

Hence, the 95% confidence interval for the difference of the two means is
$[-4.8355 \times 10^{-14}, -2.1683 \times 10^{-14}]$.

Although the confidence interval provides some knowledge about the
difference of the mean best fitness values obtained by PSO-Co and PSO-In,
it does not answer the question: 'At the 0.05 significance level, is it reasonable to conclude that the mean best fitness value generated by PSO-Co
is smaller than that by PSO-In?' To answer this question, the following
hypothesis involving a five-step testing procedure should be performed.

Step 1: State the null hypothesis and the alternate hypothesis. The null
hypothesis is that there is no difference in the mean best fitness values
for the two algorithms. In other words, the difference of -3.5019×10^{-14}
between the mean best fitness values obtained by PSO-Co and PSO-In is
yielded by chance. The alternate hypothesis is that the mean best fitness
value is smaller for PSO-Co. Let μ_1 refer to the mean best fitness value
for PSO-Co and μ_0 the mean best fitness value for PSO-In. The null and
alternate hypotheses are

$$H_0 : \mu_1 \geq \mu_0$$

$$H_1 : \mu_1 < \mu_0$$

Step 2: Select the level of significance. The significance level is the probability that the null hypothesis is rejected while it is actually true. This
likelihood is determined prior to selecting the sample or performing any

calculations. The 0.05 and 0.01 significance levels, which correspond to 95% and 99% confidence levels, respectively, are the most commonly used values, but other values, such as 0.02 and 0.10, may also be used. In theory, it is possible to select any significance level value in between 0 and 1. In the numerical test in the following, the 0.05 significance level is selected.

Step 3: Determine the test statistic. Generally, for hypothesis test of difference of the means, *t* is used as the test statistic. However, for the cases of large samples (when the sample size is larger than 30), standard normal distribution (*z*) is used. In this case, the standard normal distribution (*z*) is used as the test statistic because the samples are large.

Step 4: Formulate a decision rule. The decision rule is based on the null and the alternate hypotheses (i.e. one-tailed test or two-tailed test), the level of significance, and the test statistic used. In this case, the hypothesis test is one tailed. The significance level value is chosen as 0.05 with the *z* distribution as the test statistic. The aim here is to determine whether the mean best fitness value is smaller for PSO-Co. Suppose the alternate hypothesis is set to indicate that the mean best fitness value is smaller for PSO-Co than for PSO-In. Hence, the rejection region is in the lower tail of the standard normal distribution. To find the critical value, place 0.05 of the total area in the lower tail. This means that −0.45 (−0.5 + 0.05) of the area is located between the *z* value of 0 and the critical value. Next, obtain the data of *z* distribution using standard statistics table for a value located near −0.45. This value is obtained as −1.65, and hence the decision is to reject H_0 if the value computed from the test statistic is smaller than −1.65.

Step 5: Make the decision regarding H_0 and interpret the result. Equation 2.18 is to be used to compute the value of the test statistic.

$$z = \frac{\bar{X}_1 - \bar{X}_2}{\sqrt{(s_1^2/n_1) + (s_2^2/n_2)}} = \frac{\bar{X}_1 - \bar{X}_2}{s}. \tag{2.18}$$

The computed value of *z* is $-3.5019 \times 10^{-14}/6.08042 \times 10^{-15} = -5.1467$, which is smaller than the critical value of −1.65. The decision is to reject the null hypothesis and accept the alternate hypothesis. The difference of -3.5019×10^{-14} between the mean best fitness values by PSO-Co and PSO-In is too large to have occurred by chance. It is now possible to conclude that the mean fitness value is smaller for PSO-Co than for PSO-In on sphere function at significance level of 0.05.

Moreover, it is also useful to find the p-value for the test statistic. It is the probability of finding a value of the test statistic at this extreme when the null hypothesis is true. To calculate the p-value, the probability of a z value smaller than -5.1467 is needed. Using Microsoft Excel the computed p-value is 3.1427×10^{-4}. Therefore, the p-value is less than 0.05 and it is possible to conclude that there is very little likelihood that the null hypothesis is true!

We can further define the significance such that if the p-value is smaller than 0.05 but larger than 0.01, the fact that $\mu_1 < \mu_0$ is significant; if the p-value is smaller than 0.01 but larger than 0.0001, the fact that $\mu_1 < \mu_0$ is very significant; if the p-value is smaller than 0.0001, the fact that $\mu_1 < \mu_0$ is extremely significant. Therefore, it is concluded that the mean best fitness value is smaller for PSO-Co than for PSO-In in a very significant manner.

Following the previous procedure, statistical comparison between PSO-Co and the original PSO, or between PSO-In and the original PSO, can be carried out. The results including differences of means, standard errors, 95% confidence intervals, z values, p-values and significances are all listed in Table 2.2. It is found that for sphere function, PSO-Co has the best performance, PSO-In has the second best performance, and the original PSO performed most poorly.

2.4.3.3 Comparison of Convergence Speed

The quality of the solution or the fitness value obtained is only one of the aspects that reflect the performance of the algorithms. There are other important factors that may be employed to evaluate the algorithms, including the convergence speed of the fitness values. To trace the convergence history of the PSO algorithms, the fitness values at each iteration over 50 runs are averaged and plotted. Figure 2.7, where the ordinate is in

TABLE 2.2 Results of Statistical Test between the PSO Algorithms

Compared Algorithms	PSO-Co versus PSO-In	PSO-Co versus Original PSO	PSO-In versus Original PSO
Difference of means	-3.5019×10^{-14}	3.289×10^{-14}	3.289×10^{-14}
Standard error	6.8042×10^{-14}	529.8210	529.8210
95% Confidence interval	$[-4.84 \times 10^{-14},$ $-2.17 \times 10^{-14}]$	$[-3.38 \times 10^4,$ $-3.20 \times 10^4]$	$[-3.38 \times 10^4,$ $-3.20 \times 10^4]$
z-Value	-5.1467	-62.0776	-62.0776
p-Value	3.1427×10^{-4}	<0.0001	<0.0001
Significance	Very significant	Extremely significant	Extremely significant

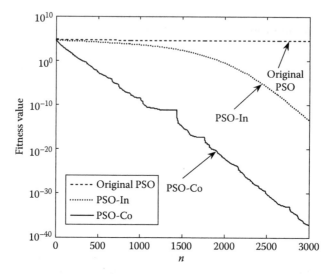

FIGURE 2.7 Convergence history of the fitness value averaged over 50 runs of each version of PSO.

logarithmic scale, traces the convergence history of each of the PSO versions for the sphere function. It can be observed that PSO-Co shows the fastest convergence speed among the three algorithms, while the original PSO is the slowest one.

2.4.3.4 Further Discussion
The sphere function is a well-known benchmark function for testing the local search ability of an optimisation algorithm. PSO-Co and PSO-In are shown to be able to find the optimum easily, but the original PSO is not able to do so. The poor performance of the original PSO may be due to either premature convergence or the slow convergence speed of the particles. Premature convergence of the algorithm is a result of excessively fast declining of the population diversity, while on the other hand, the slow convergence speed of the particles means slow decreasing of the population diversity.

The diversity of the swarm particles can be defined formally by using the average of the distances from the particles' current positions to the mean position [106], namely

$$\text{diversity } (S) = \frac{1}{|S||A|} \sum_{i=1}^{|S|} \sqrt{\sum_{j=1}^{N} (X_{i,n}^j - \bar{X}_n^j)^2}, \qquad (2.19)$$

where

S represents the swarm of particles

$|S| = M$ is the population size

$|A|$ is the length of the longest diagonal in the search space

N is the dimensionality of the problem

$X^j_{i,n}$ is the jth component of the ith particle's position at the nth iteration

X^j_n is the jth component of the mean position

In the experiment for the sphere function, $|S| = 20$ and $|A| = (X_{max} - X_{min})\sqrt{N} = 200\sqrt{30}$ (Table 2.3).
The diversity of the particle swarm at each iteration during each run of the algorithm is recorded. Table 2.3 lists the means of the initial diversity measures and the final diversity measures at the end of the search process. The decreasing process averaged over 50 runs of each PSO is depicted in Figure 2.8. Note that the diversity measure

TABLE 2.3   Initial and Final Diversity Measure in the PSO Algorithms

Algorithms	Initial Diversity	Final Diversity
Original PSO	0.2820	0.2888
PSO-In	0.2797	4.1872×10^{-10}
PSO-Co	0.2792	1.6607×10^{-22}

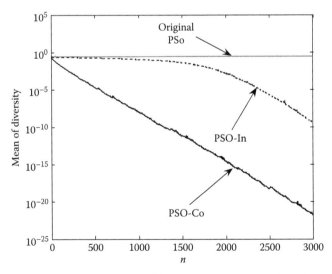

FIGURE 2.8   Convergence history of the diversity measure for sphere function averaged over 50 runs of each version of PSO.

decreased rapidly during the search process for PSO-Co or PSO-In. The initial diversity measures were 0.2797 and 0.2792, respectively for the two algorithms, and at the end of the search the diversity measures are 4.1872×10^{-10} and 1.6607×10^{-22}, respectively. Hence, PSO-Co and PSO-In have good local search ability and the iterative processes are able to reach solutions very close to the optimum. However, for the original PSO, the diversity measure in this numerical experiment did not show significant changes during the search process. The initial and final values were obtained as 0.2820 and 0.2888, respectively.

For a population-based random search technique, such as PSO, the diversity measure is a very important factor in order to evaluate the global search and local search abilities of the algorithm. When the diversity measure is large, the global search ability of the algorithm is relatively stronger and the local search ability is weak, and vice versa. Thus, the results of the diversity measure for the original PSO imply that it converges very slowly, that is, its local search ability is poor. Note also that the trade-off between local search and global search is vital for the performance of the algorithm. It seems that PSO-Co and PSO-In are able to balance the two kinds of search abilities effectively.

2.5 SOME HARDER EXAMPLES

2.5.1 Benchmark Problems

2.5.1.1 Introduction

The sphere function is a uni-modal function usually used to test the local search ability of an algorithm. It is inconclusive of the overall performance of the algorithms using the sphere function only. In the field of metaheuristics, it would be better to compare different algorithms using a large test set of functions for optimisation [107]. However, the effectiveness of an algorithm against another algorithm cannot be measured by simply taking the number of problems that are effectively solved. The 'no free lunch' theorem [108] tells that, if two searching algorithms are compared with all possible functions, the performance of any two algorithms will be the same on average. As a result, it is probably a fruitless task to attempt to design a perfect test set containing as many functions as possible in order to determine whether an algorithm is better than another for every function. Whitley et al. [109] and Salomon [110] made a comprehensive study of optimising functions in order to construct a suitable test set containing a better selection of functions. Using this test set they concluded that the performance of an algorithm depends on the type of function. Eiben and Bäck [111] has provided such test set. It has several well-characterised functions that allow us to obtain and generalise, as far as possible, the results regarding the kind of function involved. Nevertheless, most researchers in the field of evolutionary algorithms would prefer to add two further functions to their own test set aiming to balance the number of functions of each kind. These two functions are the function of Rosenbrock [112] extended to N dimensions and the function of Schwefel [113]. Both functions have been widely used in evolutionary optimisation literature.

Each function can be characterised by separability, multi-modality, regularity and dimensionality. If a function has more than one local optimum, it is known to be multi-modal. If a function of N variables can be rewritten as the sum of N functions of just one variable, it is said to be separable [114]. The separability of a function has a close relationship with the concept of epistasis or interrelation among the variables of the function. In the research area of evolutionary algorithms, the concept of epistasis is generally used to measure how much the contribution of a gene (or an individual) to the fitness of the individual depends on other genes.

If a function is non-separable, it is usually more difficult to optimise, since the accurate search direction depends on two or more genes. On the other hand, separable functions can be optimised for each variable in turn. If a function is multi-modal, the problem becomes more difficult since the search process must be able to avoid the regions around local minima so as to approximate the global optimum as far as possible. The most complex case appears when the local optima are not regularly distributed in the search space. In this case, the function is known to be irregular.

The dimensionality of the search space is, of course, another important factor in measuring the complexity of the problem. A problem with higher dimensionality is generally harder than the problem of the same form with lower dimensionality. For example, a high-dimensional sphere function is more difficult to optimise than a low-dimensional one. In [115], Friedman carried out a comprehensive study of the dimensionality problem and its features. In order to establish the same degree of difficulty in all the problems, we chose a search space of dimensionality $N = 30$ for most of the functions in the experiments relevant to this book.

2.5.1.2 Testing Functions

Eight well-known benchmark functions have been used to test the performance of the three versions of PSO algorithms. The first function is the
sphere function as described in Section 2.5.1.1. The other functions used
in the test set are described as follows.

Schwefel's Problem 1.2:

$$f_2(x) = \sum_{i=1}^{N} \left(\sum_{j=1}^{i} x_j \right)^2 \quad (-100 \le x_i \le 100, 1 \le i \le N). \qquad (2.20)$$

Schwefel's Problem 1.2 (Figure 2.9) is also known as Schwefel's double sum
function [116]. This function is peculiar because its gradient is not ori-
ented along its axis due to the epistasis among their variables. Algorithms
that use the gradient would converge slowly. The search domain for this
function in each dimension is [−100, 100]. The global minimum of the
function is $x^* = (0, 0, \ldots, 0)$ and the $f_2(x^*) = 0$.

Rosbrock function:

$$f_3(x) = \sum_{i=1}^{N-1} (100 \cdot (x_{i+1} - x_i^2)^2 + (x_i - 1)^2) \quad (-100 \le x_i \le 100, 1 \le i \le N).$$

$$(2.21)$$

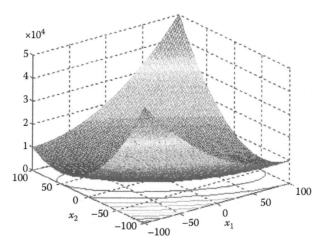

FIGURE 2.9   Schwefel's Problem 1.2 ($N = 2$).

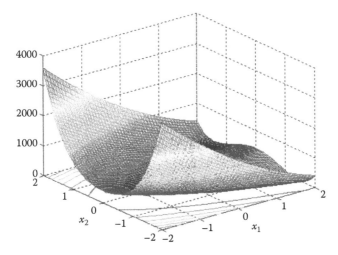

FIGURE 2.10   Rosenbrock function ($N = 2$).

Rosenbrock function [112], or De Jong's function F2, is a 2D function with a deep valley with the shape of a parabola of the form $x_1^2 = x_2$ that leads to the global minimum. Due to the non-linearity of the valley, many algorithms converge slowly because they change the direction of the search repeatedly. The extended version of this function was proposed by Spedicato [117]. Other versions have been proposed in [118,119]. It is considered by many authors as a challenge for any optimisation algorithm [120]. Its difficulty is mainly due to the non-linear interaction among its variables. The search domain for the function in each dimension is [−100, 100]. The global minimum of the function is $x^\star = (1, 1, ..., 1)$ and the $f_3(x^\star) = 0$ (Figure 2.10).

Rastrigin function:

$$f_4(X) = \sum_{i=1}^{N} (x_i^2 - 10 \cdot \cos(2\pi x_i) - 10) \quad (-5.12 \leq x_i \leq 5.12, 1 \leq i \leq N).$$

$$(2.22)$$

Rastrigin function [121] is constructed by adding a modulator term, $\alpha \cdot \cos(2\pi x_i)$, to the sphere function. Its contour is made up of a large number of local minima whose value increases with the distance to the global minimum. The search domain for the function on each dimension is [−5.12, 5.12]. The global minimum of the function is $X^\star = (0, 0, ..., 0)$ and the $f_4(X^\star) = 0$ (Figure 2.11).

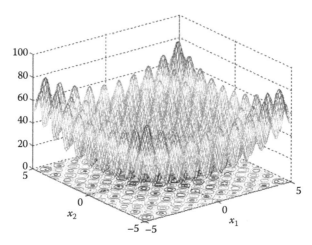

FIGURE 2.11 Rastrigin function ($N = 2$).

Ackley function:

$$f_5(x) = -20\exp\left\{-0.2\sqrt{\left(\frac{1}{N}\right)\sum_{i=1}^{N} x_i^2}\right\}$$

$$-\exp\left\{\left(\frac{1}{N}\right)\sum_{i=1}^{N}\cos(2\pi x_i)\right\} + 20 + e \quad (32 \le x_i \le 32, 1 \le i \le N).$$

(2.23)

Ackley function, originally proposed by Ackley [122] and generalised by Bäck and Schwefel [123], has an exponential term that covers its surface with numerous local minima. It should be noted that a search strategy based on a gradient or steepest descent algorithm will be trapped into a local optima, but if a search strategy can cover a wider region, it will be able to cross the valley among the optima and achieve better results. The search domain for the function on each dimension is [−32, 32]. The global minimum of the function is $x^* = (0, 0, \ldots, 0)$ and the $f_5(x^*) = 0$ (Figure 2.12).

Griewank function:

$$f_6(x) = \left(\frac{1}{4000}\right)\sum_{i=1}^{N} x_i^2 - \prod_{i=1}^{N}\cos\left(\frac{x_i}{\sqrt{i}}\right) + 1 \quad (600 \le x_i \le 600, 1 \le i \le N).$$

(2.24)

Griewank function [124] has a product term that introduces interdependence among the variables. The aim is the failure of the techniques that

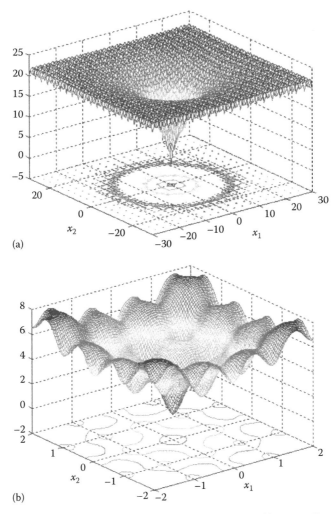

(a)

(b)

FIGURE 2.12   Ackley function ($N = 2$) (a) in large scope; (b) in small scope.

optimise each variable independently. As in Ackley function, the optima of Griewangk function are regularly distributed. The search domain for the function on each dimension is [−600, 600]. The global minimum of the function is $x^* = (0, 0, \ldots, 0)$ and the $f_6(x^*) = 0$ (Figure 2.13).

Schaffer's F6 function:

$$f_7(x) = 0.5 + \frac{\left(\sin\sqrt{x_1^2 + x_2^2}\right)^2 - 0.5}{\left(1.0 + 0.001\left(x_1^2 + x_2^2\right)\right)^2} \quad (-100 \le x_i \le 100, 1 \le i \le 2). \quad (2.25)$$

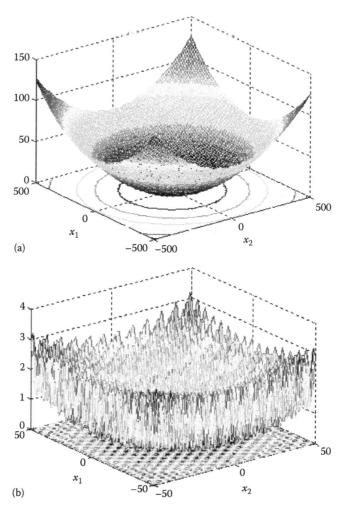

FIGURE 2.13 Griewank function ($N = 2$) (a) in large domain; (b) in small domain.

This function, proposed by Schaffer [124], is a 2D, multi-modal and non-linear function with the global minimum $X^* = (0,0)$ and $f_7(X^*) = 0$. An interesting feature of this function is that there are infinite number of local minima within the circle of radius π centred at $X^* = (0,0)$. The function value vibrates violently when the decision variable vector is close to the origin, making the function very difficult to optimise. For this problem, the search domain in each dimension is [−100, 100] (Figure 2.14).

Schwefel function:

$$f_8(x) = 418.9829 \cdot N + \sum_{i=1}^{N} x_i \sin(\sqrt{|x_i|}) \quad (600 \le x_i \le 600, 1 \le i \le N). \quad (2.26)$$

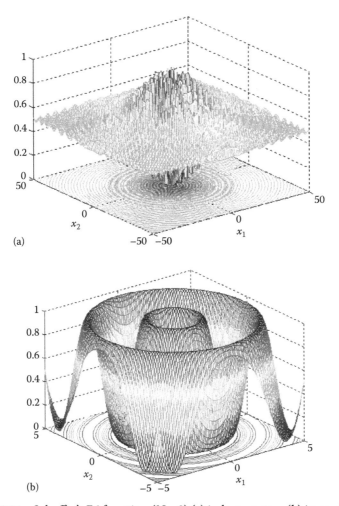

FIGURE 2.14 Schaffer's F6 function ($N = 2$) (a) in large scope; (b) in small scope.

The surface of Schwefel function [113] is composed of a large number of peaks and valleys. The function has a second best minimum far from the global minimum at $X^\star = (-420.9687, \ldots, -420.9687)$, where many search algorithms are trapped. Moreover, the global minimum is near the bounds of the domain which is the interval [600, 600] on each dimension (Figure 2.15).

The features of each benchmark function are summarised in Table 2.4.

2.5.2 Results on Benchmark Problems

The three versions of PSO were tested on the eight benchmark functions with the algorithmic parameters configured as those for the sphere

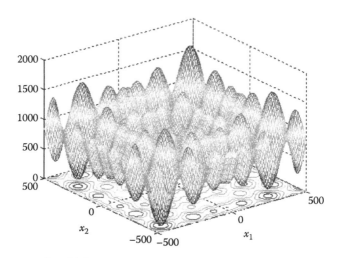

FIGURE 2.15   Schwefel function.

TABLE 2.4   The Features of Each Function

Benchmark Functions	Multi-Modality	Separability	Regularity
Sphere function (f_1)	No	Yes	N/A
Schwefel function 1.2 (f_2)	No	No	N/A
Rosenbrock function (f_3)	No	No	N/A
Rastrigin function (f_4)	Yes	Yes	N/A
Ackley function (f_5)	Yes	No	Yes
Griewank function (f_6)	Yes	Yes	Yes
Schaffer's F6 function (f_7)	Yes	No	N/A
Schwefel function (f_8)	Yes	Yes	N/A

function. The mean best fitness values and the standard deviation of the best fitness values over 50 independent runs of each algorithm are listed in Table 2.5. The results for the sphere function are recorded from the previous section for the purpose of overall performance comparison. It is demonstrated from the results that the poorest performing algorithm on all benchmark function is the original PSO. For the sphere function (f_1), PSO-Co generated better results than the original PSO and PSO-In. For Schwefel's Problem 1.2 (f_2), PSO-Co also performed better than the other variants. For Rosenbrock function (f_3), PSO-Co was also the winner. For Rastrigin function (f_4), PSO-In yielded the best results. For Ackley function (f_5), PSO-In obtained better quality of the average optimum solution than any other algorithm. For Griewank function (f_6), the performance of PSO-Co was superior to those of the other PSO algorithms. The results for Schaffer's F6 function (f_7) show that PSO-In outperformed the original

TABLE 2.5   Results for Each Benchmark Function Obtained by Each Algorithm

Benchmark Functions	Algorithms	Mean Best Fitness	Standard Deviation
Sphere function (f_1)	Original PSO	3.2890×10^4	3.7464×10^3
	PSO-In	3.5019×10^{-14}	4.8113×10^{-14}
	PSO-Co	8.6167×10^{-38}	3.1475×10^{-37}
Schwefel's Problem 1.2 (f_2)	Original PSO	6.0089×10^4	9.4277×10^3
	PSO-In	419.2338	241.8106
	PSO-Co	0.0158	0.0404
Rosenbrock function (f_3)	Original PSO	8.3530×10^9	2.1263×10^9
	PSO-In	201.1325	423.1043
	PSO-Co	57.9633	91.3910
Rastrigin function (f_4)	Original PSO	333.2886	20.7637
	PSO-In	39.8305	11.5816
	PSO-Co	73.5472	19.4169
Ackley function (f_5)	Original PSO	19.2673	0.3151
	PSO-In	0.1096	0.3790
	PSO-Co	2.9731	1.9449
Griewank function (f_6)	Original PSO	297.0924	33.8384
	PSO-In	0.0139	0.0151
	PSO-Co	0.0479	0.0741
Schaffer's F6 function (f_7)	Original PSO	0.0093	0.0019
	PSO-In	1.9432×10^{-4}	0.0014
	PSO-Co	0.0035	0.0047
Schwefel function (f_8)	Original PSO	6.1959×10^3	454.1043
	PSO-In	4.0872×10^3	743.9391
	PSO-Co	3.7826×10^3	620.7381

PSO and PSO-Co, since it was able to find the solutions close to the optima more frequently. For Schwefel function (f_8), PSO-Co beat the original PSO and PSO-Co.

To compare the algorithms in a statistical manner, for each benchmark function, the pairwise comparisons were undertaken between the results of the algorithms listed in Table 2.5, as the statistical comparison for the sphere function. Then the performance of each algorithm on each function was ranked. For the sphere function, as indicated by the statistical test, PSO-Co, PSO-In and the original PSO had the best, second best and worst performances, respectively. Correspondingly, the ranks of PSO-Co, PSO-In and the original PSO are 1, 2 and 3, respectively. In a similar way, the ranking of the algorithms' performance on other benchmark function were performed and their ranks are listed in Table 2.6. It is shown

TABLE 2.6 Ranking by Algorithms and Problems

Algorithms	f_1	f_2	f_3	f_4	f_5	f_6	f_7	f_8	Total Rank
Original PSO	3	3	3	3	3	3	3	3	32
PSO-In	2	2	2	1	1	1	1	2	12
PSO-Co	1	1	1	2	2	2	2	1	12

that the total ranks of the original PSO, PSO-In and PSO-Co are 32, 12 and 12, respectively, implying that PSO-Co and PSO-In have the comparable overall performances, while the original PSO has the worst overall performance.

The convergence histories of each algorithm averaged over 50 runs for all the benchmark functions, except those for the sphere function as shown in the previous section, are depicted in Figures 2.16 through 2.22. For Schwefel's Problem 1.2 (f_2), the fastest converging algorithm during the whole search process is PSO-Co, and, consequently, the best solutions at the end of the search process were found. The convergence speed of PSO-In, though faster than that of the original PSO, was much slower in the early stage of the search process that failed to reach the solution close to the optima at the end of the running. The original PSO, for this function, exhibited the slowest convergence speed which means that its local search ability was poor.

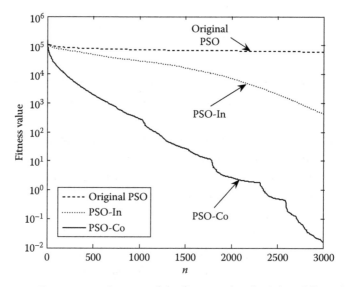

FIGURE 2.16 Convergence history of the fitness value for Schwefel's Problem 1.2 averaged over 50 runs of each version of PSO.

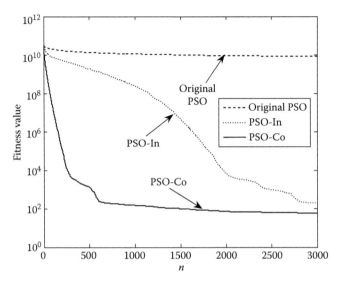

FIGURE 2.17   Convergence history of the fitness value for Rosenbrock function averaged over 50 runs of each version of PSO.

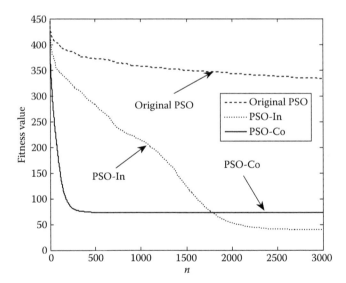

FIGURE 2.18   Convergence history of the fitness value for Rastrigin function averaged over 50 runs of each version of PSO.

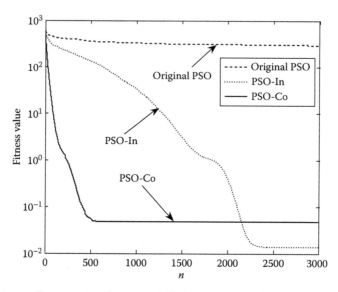

FIGURE 2.19 Convergence history of the fitness value for Griewank function averaged over 50 runs of each version of PSO.

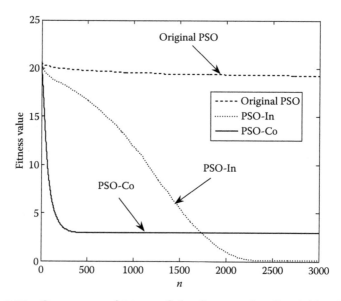

FIGURE 2.20 Convergence history of the fitness value for Ackley function averaged over 50 runs of each version of PSO.

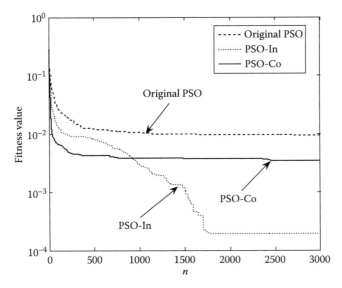

FIGURE 2.21 Convergence history of the fitness value for Schaffer's F6 function averaged over 50 runs of each version of PSO.

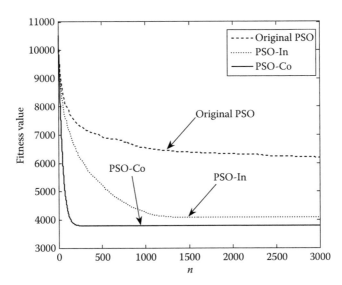

FIGURE 2.22 Convergence history of the fitness value for Schwefel function averaged over 50 runs of each version of PSO.

For Rosenbrock function (f_3), PSO-Co demonstrated the fastest convergence speed among the three versions of PSO, particularly in the early stage of the search. However, in the later stage, its convergence speed slowed down due to premature convergence which results in the sub-optimal solution. PSO-In converged relatively slowly during the early stage, and the convergence was accelerated in the later stage, but consequently found a worse result compared to PSO-In. For Rastrigin function (f_4), although PSO-Co converged fastest in the early stage of the search process, it encountered premature convergence during the middle and later stages. PSO-In appeared to converge slower during the early and middle stages of the search process, but in the later stage, it could make a further search and found a better solution. A similar result was seen in Griewank function (f_5), Ackley function (f_6) and Schaffer's F6 function (f_7). For Schwefel function (f_8), PSO-Co had the fastest convergence speed over the whole search process while the original PSO had the slowest convergence speed.

REFERENCES

1. G. Beni, J. Wang. Swarm intelligence in cellular robotic systems. In *Proceedings of the NATO Advanced Workshop on Robots and Biological Systems*, Tuscany, Italy, June 26–30, 1989.
2. J. Kennedy, R.C. Eberhart. Particle swarm optimization. In *Proceedings of the IEEE International Conference on Neural Networks*, Piscataway, NJ, 1995, vol. IV, pp. 1942–1948.
3. P.J. Angeline. Evolutionary optimization versus particle swarm optimization: Philosophy and performance differences. *Evolutionary Programming VII*, Lecture Notes in Computer Science, Springer-Verlag, Berlin, Germany, 1998, vol. 1447, pp. 601–610.
4. C.A. Coello, M.S. Lechuga. MOPSO: A proposal for multiple objective particle swarm optimization. In *Proceedings of the 2002 IEEE Congress on Evolutionary Computation*, Honolulu, HI, 2002, pp. 1051–1056.
5. J.E. Fieldsen, S. Singh. A multi-objective algorithm based upon particle swarm optimization, an efficient data structure and turbulence. In *Proceedings of the 2002 U.K. Workshop on Computational Intelligence*, Birmingham, U.K., 2002, pp. 34–44.
6. X. Hu. Multiobjective optimization using dynamic neighbourhood particle swarm optimization. In *Proceedings of the 2002 IEEE Congress on Evolutionary Computation*, Honolulu, HI, 2002, pp. 1677–1682.
7. K.E. Parsopoulos, M.N. Vrahtis. Particle swarm optimization method in multiobjective problems. In *Proceedings of the 2002 ACM Symposium on Applied Computing*, Madrid, Spain, 2002, pp. 603–607.

8. S. Mostaghim, J. Teich. Strategies for finding good local guides in multi-objective particle swarm optimization (MOPSO). In *Proceedings of the IEEE 2003 Swarm Intelligence Symposium*, Indianapolis, IN, 2003, pp. 26–33.

9. X. Li. A non-dominated sorting particle swarm optimizer for multiobjective optimization. In *Proceedings of the 2003 Genetic and Evolutionary Computation Conference*, Lecture Notes in Computer Science, Chicago, IL, 2003, vol. 2723, pp. 1611–3349.

10. E.C. Laskari, K.E. Parsopoulos, M.N. Vrahatis. Particle swarm optimization for minimax problems. In *Proceedings of the IEEE 2002 Congress on Evolutionary Computation*, Honolulu, HI, 2002, pp. 1582–1587.

11. Y. Shi, R.A. Krohling. Co-evolutionary particle swarm optimization to solve min-max problems. In *Proceedings of the IEEE 2002 Congress on Evolutionary Computation*, Honolulu, HI, 2002, pp. 1682–1687.

12. E.C. Laskari, K.E. Parsopoulos, M.N. Vrahatis. Particle swarm optimization for integer programming. In *Proceedings of the IEEE 2002 Congress on Evolutionary Computation*, Honolulu, HI, 2002, pp. 1576–1581.

13. X. Hu, R.C. Eberhart, Y. Shi. Swarm intelligence for permutation optimization: A case study of N-queens problem. In *Proceedings of the IEEE Swarm Intelligence Symposium*, Indianapolis, IN, 2003, pp. 243–246.

14. H.S. Lope, L.S. Coelho. Particle swarm optimization with fast local search for the blind traveling salesman problem. In *Proceedings of the Fifth International Conference on Hybrid Intelligent Systems*, Shenyang, China, 2005, pp. 245–250.

15. D.W. van der Merwe, A.P. Engelbrecht. Data clustering using particle swarm optimization. In *Proceedings of the IEEE Congress on Evolutionary Computation*, Barcelona, Spain, 2003, vol. 1, pp. 215–220.

16. S.C.M. Cohen, L.N. de Castro. Data clustering with particle swarms. In *Proceedings of the IEEE Congress on Evolutionary Computation*, Barcelona, Spain, 2006, pp. 1792–1798.

17. L.S. Oliveira, A.S. Britto, R. Sabourin. Improving cascading classifiers with particle swarm optimization. In *Proceedings of the Eighth International Conference on Document Analysis and Recognition*, Barcelona, Spain, 2005, vol. 2, pp. 570–574.

18. J. Nanbo, Y. Rahmat-Samii. Advances in particle swarm optimization for antenna designs: Real-number, binary, single-objective and multi-objective implementations. *IEEE Transactions on Antennas and Propagation*, 2007, 55(3): 556–567.

19. M. Donelli, R. Azaro, F.G.B. De Natale. An innovative computational approach based on a particle swarm strategy for adaptive phased-arrays control. *IEEE Transactions on Antennas and Propagation*, 2006, 54(3): 888–898.

20. M.M. Khodier, C.G. Christodoulou. Linear array geometry synthesis with minimum sidelobe level and null control using particle swarm optimization. *IEEE Transactions on Antennas and Propagation*, 2005, 53(8): 2674–2679.

21. K. Youngwook, L. Hao. Equivalent circuit modeling of broadband antennas using vector fitting and particle swarm optimization. In *Proceedings of the IEEE Antennas and Propagation International Symposium*, Albuquerque, NM, 2006, pp. 3555–3558.

22. J. Nanbo, Y. RahmatSamii. Real-number and binary multi-objective particle swarm optimizations: Aperiodic antenna array designs. In *Proceedings of the IEEE Antennas and Propagation Society International Symposium*, Honolulu, HI, 2006, pp. 3523–3526.

23. J.R. Perez, J. Basterrechea. Comparison of different heuristic optimization methods for near-field antenna measurements. *IEEE Transactions on Antennas and Propagation*, 2007, 55(3): 549–555.

24. R.C. Eberhart, X. Hu. Human tremor analysis using particle swarm optimization. In *Proceedings of the Congress on Evolutionary Computation*, Washington, DC, 1999, vol. 3, pp. 1927–1930.

25. L. Messerschmidt, A.P. Engelbrecht. Learning to play games using a PSO-based competitive learning approach. *IEEE Transactions on Evolutionary Computation*, 2004, 8(3): 280–288.

26. N. Franken, A.P. Engelbrecht. Particle swarm optimization approaches to coevolve strategies for the iterated prisoner's dilemma. *IEEE Transactions on Evolutionary Computation*, 2005, 9(6): 562–579.

27. N. Khemka, C. Jacob, G. Cole. Making soccer kicks better: A study in particle swarm optimization and evolution strategies. In *Proceedings of the IEEE Congress on Evolutionary Computation*, Barcelona, Spain, 2005, vol. 1, pp. 735–742.

28. A. Elgallad, M. El-Hawary, W. Phillips. PSO-based neural network for dynamic bandwidth re-allocation power system communication. In *Proceedings of the Large Engineering Systems Conference on Power Engineering*, Halifax, NS, 2002, pp. 98–102.

29. W. Mohemmed, N. Kamel. Particle swarm optimization for bluetooth scatternet formation. In *Proceedings of the Second International Conference on Mobile Technology, Applications and Systems*, Seattle, WA, 2005, pp. 162–166.

30. G. Zwe-Lee. A particle swarm optimization approach for optimum design of PID controller in AVR system. *IEEE Transactions on Energy Conversion*, 2004, 19(2): 384–391.

31. J.-G. Juang, B.-S. Lin, K.-C. Chin. Automatic landing control using particle swarm optimization. In *Proceedings of the IEEE International Conference on Mechatronics*, Keelung, Taiwan, 2005, pp. 721–726.

32. R.C. Eberhart, Y. Shi. Particle swarm optimization and its applications to VLSI design and video technology. In *Proceedings of the IEEE International Workshop on VLSI Design and Video Technology*, Suzhou, China, 2005, pp. xxiii–xxiii.

33. E.H. Luna, C.A. Coello, A.H. Aguirre. On the use of a population-based particle swarm optimizer to design combinational logic circuits. In *Proceedings of the NASA/DoD Conference on Evolvable Hardware*, Seattle, WA, 2004, pp. 183–190.

34. S. Baskar, A. Alphones, P.N. Suganthan. Design of Yagi-Uda antennas using comprehensive learning particle swarm optimization. *IEE Proceedings—Microwaves, Antennas and Propagation*, 2005, 152: 340–346.

35. W. Wang, Y. Lu, J.S. Fu. Particle swarm optimization and finite-element based approach for microwave filter design. *IEEE Transactions on Magnetics*, 2005, 41(5): 1800–1803.

36. M.A. Abido. Optimal design of power-system stabilizers using particle swarm optimization. *IEEE Transactions on Energy Conversion*, 2002, 17(3): 406–413.

37. S. Cui, D.S. Weile. Application of a parallel particle swarm optimization scheme to the design of electromagnetic absorbers. *IEEE Transactions on Antennas and Propagation*, 2005, 53(11): 3616–3624.

38. F.-Y. Huang, R.-J. Li, H.-X. Liu, R. Li. A modified particle swarm algorithm combined with fuzzy neural network with application to financial risk early warning. In *Proceedings of the IEEE Asia-Pacific Conference on Services Computing*, Guangzhou, China, 2006, pp. 168–173.

39. J. Nenortaite, R. Simutis. Adapting particle swarm optimization to stock markets. In *Proceedings of the Fifth International Conference on Intelligent Systems Design and Applications*, Washington, DC, 2005, pp. 520–525.

40. R. Senaratne, S. Halgamuge. Optimised landmark model matching for face recognition. In *Proceedings of the International Conference on Automatic Face and Gesture Recognition*, Southampton, U.K., 2006, pp. 120–125.

41. F. Juang. A hybrid of genetic algorithm and particle swarm optimization for recurrent network design. *IEEE Transactions on Systems, Man and Cybernetics, Part B*, 2004, 34(2): 997–1006.

42. S. Chandrasekaran, S.G. Ponnambalam, R.K. Suresh. A hybrid discrete particle swarm optimization algorithm to solve flow shop scheduling problems. In *Proceedings of the IEEE Conference on Cybernetics and Intelligent Systems*, Coimbatore, India, 2006, pp. 1–6.

43. Y. Yong, H. Zhihai, C. Min. Virtual MIMO-based cross-layer design for wireless sensor networks. *IEEE Transactions on Vehicular Technology*, 2005, 55(3): 856–864.

44. S. Janson, D. Merkle, M. Middendor. Molecular docking with multi-objective particle swarm optimization. *Applied Soft Computing Journal*, 2008, 8(1): 666–675.

45. J. Kennedy. The behavior of particle. In *Proceedings of the Seventh Annual Conference on Evolutionary Programming VII*, San Diego, CA, 1998, pp. 581–589.

46. E. Ozcan, C.K. Mohan. Particle swam optimization: Surfing the waves. In *Proceedings of the IEEE Congress Evolutionary Computation*, Barcelona, Spain, 1999, pp. 1939–1944.

47. M. Clerc, J. Kennedy. The particle swarm-explosion, stability and convergence in a multidimensional complex space. *IEEE Transactions on Evolutionary Computation*, 2002, 6(2): 58–73.

48. F. van den Bergh. An analysis of particle swarm optimizers. PhD dissertation, University of Pretoria, Pretoria, South Africa, 2002.

49. R.C. Eberhart, Y. Shi. Parameter selection in particle swarm optimization. In *Proceedings of the Seventh Conference on Evolutionary Programming*, San Diego, CA, 1998, pp. 591–600.

50. C. Trelea. The particle swarm optimization algorithm: Convergence analysis and parameter selection. *Information Processing Letters*, 2003, 85: 317–325.

51. H.M. Emara, H.A.A. Fattah. Continuous swarm optimization technique with stability analysis. In *Proceedings of the American Control Conference*, Cairo, Egypt, 2004, pp. 2811–2817.

52. V. Gavi, K.M. Passino. Stability analysis of social foraging swarms. *IEEE Transactions on Systems, Man and Cybernetics*, 2003, 34(1): 539–557.

53. V. Kadirkamanathan, K. Selvarajah, P.J. Fleming. Stability analysis of the particle dynamics in particle swarm optimizer. *IEEE Transactions on Evolutionary Computation*, 2006, 10(3): 245–255.

54. M. Jiang, Y.P. Luo, S.Y. Yang. Stochastic convergence analysis and parameter selection of the standard particle swarm optimization algorithm. *Information Processing Letters*, 2007, 102: 8–16.

55. F.J. Solis, R.J.-B. Wets. Minimization by random search techniques. *Mathematics of Operations Research*, 1981, 6(1): 19–30.

56. Y. Shi, R.C. Eberhart. A modified particle swarm optimizer. In *Proceedings of the IEEE International Conference on Evolutionary Computation*, Anchorage, AK, 1998, pp. 69–73.

57. M. Clerc. The swarm and the queen: Towards a deterministic and adaptive particle swarm optimization. In *Proceedings of the Congress on Evolutionary Computation*, Washington, DC, 1999, vol. 3, pp. 1951–1957.

58. P.J. Angeline. Using selection to improve particle swarm optimization. In *Proceedings of the IEEE International Conference on Evolutionary Computation*, Anchorage, AK, 1998, pp. 84–89.

59. P.N. Suganthan. Particle swarm optimizer with neighborhood operator. In *Proceedings of the Congress on Evolutionary Computation*, Washington, DC, 1999, vol. 3, pp. 1962–1967.

60. J. Kennedy. Stereotyping: Improving particle swarm performance with cluster analysis. In *Proceedings of the Congress on Computational Intelligence*, Honolulu, HI, 2002, pp. 1671–1676.

61. J. Kennedy. Small worlds and mega-minds: Effects of neighborhood topology on particle swarm performance. In *Proceedings of the Congress on Evolutionary Computation*, Washington, DC, 1999, vol. 3, pp. 1931–1938.

62. M. Clerc. *Particle Swarm Optimization*. ISTE Publishing Company, London, U.K., 2006.

63. J.J. Liang, P.N. Suganthan. Dynamic multiswarm particle swarm optimizer (DMS-PSO). In *Proceedings of the IEEE Swarm Intelligence Symposium*, Pasadena, CA, 2005, pp. 124–129.

64. A. Mohais, R. Mendes, C. Ward, C. Postoff. Neighborhood re-structuring in particle swarm optimization. In *Proceedings of the 18th Australian Joint Conference Artificial Intelligence*, Sydney, Australia, 2005, pp. 776–785.

65. F. van den Bergh, A.P. Engelbrecht. A cooperative approach to particle swarm optimization. *IEEE Transactions on Evolutionary Computation*, 2004, 8(3): 225–239.

66. R. Mendes, J. Kennedy, J. Neves. The fully informed particle swarm: Simpler, maybe better. *IEEE Transactions on Evolutionary Computation*, 2004, 8(3): 204–210.

67. S. Janson, M. Middendorf. A hierarchical particle swarm optimizer and its adaptive variant. *IEEE Transactions on Systems, Man, and Cybernetics, Part B: Cybernetics*, 2005, 35(6): 1272–1282.
68. A. Parrott, X. Li. Locating and tracking multiple dynamic optima by a particle swarm model using speciation. *IEEE Transactions on Evolutionary Computation*, 2006, 10(4): 440–458.
69. R.A. Krohling. Gaussian swarm: A novel particle swarm optimization algorithm. In *Proceedings of the IEEE Conference on Cybernetics and Intelligent Systems*, Singapore, 2004, pp. 372–376.
70. B. Secrest, G. Lamont. Visualizing particle swarm optimization-gaussian particle swarm optimization. In *Proceedings of the IEEE Swarm Intelligence Symposium*, Indianapolis, IN, 2003, pp. 198–204.
71. T.J. Richer, T.M. Blackwell. The Levy particle swarm. In *Proceedings of the Congress on Evolutionary Computation*, New York, 2006, pp. 808–815.
72. J. Kennedy. Bare bones particle swarms. In *Proceedings of the 2003 IEEE Swarm Intelligence Symposium*, Indianapolis, IN, 2003, pp. 80–87.
73. J. Kennedy. Probability and dynamics in the particle swarm. In *Proceedings of the 2004 Congress on Evolutionary Computation*, Portland, OR, 2004, vol. 1, pp. 340–347.
74. J. Kennedy. In search of the essential particle swarm. In *Proceedings of the 2006 IEEE World Congress on Computational Intelligence*, Los Alamitos, CA, 2006, pp. 1694–1701.
75. J. Sun, B. Feng, W.B. Xu. Particle swarm optimization with particles having quantum behavior. In *Proceedings of the Congress on Evolutionary Computation*, Portland, OR, 2004, vol. 1, pp. 326–331.
76. J. Sun, W.B. Xu, B. Feng. A global search strategy of quantum-behaved particle swarm optimization. In *Proceedings of the IEEE Conference on Cybernetics and Intelligent Systems*, Singapore, 2004, pp. 111–116.
77. J. Sun, W.B. Xu, B. Feng. Adaptive parameter control for quantum-behaved particle swarm optimization on individual level. In *Proceedings of the 2005 IEEE International Conference on Systems, Man and Cybernetics*, Honolulu, HI, 2005, vol. 4, pp. 3049–3054.
78. A. Banks, J. Vincent, C. Anyakoha. A review of particle swarm optimization. Part I: Background and development. *Natural Computing*, 2007, 6(4): 467–484.
79. A. Banks, J. Vincent, C. Anyakoha. A review of particle swarm optimization. Part II: Hybridisation, combinatorial, multicriteria and constrained optimization, and indicative applications. *Natural Computing*, 2008, 7(1): 109–124.
80. I. Giardina. Collective behavior in animal groups: Theoretical models and empirical studies. *HFSP Journal*, 2008, 2(4): 205–219.
81. J.T. Emlen. Flocking behaviour in birds. *The Auk*, 1952, 69(2): 160–170.
82. D.V. Radakov. *Schooling in the Ecology of Fish*. Israeli Scientific Translation Series, Wiley, New York, 1973.
83. T.J. Pitcher, A.E. Magurran. Shoal size, patch profitability, and information exchange in foraging goldfish. *Animal Behavior*, 1983, 31: 546–555.

84. J.S. Kennedy. The migration of desert locust (*Schistocerca gregaria*). *Philosophical Transactions of the Royal Society of London. Series B, Biological Sciences*, 1951, 235(625): 163–290.

85. E.O. Wilson. *The Insect Societies*. Belknap Press of Harvard University, Cambridge, MA, 1971.

86. A.R.E. Sinclair. *The African Buffalo: A Study of Resource Limitation of Population*. University Press of Chicago, Chicago, IL, 1977.

87. G. Nicolis, I. Prigogine. *Self-Organization in Non-Equilibrium Systems*. Wiley, New York, 1977.

88. S. Camazine, J.L. Deneubourg, N.R. Franks, J. Sneyd, G. Theraulaz, E. Bonabeau. *Self-Organization in Biological Systems*. Princeton Studies in Complexity, Princeton University Press, Princeton, NJ, 2001.

89. D.J.T. Sumpter. The principles of collective animal behaviour. *Philosophical Transactions of the Royal Society of London. Series B, Biological Sciences*, 2006, 361(1465): 5–22.

90. S. Garnier, J. Gautrais, G. Theraulaz. The biological principles of swarm intelligence. *Swarm Intelligence*, 2007, 1: 3–31.

91. I. Aoki. A simulation study on the schooling mechanism in fish. *Bulletin of the Japanese Society of Scientific Fisheries*, 1982, 48(8): 1081–1088.

92. A. Okubo. Dynamical aspects of animal grouping: Swarms, schools, flocks, and herds. *Advances in Biophysics*, 1986, 22: 1–94.

93. A. Huth, C. Wissel. The simulation of the movement of fish schools. *Journal of Theoretical Biology*, 1992, 156: 365–385.

94. C.W. Reynolds. Flocks, herds and schools: A distributed behavioral model. *Computer Graphics*, 1987, 21(4): 25–34.

95. F. Heppner, U. Grenander. *A Stochastic Nonlinear Model for Coordinated Bird Flocks*. AAAS Publications, Washington, DC, 1990.

96. S. Gueron, S.A. Levin, D.I. Rubenstein. The dynamics of herds: From individuals to aggregations. *Journal of Theoretical Biology*, 1996, 182: 85–98.

97. I.D. Couzin, J. Krause. Self-organization and collective behaviour in vertebrates. *Advances in Study of Behavior*, 2003, 32: 1–75.

98. I.D. Couzin, J. Krause, R. James, G.D. Ruxton, N.R. Franks. Collective memory and spatial sorting in animal groups. *Journal of Theoretical Biology*, 2002, 218(1): 1–11.

99. K. Warburton, J. Lazarus. Tendency-distance models of social cohesion in animal groups. *Journal of Theoretical Biology*, 1991, 150(4): 473–488.

100. T. Vicsek, A. Czirók, E. Ben-Jacob, I.I. Cohen, O. Shochet. Novel type of phase transition in a system of self-driven particles. *Physical Review Letters*, 1995, 75(6): 1226–1229.

101. G. Gregoire, H. Chate, Y.H. Tu. Moving and staying together without a leader. *Physica D*, 2003, 181: 157–170.

102. A. Mogilner, L. Edelstein-Keshet, L. Bent, A. Spiros. Mutual interactions, potentials, and individual distance in a social aggregation. *Journal of Theoretical Biology*, 2003, 47: 353–389.

103. O.E. Wilson. *Sociobiology: They New Synthesis*. Belknap Press, Cambridge, MA, 1975.

104. I. Rechenberg. Evolutionsstrategie-optimierum technischer systeme nach prinzipien der biologischen evolution. PhD thesis, Frommann-Holzboog, Stuttgart, Germany, 1973.
105. K.A. De Jong. An analysis of the behavior of a glass of genetic adaptive systems. PhD thesis, University of Michigan, Ann Arbor, MI, 1975.
106. R.K. Ursem. Diversity-guided evolutionary algorithms. In *Proceedings of the Parallel Problem Solving from Nature Conference*, Berlin, Germany, 2001, pp. 462–471.
107. V.S. Gordon, D. Whitley. Serial and parallel genetic algorithms as function optimizers. In *Proceedings of the Fifth International Conference on Genetic Algorithms*, San Mateo, CA, 1993, pp. 177–183.
108. D.H. Wolpert, W.G. Macready. No free-lunch theorems for search, Santa Fe Institute, Technical Report 95-02-010, 1995.
109. L.D. Whitley, K.E. Mathias, S.B. Rana, J. Dzubera. Building better test functions. In *Proceedings of the Sixth International Conference on Genetic Algorithms*, Pittsburgh, PA, 1995, pp. 239–246.
110. R. Salomon. Reevaluating genetic algorithm performance under coordinate rotation of benchmark functions. *BioSystems*, 1996, 39(3): 263–278.
111. A.E. Eiben, T. Bäck. Empirical investigation of multi-parent recombination operators in evolution strategies. *Evolutionary Computation*, 1997, 5(3): 347–365.
112. H.H. Rosenbrock. An automatic method for finding the greatest or least value of a function. *Computer Journal*, 1960, 3(3): 175–184.
113. H.P. Schwefel. *Numerical Optimization of Computer Models*. John Wiley & Sons, Chichester, U.K., 1981. English translation of Numerische Optimierung von Computer-Modellen mittels der Evolutionsstrategie, 1977.
114. G. Hadley. *Nonlinear and Dynamics Programming*. Addison Wesley, Reading, MA, 1964.
115. J.H. Friedman. An overview of predictive learning and function approximation. In V. Cherkassky, J.H. Friedman, H. Wechsler (Eds.), *From Statistics to Neural Networks, Theory and Pattern Recognition Applications*, vol. 136 of NATO ASI Series F. Springer-Verlag, Berlin, Germany, 1994, pp. 1–61.
116. H.P. Schwefel. *Evolution and Optimum Seeking*. John Wiley & Sons, Chichester, U.K., 1995.
117. E. Spedicato. Computational experience with quasi-newton algorithms for minimization problems of moderately large size, Centro Informazioni Studi Esperienze, Segrate (Milano), Italy, Technical Report CISE-N-175, 1975.
118. S.S. Oren. On the selection of parameters in self scaling variable metric algorithms. *Mathematical Programming*, 1974, 7(1): 351–367.
119. L.C.W. Dixon. Nonlinear optimization: A survey of the state of the art. In *Software for Numerical Mathematics*, Academic Press, London, U.K., 1974, pp. 193–216.
120. D. Schlierkamp-Voosen. Strategy adaptation by competition. In *Proceedings of the Second European Congress on Intelligent Techniques and Soft Computing*, Aachen, Germany, 1994, pp. 1270–1274.
121. L.A. Rastrigin. Extremal control systems. In *Theoretical Foundations of Engineering Cybernetics Series*. Nauka, Moscow, Russia, 1974.

122. D. Ackley. An empirical study of bit vector function optimization. In *Genetic Algorithms and Simulated Annealing*. Morgan Kaufmann, Los Altos, CA, 1987, pp. 170–215.

123. T. Bäck, H.P. Schwefel. An overview of evolutionary algorithms for parameter optimization. *Evolutionary Computation*, 1993, 1(1): 1–23.

124. T. Bäck, D. Fogel, Z. Michalewicz. *Handbook of Evolutionary Computation*. Institute of Physics Publishing Ltd, Bristol, U.K., 1997.

CHAPTER 3

Some Variants of Particle Swarm Optimisation

3.1 WHY DOES THE PSO ALGORITHM NEED TO BE IMPROVED?

Recall briefly the comparison of performance in Chapter 2 that the original PSO performed unsatisfactorily on some simple examples, such as the sphere function. The reason is because the original PSO algorithm has weak local search ability. It should be noted here that the trade-off between local search (exploitation) and global search (exploration) is vital for the performance of the algorithm.

The original PSO needs to accelerate the convergence speed of the particles in order to achieve a better balance between exploitation and exploration. Work in this area, first carried out by Shi and Eberhart [1], involves introducing an inertia weight into the update equation for velocities. Clerc [2] proposed another acceleration method by adding a constriction factor in the velocity update equation to avoid any velocity restriction during the convergence history.

The acceleration techniques were shown to work well. The previous two variants of PSO have laid the foundation for further enhancement of PSO. However, there is an issue of the topological structure of particle swarm as described in the following. In the original PSO, PSO-In and PSO-Co, the search is guided by the global best position and its personal best position. This neighbourhood topology is known as the global best model. Although the algorithm with this model is able to obtain the best approximate solutions efficiently for many problems, it is more prone to encounter premature convergence when solving harder problems. If the global best particle sticks to a local or sub-optimal point, it would mislead the other particles to move towards that point. In other words, the other promising search areas might be missed. This had led to the investigation of another neighbourhood known as the local best model first studied by Eberhart and Kennedy [3] and, subsequently, in depth by many other researchers. The objective there was to find other possible topologies to improve the performance of PSO.

In PSO, the particles essentially follow a deterministic trajectory defined by a velocity update formula with two random acceleration coefficients. This is a semi-deterministic search which restricts the search domain of each particle and may weaken the global search ability of the algorithm, particularly, at the later stage of search process. In view of this limitation, several probabilistic PSO algorithms simulate the particle trajectories by direct sampling, using a random number generator, from a distribution of practical interests. The Bare Bones PSO (BBPSO) family are typical probabilistic PSO algorithms. In BBPSO, each particle does not have a velocity vector, but its new position is sampled 'around' a supposed good one according to certain probability distribution, such as Gaussian distribution in the original development [4]. Several other new BBPSO variants used other distributions that seem to generate better results. The quantum-behaved particle swarm optimisation (QPSO) algorithm, which is addressed in the second half of this book, is also a probabilistic PSO.

Note also that the *no free lunch* (NFL) theorem says that there are other random search techniques that may well be more efficient than PSO on many other problems. As such, researchers also turn to hybrid algorithms that incorporate other search methods into the PSO algorithms.

Throughout the discussion of various methods in the following sections, readers should refer to the original publication for a complete description of the numerical experiments and results comparison.

It should be noted that normal distribution and Gaussian distribution are frequently used in this chapter and Chapters 4 through 6. $N(a, b)$ and $G(a, b)$ denote the normal distribution and Gaussian distribution with mean a and standard deviation b, respectively. In essence, they refer to the same probability distribution and their usage may be changeable in the text. In addition, $U(a, b)$ denotes the uniform distribution on the range (a, b).

3.2 INERTIA AND CONSTRICTION—ACCELERATION TECHNIQUES FOR PSO

3.2.1 PSO with Inertia Weights (PSO-In)

3.2.1.1 Qualitative Analysis of the Velocity Updating Equation

Recall that the velocity and position of a particle in the original PSO are updated according to the equations

$$V_{i,n+1}^{j} = V_{i,n}^{j} + c_1 r_{i,n}^{j} \left(P_{i,n}^{j} - X_{i,n}^{j} \right) + c_2 R_{i,n}^{j} \left(G_n^{j} - X_{i,n}^{j} \right), \tag{3.1}$$

$$X_{i,n+1}^{j} = X_{i,n}^{j} + V_{i,n+1}^{j} \tag{3.2}$$

for $i = 1, 2, \ldots, M$, $j = 1, 2, \ldots, N$. The parameters c_1 and c_2 are known as the acceleration coefficients which are two positive constants; $r_{i,n}^{j}$ and $R_{i,n}^{j}$ are two random numbers uniformly distributed on $(0,1)$. $P_{i,n}$ is the personal best position of the ith particle and G_n is the global best position at the nth iteration. The global best position is selected from the personal best positions of all particles in the swarm at each iteration. From Equation 3.1, it is clear that the updating velocity of each particle is composed of three parts as listed in the following:

1. The previous velocity, $V_{i,n}^{j}$, known as the 'inertia part', providing the necessary momentum for the particles to fly in the search space

2. The 'cognition' part, $c_1 r_{i,n}^{j} \left(P_{i,n}^{j} - X_{i,n}^{j} \right)$, representing the particle's own experience

3. The 'social' part, $c_2 R_{i,n}^{j} \left(G_n^{j} - X_{i,n}^{j} \right)$, reflecting the information shared among all the particles

These three parts play different roles in the search process of the particle. Shi and Eberhart [1] provided an analysis of these three parts which is briefly discussed as follows.

When $c_1 = c_2 = 0$, there are no 'cognition' and 'social' parts in the velocity update. The particle flies uniformly in a straight line until it reaches the boundary of the search space. No optimal solution can be found unless the solution is on its trajectory.

When $c_1 = 0$, there is no 'cognition' part in Equation 3.1. The resulting velocity update equation is known as the 'social-only' model. With the interaction among particles, the PSO algorithm has the ability to reach a new search space. However, the particle swarm converges at a relatively fast speed. The algorithm can be easily trapped into the local optima for complex problems, but it may have good performance in solving some specific problems.

When $c_2 = 0$, there is no 'social part' in the velocity update. This means that there is no information sharing between the particles. The PSO algorithm is known as running in 'cognition-only' model. Without any interaction amongst the particles, running a swarm with population size M is essentially equivalent to running M individuals independently. As such, the PSO algorithm becomes a multi-start random search algorithm and has slower convergence rate than the algorithm in 'social-only' model.

When there is no inertia part, the particle velocity is determined by its personal best position and the global best position. In this model, the velocity of the particle is said to be memory-less and the particle has better local search ability. Hence, the algorithm can converge more rapidly than the original PSO. However, without the inertia part, particles can only sample their positions around their personal best and global best positions, and have less probability of searching other areas.

3.2.1.2 PSO with Inertia Weight

In order to balance the local search and global search during an optimisation process in the original PSO, the concept of an inertia weight [1], w, is introduced into Equation 3.1, leading to the modified velocity update:

$$V_{i,n+1}^{j} = wV_{i,n}^{j} + c_1 r_{i,n}^{j}\left(P_{i,n}^{j} - X_{i,n}^{j}\right) + c_2 R_{i,n}^{j}\left(G_n^{j} - X_{i,n}^{j}\right). \qquad (3.3)$$

The PSO algorithm supplemented with (3.3) is known as PSO with inertia weight (PSO-In). The inertia weight w can be a positive value chosen according to experience or from a linear or non-linear function of the iteration number. When w is 1, the PSO-In is equivalent to the original PSO. The values of c_1 and c_2 in Equation 3.1 are generally set to be 2 as recommended by Kennedy and Eberhart [5], which implies that the 'social' and 'cognition' parts have the same influence on the velocity update.

Experiences on the choice of w for the benchmark problem Schaffer's F6 function [1,6] are given here. When w is smaller than 0.8, PSO could find the global optimum efficiently provided the optimal solution is in the initial search space. When w is larger than 1.2, the particles have stronger exploration ability and, hence, better chances to locate the global optimum area. However, its exploitation ability is worse than the original PSO.

When w is between 0.8 and 1.2, the PSO algorithm has better opportunities to find the global optimum in a reasonable number of iterations. Within the range of 0.9–1.2, Shi and Eberhart [6] introduced a significant improvement in the performance of the PSO method with a time-decreasing inertia weight over the generations. This time-decreasing strategy for w can be written as

$$w_n = \frac{(w_{initial} - w_{final}) \cdot (n_{max} - n)}{n_{max} + w_{final}},$$ (3.4)

where

w_n is the value of the inertial weight at the nth iteration

$w_{initial}$ and w_{final} are the initial and final values of the inertia weight, respectively

n is the current iteration number

n_{max} is the maximum number of iterations

Therefore, at the beginning of the search within a run, the PSO with a large inertia weight can provide high diversity in order to use the full range of the search space; and at the end of the search, a small inertia weight can help the PSO converge to the optimal solution with fine-tuning.

Shi and Eberhart provided an extensive investigation on the linear-decreasing inertia weight [7]. Simulation tests there took the value of w from 0.9 at the beginning of the search to 0.4 at the end of the search. Shi et al. were able to demonstrate that with the linear-decreasing inertia weight in the range [0.4, 0.9], the PSO algorithm would converge efficiently and its performance was not sensitive to the other parameters such as population size and the maximum number of iterations. However, they highlighted two problems with the PSO algorithm, one being that the PSO may lack of global search ability ultimately as the value of the inertia weight is small, and the other being that PSO may experience difficulties in dealing with complicated and complex problems.

3.2.1.3 Modifications of PSO Using Various Inertia Weights

3.2.1.3.1 Random Inertia Weight   In order to solve the dynamic nature of real-world applications effectively by PSO, the following random inertia weight was proposed [8]:

$$w_n = \frac{0.5 + rand}{2.0},$$ (3.5)

where *rand* is a random number uniformly distributed within the range (0,1). The value of the inertia weight ranges between 0.5 and 1.0 with a mean value of 0.75. Several benchmark functions were examined and the results showed that it has rapid convergence in the early stage of the optimisation process and can find a reasonably good solution for most of the functions.

In [9], Pant et al. proposed a new dynamic inertia weight by using Gaussian distribution (i.e. normal distribution). This kind of random inertia weight is defined by the equation

$$w_n = \frac{|randn|}{2}, \tag{3.6}$$

where *randn* is a random number with the standard normal distribution. Different probability distributions, such as Gaussian distribution and exponential probability distribution, may be used to initialise the velocities and positions of the particles. The Gaussian distribution used in their study has the probability density function

$$f(x) = \frac{1}{\sqrt{2\pi}} e^{-x^2/2}, \quad -\infty < x < \infty, \tag{3.7}$$

with mean 0 and standard deviation 1. The exponential distribution used for the generation of the initial swarm has the probability function

$$f(x) = \frac{1}{2b} \exp\left(-\frac{|x-a|}{b}\right), \quad -\infty < x < \infty, \tag{3.8}$$

where $a, b > 0$ are two parameters that can be changed to control the mean and the variance. In [9], the values of a and b were chosen to be $a = 0.3$ and $b = 1$, which provides a mean value 0.78 with standard deviation 0.93 for the random number with the exponential distribution. Pant et al. found that the PSO with Gaussian inertia weight was able to find better solutions to most of the tested benchmark problems than the PSO-In algorithm when uniform distribution is used to initialise the particle swarm. In addition, using Gaussian distribution or exponential probability distribution for initialisation of the swarm could improve the effectiveness of the algorithm.

3.2.1.3.2 Linearly Increasing Inertia Weight In contrast to the PSO with decreasing inertia weight, a PSO with increasing inertia weight was proposed by Zheng et al. [10]. The inertia weight of PSO as shown in [10] took values from 0.4 to 0.9 linearly and numerical tests were performed for four benchmark functions. A significant characteristic of their proposed inertia weight is that it has the same computational load as the PSO with decreasing inertia weight, but the experimental results show that the former outperforms the latter both in convergence speed and accuracy.

3.2.1.3.3 Non-Linear-Decreasing Inertia Weight To adjust the search behaviour during the initial iterations and later iterations of the search, Chatterjee and Siarry [11] proposed a non-linear adaptive inertia weight for PSO. The non-linear variation of the inertia weight multiplying the old velocity of the particle can improve the speed of convergence and be able to fine-tune the search in the multi-dimensional space. This non-linear inertia weight is given by

$$w_n = \frac{(n_{max} - n)^r}{(n_{max})^r} (w_{initial} - w_{final}) + w_{final},$$ (3.9)

where

$w_{initial}$ and w_{final} are the initial and final values of the inertia weight respectively

n is the current iteration number

n_{max} is the maximum number of iterations

r is the non-linear modulation index

It is clear that when $r = 1$, this inertia weight is equivalent to the linearly decreasing inertia weight. Figure 3.1 is reproduced by Chatterjee and Siarry [11], which provides the shapes of w_n developing with iteration number n for different r values.

Similar to the ideas of decreasing inertia weight in PSO, two types of natural exponential functions used for determining the value of inertia weight were proposed in [12]. The idea is to provide a very fast change of the inertia weights during the first part and a very slow change of the inertia weights during the final part of the search process. The two exponential functions used in [12] are defined by these two equations:

$$w_n = (w_{initial} - w_{final})e^{-n/(n_{max}/10)} + w_{final},$$ (3.10)

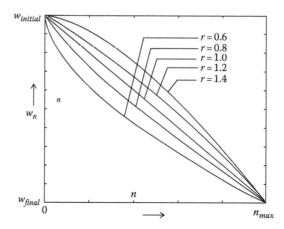

FIGURE 3.1 Variations of inertia weights with iterations for different values of non-linear modulation index (r). (From Chatterjee, A. and Siarry, P., *Comput. Oper. Res.*, 33(3), 859, 2006.)

$$w_n = (w_{initial} - w_{final})e^{-[n/(n_{max}/4)]^2} + w_{final}, \qquad (3.11)$$

where $w_{initial}$, w_{final}, n and n_{max} share the same meaning with those in Equation 3.4. Figure 3.2 depicts the exponential inertia weights [12]. Four benchmark functions were used to test the performance of these two inertia weights. The current two variants of PSO converge faster during the early stage of the search process. For most of the continuous optimisation problems the two PSO algorithms performed better.

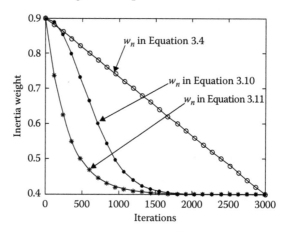

FIGURE 3.2 Comparison of the three decreasing strategies. (From Chen, G. et al., Natural exponential inertia weight strategy in particle swarm optimization, in *Proceedings of the Regional Postgraduate Conference on Engineering and Science*, Johor, Bahru, Malaysia, 2006, pp. 247–252.)

A dynamic inertia weight decreases according to certain inverse power law [13] was proposed:

$$w_n = au^{-n}, \tag{3.12}$$

where

a and u are user-defined values in the range of [0,1] and [1.00001, 1.005], respectively

n is the current iteration number

Some different values of a were tested while the value of u was set to 1.00002, and it was found that the modified PSO had the best performance when $a \in [0.3, 0.4]$.

3.2.1.3.4 Sigmoid Variations of Inertia Weight   A sigmoid-decreasing inertia weight for PSO was proposed by Adriansyah and Amin [14]. The concept behind this inertia weight depends on the basic sigmoid-decreasing function which begins at a maximum value and ends at a minimum value, where the change from the maximum to the minimum takes place in a very small region instead of a sharp discontinuity or a gradual change. Typical shapes of this function are depicted in Figure 3.3. The idea is to allow the first few iterations of the search being concentrated on the global search and towards the final few iterations of the search being concentrated on a fine-tune of the approximate solution around optimal solution.

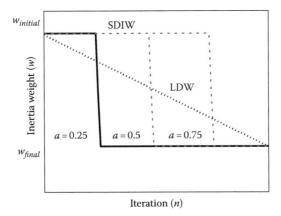

Iteration (n)

FIGURE 3.3   Sigmoid and linearly decreasing inertia weight.

This method becomes an effective exploration–exploitation trade-off in PSO-In. The sigmoid function for inertia weight variation is given by

$$w_n = \frac{w_{initial} - w_{final}}{1 + e^{-u(n - an_{max})}} + w_{final}, \tag{3.13}$$

$$u = 10^{(\log(n_{max}) - 2)}, \tag{3.14}$$

where

w_n is the inertia weight at the nth iteration

$w_{initial}$ and w_{final} are the initial and final values of the inertia weight, respectively

u is a constant used to adjust the sharpness of the sigmoid function

n_{max} is the maximum number of iterations

a is a constant governing the abrupt changes of the sigmoid function

Solutions to four benchmark functions, using the PSO with sigmoid-decreasing inertia weight, were shown to have faster convergence and better solution accuracy than those results obtained by using the PSO with linear-decreasing inertia weight.

In contrast to the sigmoid-decreasing inertia weight, a new inertia weight using the sigmoid-increasing function was proposed by Malik et al. [15]. The function takes a small value in the first few iterations and increases abruptly to maximum value during the last few iterations. The increasing inertia weight is given by

$$w_n = \frac{w_{initial} - w_{final}}{1 + e^{u \cdot (n - a \cdot n_{max})}} + w_{final}, \tag{3.15}$$

where w_n, $w_{initial}$, w_{final}, u and a have the same meanings with those in Equation 3.13. The shape of the sigmoid-increasing function is depicted in Figure 3.4.

3.2.1.3.5 Fuzzy Inertia Weight   The search ability of the PSO with linearly decreasing inertia weight is a linear transition from global to local search. In order to better resemble the actual search process, which is usually non-linear and complicated, a fuzzy system was implemented to dynamically

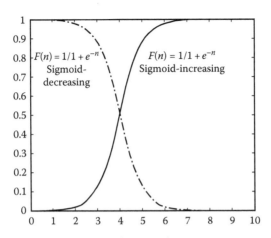

FIGURE 3.4   Sigmoid-decreasing and sigmoid-increasing inertia weight.

adapt the inertia weight of PSO [16]. The aim was to balance the global and local search abilities. It has two inputs and one output. The inputs are the current best performance evaluation (CBPE) and the current inertia weight, and the output is the change of the inertia weight. The CBPE measures the performance of the best candidate solution found so far by PSO. In order to design the fuzzy system with the CBPE as one of the inputs that is applicable to a wide range of optimisation problems, the CBPE needs to be normalised as

$$CBPE = \frac{CBPE - CBPE_{min}}{CBPE_{max} - CBPE_{min}}, \tag{3.16}$$

where
   $CBPE_{min}$ denotes the estimated minimum
   $CBPE_{max}$ denotes the non-optimal CBPE representing any solution with
      CBPE greater than or equal to $CBPE_{max}$ that is not acceptable to the
      optimisation problem

The three fuzzy variables, including two input variables and one output variables, have three fuzzy sets: *low*, *medium* and *high*, with the associated membership functions as *left_Triangle*, *Triangle* and *right_Triangle*, respectively. These three membership functions are defined as follows.

Left_Triangle membership function:

$$f_{left\_Triangle} = \begin{cases} 1 & \text{if } x < x_1 \\ \dfrac{x_2 - x}{x_2 - x_1} & \text{if } x_1 \leq x \leq x_2 . \\ 0 & \text{if } x > x_2 \end{cases} \qquad (3.17)$$

Triangle membership function:

$$f_{Triangle} = \begin{cases} 0 & \text{if } x < x_1 \\ 2\dfrac{x - x_1}{x_2 - x_1} & \text{if } x_1 \leq x \leq \dfrac{x_2 + x_1}{2} \\ 2\dfrac{x_2 - x}{x_2 - x_1} & \text{if } \dfrac{x_2 + x_1}{2} < x \leq x_2 \\ 0 & \text{if } x > x_2 \end{cases} \qquad (3.18)$$

Right_Triangle member function:

$$f_{right\_Triangle} = \begin{cases} 0 & \text{if } x < x_1 \\ \dfrac{x - x_1}{x_2 - x_1} & \text{if } x_1 \leq x \leq x_2 . \\ 1 & \text{if } x > x_2 \end{cases} \qquad (3.19)$$

where x_1 and x_2 are critical parameters that determine the shape and location of the functions.

Based on the membership functions defined earlier, Shi provided the fuzzy system for dynamically adjusting the inertia weight. Although other forms of membership functions are possible, the previous ones are useful for a variety of problems and easy to be implemented in microcontrollers and microprocessors.

3.2.1.3.6 Adaptive Inertia Weights Since each particle in a swarm has inherited the differences of the search state in each iteration, it is logical to consider an adaptive inertia weight at the individual level. In [17],

Feng et al. analysed the particle behaviour according to the principle of mechanics by partitioning the right-hand side of the velocity update Equation 3.3 into two parts as given in the following:

$$V_{i,n+1}^j = wV_{i,n}^j + F_{i,n}^j,$$ (3.20)

where

$$F_{i,n}^j = c_1 r_{i,n}^j \left(P_{i,n}^j - X_{i,n}^j \right) + c_2 R_{i,n}^j \left(G_n^j - X_{i,n}^j \right).$$ (3.21)

$F_{i,n}^j$ represents the acceleration of the particle or the external force pulling the particle towards $P_{i,n}^j$ and G_n^j on the jth component. If $V_{i,n}^j$ and $F_{i,n}^j$ are on different directions, relatively large $|F_{i,n}^j|/|V_{i,n}^j|$ means that the particle is far from the optimal region around $P_{i,n}^j$ and G_n^j. As a result, the particle needs to change the jth component of its velocity and the jth component of the inertia weight $w_{i,n}^j$ needs to be set as a smaller value. On the other hand, relatively small $|F_{i,n}^j|/|V_{i,n}^j|$ implies that it is not urgent to change the jth component of the particle velocity and $w_{i,n}^j$ could be set as a larger value. If $V_{i,n}^j$ and $F_{i,n}^j$ are in the same direction, relatively large $|F_{i,n}^j|/|V_{i,n}^j|$ means that the jth component of the particle is in the right direction, but the velocity is so small that its jth component needs to speed up and the inertia weight needs to be set at a larger value. On the other hand, relatively small $|F_{i,n}^j|/|V_{i,n}^j|$ means that the particle is near the optimal region around $P_{i,n}^j$ and G_n^j. As a result, the particle needs to slow down on the jth component and $w_{i,n}^j$ should be set to a smaller value. With this in mind, Feng et al. concluded that the value of $w_{i,n}^j$ is a function of $|F_{i,n}^j|/|V_{i,n}^j|$ and thus suggested that $w_{i,n}^j$ ($1 \leq i \leq M$, $1 \leq j \leq N$) change adaptively according to

$$w_{i,n}^j = w_{min} + (w_{max} - w_{min})\alpha_{i,n}^j,$$ (3.22)

where
n is the current iteration number
$w_{i,n}^j$ is the inertia weight for the jth component of particle i at the nth iteration
w_{min} and w_{max} are user-defined limits of the inertia weight $w_{i,n}^j$

Here $\alpha_{i,n}^{j}$ is calculated by the equation

$$
\alpha_{i,n}^{j} = \begin{cases} \dfrac{1}{\sqrt{1+\left(\tilde{F}_{i,n}^{j}/F\right)^{4}}} & \tilde{F}_{i,n}^{j} < 0, \\[3mm] 1-\dfrac{1}{\sqrt{1+\left(\tilde{F}_{i,n}^{j}/F\right)^{4}}} & \tilde{F}_{i,n}^{j} \geq 0, \end{cases} \tag{3.23}
$$

where

F is defined as the cut-off constant specified by users to control the rate of change for the inertia weight

$\tilde{F}_{i,n}^{j}$ is determined by

$$
\tilde{F}_{i,n}^{j} = \frac{F_{i,n}^{j}}{V_{i,n}^{j}}. \tag{3.24}
$$

Figure 3.5 shows the adaptive inertia weights used in [17]. Readers should refer to the reference for further details of the numerical experiments.

Another adaptive inertia weight based on the influence of a particle on the others according to the effect of attraction towards the global best position, that is, the distance from the particles to the global best position, was proposed by Suresh et al. [18]. At the nth iteration, the value of inertia weight for particle i is determined by

$$
w_{i,n} = w_0\left(1 - \frac{d_{i,n}}{D_n}\right), \tag{3.25}
$$

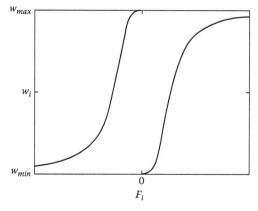

FIGURE 3.5 The adaptive inertia weight as proposed in [13].

where

w_0 is a random number uniformly distributed in the range of $[0.5, 1]$

$d_{i,n}$ is the current Euclidean distance from particle i to the global best position

D_n is the maximum distance from a particle to the global best position at iteration n

That is,

$$D_n = \arg\max_i(d_{i,n}). \tag{3.26}$$

The Euclidean distance $d_{i,n}$ is calculated as follows:

$$d_{i,n} = \sqrt{\sum_{j=1}^{N}\left(G_n^j - X_{i,n}^j\right)^2}. \tag{3.27}$$

In order to avoid premature convergence in the final stages of the search, Suresh et al. also changed the position update equation to the one given as follows:

$$X_{i,n+1}^j = V_{i,n+1}^j + (1-\rho)X_{i,n}^j, \tag{3.28}$$

where ρ is a random number uniformly distributed on the range of $(-0.25, 0.25)$.

From experience, it is well known that the convergence of PSO depends on the evolution speed and the fitness of the particles. Therefore, in order to handle a complex and non-linear optimisation process, it would be helpful to include such factor into the inertia weight design. Zhang et al. [19] proposed a novel dynamically varying inertia weight for PSO involving two factors, namely, the evolution speed factor and the aggregation degree factor. The evolution speed factor at the nth iteration is given by

$$h_n = \frac{\min(f(G_{n-1}), f(G_n))}{\max(f(G_{n-1}), f(G_n))}, \tag{3.29}$$

where

$f(G_{n-1})$ represents the fitness value of G_{n-1}, the global best position at the $(n-1)$th iteration

n is the current iteration number

Note that the evolutionary speed factor lies in [0,1]. This parameter reflects the run-time history of the algorithm and the evolution speed. The smaller the value of h_n, the faster the evolution speed. After certain number of iterations, the value of h_n attains its maximum value, indicating that the algorithm stagnates or finds the optimum solution. The aggregation degree factor s is defined by

$$s_n = \frac{\min\left(f(G_{n-1}), \bar{f}_n\right)}{\max\left(f(G_{n-1}), \bar{f}_n\right)},\tag{3.30}$$

where $\bar{f}_n$ is the average fitness value of all the particles at the nth iteration, that is,

$$\bar{f}_n = \frac{1}{M}\sum_{i=1}^{M} f(X_{i,n}),\tag{3.31}$$

where M is the swarm size. It is obvious that the value of s_n falls in [0,1]. The aggregation can reflect not only the aggregation degree but also the diversity of the swarm. The larger the value of s_n, the more aggregation or the smaller the swarm diversity. In the case when s_n is equal to 1, all the particles have the same identity with each other. The new inertia weight of Zhang et al. is given by

$$w_n = w_{initial} - h_n w_h + s_n w_s,\tag{3.32}$$

where
  $w_{initial}$ is the initial value of inertia weight, usually set to 1
  w_h and w_s are two user-defined weights that lie between 0 and 1

Their numerical experiments show that the adaptive PSO with dynamically changing inertia weight could achieve good performances when w_h lies between 0.4 and 0.6 and w_s between 0.05 and 0.2. When $0 < h_n \leq 1$ and $0 < s_n \leq 1$, the value of inertia weight must lie between $w_{initial} - w_h$ and $w_{initial} + w_s$. The modified PSO algorithm was compared with the PSO with linearly decreasing inertia weight by using three benchmark functions. The experimental results showed that the algorithm was superior to its competitor in the convergence speed and the solution accuracy.

However, the modified PSO algorithm had weak ability to jump out of local search than the PSO with linearly decreasing inertia weight.

With the aforementioned shortcomings, Yang et al. examined two new definitions for the evolution speed factor and the aggregation degree factor [20] taking care of the individual particle. These are given as follows with the additional subscript i denoting the ith particle:

$$h_{i,n} = \frac{\min\left(f(P_{i,n-1}), f(P_{i,n})\right)}{\max\left(f(P_{i,n-1}), f(P_{i,n})\right)}, \tag{3.33}$$

where $f(P_{i,n})$ is the fitness value of the personal best position of particle i at iteration n. The aggregation degree factor for particle i is defined as

$$s_{i,n} = \frac{\min\left(f(B_n), \overline{f_n}\right)}{\max\left(f(B_n), \overline{f_n}\right)}, \tag{3.34}$$

where $f(B_n)$ represents the optimal value that the swarm found in the nth iteration. Note that $f(B_n)$ cannot be replaced by $f(G_n)$ since $f(G_n)$ denotes the optimal value that the entire swarm has found up to the nth iteration. It was also pointed out by Yang et al. that the aggregation degree factor in Equation 3.34 can respond faster to the evolving state than the one used in Equation 3.30. The new inertia weight for each particle can be written as

$$w_{i,n} = w_{initial} - (1 - h_{i,n})\alpha + s_{i,n}\beta, \tag{3.35}$$

where

$w_{initial}$ has the same meaning as that in Equation 3.32
α and β typically lie in the range [0,1]

It is obvious that the value of inertia weight lies in between $(1 - \alpha)$ and $(1 + \beta)$.

Six benchmark functions were used to test the performance of the aforementioned inertia weights [20]. The experimental results show that the PSO with the inertia weight has a stronger ability to jump out of the local optima and is able to prevent the premature convergence effectively. These results also suggest that the performance of the algorithm is not strongly dependent on parameters α and β.

Other factors were employed to update the value of inertia weight dynamically in order to accelerate the evolutionary state of the PSO [21]. An example is given here involving two factors, that is, the dispersion degree and the advance degree factors. The dispersion degree is used to reflect the scattered condition of the particles and is defined by the difference of the fitness values of the particles at the current iteration:

$$D_n = \frac{\sqrt{\left[f(X_{1,n})-\bar{f}_n\right]^2+\left[f(X_{2,n})-\bar{f}_n\right]^2+\cdots+\left[f(X_{M,n})-\bar{f}_n\right]^2/n}}{\sqrt{\left[f_{max,n}-\bar{f}_n\right]^2+\left[f_{min,n}-\bar{f}_n\right]^2/2}},$$

(3.36)

where
  $\bar{f}_n$ is the mean fitness of all the particles at the nth iteration
  $f(X_{i,n})$ is the fitness of the ith particle at the nth iteration
  $f_{max,n}$ and $f_{min,n}$ denote the maximum and minimum fitness at the nth iteration, respectively

It can be seen that the larger the value of D_n, the more scattered are the particles in the swarm and vice versa. The advance degree factor is defined by the difference of the fitness values of the particles between the current iteration and the previous iteration, that is,

$$A_{i,n} = \frac{f(X_{i,n+1})-f(X_{i,n})}{f(X_{i,n})},$$

(3.37)

$$A_n = \frac{f(G_{n+1})-f(G_n)}{f(G_n)},$$

(3.38)

where
  $f(X_{i,n+1})$ is the fitness of particle i at the nth iteration
  $f(G_n)$ is the global best fitness value of the swarm at the nth iteration

The inertia weight for each particle is defined as follows:

$$w_{i,n} = w_{initial} - \alpha D_n + \beta A_{i,n} + \gamma A_n,$$

(3.39)

where
  $w_{initial}$ is the initial value of w
  α, β and γ are the uniformly distributed random numbers ranging in (0,1)

Experiments as shown in [21] provide a comparison between this version of PSO-In and PSO with linearly decreasing inertia weight and the PSO with dynamical inertia weight using eight benchmark functions.

3.2.2 PSO with Constriction Factors (PSO-Co)

In order to ensure the convergence of the PSO without imposing any restriction on velocities, Clerc incorporated the constriction factor [2] into the velocity update formula as given in the following:

$$V_{i,n+1}^{j} = \chi \left[V_{i,n}^{j} + c_1 r_{i,n}^{j} \left(P_{i,n}^{j} - X_{i,n}^{j} \right) + c_2 R_{i,n}^{j} \left(G_n^j - X_{i,n}^{j} \right) \right] \qquad (3.40)$$

where the constant χ is known as the constriction factor and is given by

$$\chi = \frac{2}{\left| 2 - \varphi - \sqrt{\varphi^2 - 4\varphi} \right|}, \quad \varphi = c_1 + c_2. \qquad (3.41)$$

It was shown by Clerc and Kennedy [2,22] that the swarm shows stable convergence if $\varphi \geq 4$. If $\chi \in [0,1]$, the approach is very similar to the concept of inertia weight with $w = \chi$, $c_1' = \chi c_1$, $c_2' = \chi c_2$. Clerc and Kennedy recommended a value of 4.1 for the sum of c_1 and c_2, which leads to $\chi = 0.7298$ and $c_1 = c_2 = 2.05$.

Eberhart and Shi compared the performance of the inertia weight and constriction factors in PSO [23]. Five benchmark functions were used in the comparison tests. Experimental results show that PSO-Co has better convergence rate than the PSO-In. However, PSO-Co does not reach the global optimal area in the pre-assigned number of iterations possibly caused by the distance of the particle from the search space. In order to avoid such situation, Eberhart and Shi suggested limiting the maximum [23] velocity V_{max} to the dynamic range of the variable X_{max} in each dimension when using PSO-Co.

In addition to the method of using a fixed constriction factor, an adaptive constriction factor may be used during the search in order to speed up the optimisation process. As shown in [24], both inertia weight and the constriction factor can be represented by the following equation:

$$V_{i,n+1}^{j} = \chi \left[w V_{i,n}^{j} + c_1 r_{i,n}^{j} \left(P_{i,n}^{j} - X_{i,n}^{j} \right) + c_2 R_{i,n}^{j} \left(G_n^j - X_{i,n}^{j} \right) \right] \qquad (3.42)$$

where the inertia weight w may be chosen as 0.729 [25] and the constriction factor may be set as 1.0 initially and adapted through decaying its value over time till 0.0 at the end. With this idea in mind, the inertia weight is responsible for keeping the algorithm exploring at the early stage of the search and the constriction factor takes over the role of the inertia weight as the value of the constriction factor decreases. There are two adaptive rules for the constriction factor:

$$\chi_n = 1 - \left(\frac{n}{n_{max}}\right)^h,$$ (3.43)

$$\chi_n = \left[1 - \left(\frac{n}{n_{max}}\right)^h\right]\sigma,$$ (3.44)

where

h is a positive number used to adjust the reduction rate of the velocity

n is the current iteration number

n_{max} is the maximum number of iterations

The constriction factor in Equation 3.43 is annealed by the factor $(n/n_{max})^h$ at each iteration. In Equation 3.44, a small random number, σ, uniformly distributed on $(0,1)$ is multiplied to Equation 3.43. From Equations 3.43 and 3.44, it is obvious that if h increases, the reduction of χ slows down and, therefore, reduces the speed of convergence.

For convenience, in this chapter and the rest of this book PSO-In and PSO-Co are collectively referred to as PSO or the canonical PSO frequently, except, for example, in Chapter 2, when these two versions need to be distinguished.

3.3 LOCAL BEST MODEL

Since the introduction of the algorithm, two types of PSO model with different neighbourhood topologies are commonly used, that is, the global best model and the local best model. In the global best model, there exists a global best position which may be found by any particle of the swarm and all the particles are attracted by the global best position as well as by their own personal best positions. In the local best model, each particle is attracted by the best position found by its K immediate neighbours and their personal best positions, where K depends on the neighbourhood topology employed.

3.3.1 Neighbourhood Topologies

The local best model has been studied in depth by many other researchers in order to find the topologies that may be able to improve the performance of PSO. In 1999, Kennedy studied the effects of neighbourhood topology on PSO [26] using four types of neighbourhood topology, including circles (also known as rings), wheels, stars and random edges. In the circle topology, each particle is only connected to its K immediate neighbours. The parameter K is defined at run-time and may affect the performance of the algorithm. In the wheel topology, one particle is connected to all of the other particles. In the star topology, each particle is connected to every other particle and all the particles are fully connected. If a certain amount of random symmetrical connections are assigned to the same number of particles between pairs of particles, the topology is known as random edges topology.

Bratton and Kennedy defined a standard PSO [27], based on an extension of the PSO algorithm with constriction factors defined in Section 3.2.2, which improves the performance of the algorithm. They found that combining PSO-Co with ring topology could perform well on most of the benchmark functions.

3.3.2 PSO with a Neighbourhood Operator

In order to avoid particles being trapped in local minima, Suganthan proposed to divide the swarm into multiple 'neighbourhoods' [28], where each neighbourhood maintained its own local best solution. The local best solution is defined as the best solution within one neighbourhood and it can be discovered by inspecting all personal best solutions within that neighbourhood. The size of a neighbourhood is dynamically increased as the number of iterations increases in order to change the local best solution from the personal best solution to the global best solution. However, this approach has slower convergence compared to the PSO with global best model. The velocity update is given as

$$V_{i,n+1}^j = wV_{i,n}^j + c_{1,n}r_{i,n}^j \left(P_{i,n}^j - X_{i,n}^j \right) + c_{2,n}R_{i,n}^j \left(L_n^j - X_{i,n}^j \right), \qquad (3.45)$$

where the global best position G_n^j is replaced by the local best position L_n^j, and both the parameters $c_{1,n}$ and $c_{2,n}$ decrease as the iteration number increases according to the equations

$$c_{1,n} = \frac{\left(c_1^{begin} - c_1^{end} \right)(n_{max} - n)}{n_{max}} + c_1^{end}, \qquad (3.46)$$

$$c_{2,n} = \frac{\left(c_2^{begin} - c_2^{end}\right)(n_{max} - n)}{n_{max}} + c_2^{end},$$

(3.47)

where

c_1^{begin} and c_1^{end} are the initial value and the end value of c_1, c_2^{begin}
c_2^{end} are the initial value and end value of c_2

3.3.3 Fully Informed PSO

In all of the previous methods, the own history of the particles is used. It is possible to use history information from the neighbours as developed by Mendes et al. [29]. The method is now known as the fully informed particle swarm (FIPS). With history information provided and using an appropriate topology, the method performs better than the traditional versions on a number of test functions. The FIPS model is given as

$$V_{i,n+1}^j = \chi\left(V_{i,n}^j + \sum_{m=1}^{M_i} \frac{\sigma\left(P_{\Omega,m,n}^j - X_{i,n}^j\right)}{M_i}\right),$$

(3.48)

where

χ shares the same meaning with that in Equation 3.40
M_i is the number of neighbours particle i has
$P_{\Omega,m,n}^j$ is the jth component of the personal best position of the mth neighbour of particle i
σ is a number randomly generated by a uniform distribution between 0 and φ

It can be seen from the previous equation that the neighbourhood size M_i is an important parameter in the FIPS model, which determines the diversity of the model. In the studies performed by Mendes and Kennedy, five different neighbourhood topologies were examined, including the all topology, the ring topology, the four clusters topology, the pyramid topology and the square topology, as shown in Figure 3.6.

3.3.4 Von Neumann Topology

The von Neumann topology was separately proposed by Kennedy and Mendes [30] and Hamdan [31]. It is used in the local best model. In this

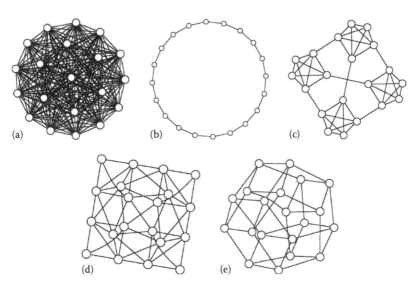

FIGURE 3.6 Different neighbourhood topologies used in FIPS: (a) all, (b) ring, (c) four clusters, (d) pyramid and (e) square.

topology, each particle is connected to its four neighbouring particles lying in the north, south, east and west directions. The key property of the PSO with von Neumann topology is to enable the particles to explore more areas in the search space in order to avoid premature convergence. As a result, the method converges slower compared to other methods. Readers interested in further details about the von Neumann topology may refer to references such as [30,31].

3.3.5 Some Dynamic Topologies

In order to locate the Pareto front of multi-objective optimisation problems, Hu and Eberhart modified the PSO by using a dynamic neighbourhood strategy [32]. In this technique, a particle finds the nearest m particles as the neighbours based on the distances between the fitness values of that particle and other particles, leading to different neighbours at each iteration.

Dynamics of particles may be changed to ensure that each particle is moving towards its neighbouring particles that have better fitness than itself. This way one requires the particles to be attracted by the best position located in the current iteration. Peram et al. proposed a modification of the direction of each particle determined by the ratio of the relative fitness and the distance of the other particles [33]. Each particle would learn from its own experience, the experience of the best particle and the experience of the neighbouring particles that have a better fitness than itself. The method is now known as the fitness–distance ratio–based PSO (FDR-PSO) and the modified velocity update is given as

$$
V_{i,n+1}^{j} = wV_{i,n}^{j} + c_1 r_{i,n}^{j}\left(P_{i,n}^{j} - X_{i,n}^{j}\right) + c_2 R_{i,n}^{j}\left(G_n^{j} - X_{i,n}^{j}\right) + c_3\left(P_{nbest,i,n}^{j} - X_{i,n}^{j}\right),
$$

(3.49)

where $P_{nbest,i,n}^{j}$ represents the jth component of the best neighbour of particle i at iteration n. In essence, this is the particle k that maximises

$$
\frac{f(P_{k,n}) - f(X_{i,n})}{P_{k,n}^{j} - X_{i,n}^{j}}.
$$

(3.50)

Note that the swarms of some local best versions of PSO with different neighbourhood structures and the multi-swarm PSOs need to be predefined or dynamically adjusted according to the distance. One such technique proposed by Liang and Suganthan is now known as the dynamic multi-swarm particle swarm optimiser (DMS-PSO) [34]. The swarm involved in this method is divided into many sub-swarms. During the search process, these sub-swarms can be regrouped as frequently as possible by using certain regrouping schedules. One typical regrouping technique used in [34] is to regroup after a given number of iterations in a random fashion and the search restarts with the new regrouped structure. After regrouping, the information of each of the former sub-swarms is exchanged among themselves. The entire swarm keeps the diversity which is helpful to avoid particles being trapped in the local minima. This is known as the randomised regrouping schedule.

The main difference between the global best model and the local best model is the size of the neighbourhood. PSO with the global best model converges faster than the local best model but is more susceptible to becoming trapped in local optima. On the other hand, PSO with the local best model search multiple regions around the local best position has better global search ability, but its convergence speed is slower. In order to inherit both the fast convergence in global best model and be able to 'flow around' local optima in the local best model, a variable neighbourhood model for PSO [35] as proposed by Liu et al. can be adopted. The concept is to make the method able to migrate from the local best model used at the beginning of the iterations to the global best model as the number of iterations increases. In the variable neighbourhood model, the number of neighbours is 2 at the beginning of the search process and increases gradually to the one less than the swarm size as the number of iterations increases.

Another modification is based on the concept that all particles dynamically share the best information of the local best particle, the global best particle and group particles as proposed by Jiao et al. [36]. Based on this idea, the velocity and position update equations are designed as

$$V_{i,n+1}^{j} = wV_{i,n}^{j} + r_{i,n}^{j}\left(\frac{a+1}{(n_{max}+1-n)}\right)\left(P_{i,n}^{j} - X_{i,n}^{j}\right) + \left(\frac{b-1}{(n_{max}+1-n)}\right)$$
$$\times \left(L_{i,n}^{j} - X_{i,n}^{j}\right) + cR_{i,n}^{j}\left(G_{n}^{j} - X_{i,n}^{j}\right), \tag{3.51}$$

$$X_{i,n+1}^{j} = X_{i,n}^{j} + V_{i,n+1}^{j}, \tag{3.52}$$

where

n_{max} is the maximum number of iterations

c is the acceleration coefficient

$L_{i,n}^{j}$ is the best position of the local neighbourhood of particle i

a and b are weights chosen between 0.6 and 1.2

The weights a and b are chosen according to the problem and are designed to reflect the degree of importance of the personal best position of and the local best position of the ith particle at the nth iteration. The remaining parameters are the same as those in Equations 3.1 and 3.2. One can easily observe from Equations 3.51 and 3.52 that each particle shares its information globally and benefits from the useful information obtained from all other particles during the search process. Experimental results of several uni-modal or multi-modal benchmark functions can be found in the original publication. Performance of this method as compared to PSO-In shows the effectiveness of the algorithm [36].

It is important not to collect useless information and lose any useful information. Using the concept of the Voronoi neighbourhood to create a geometric dynamic neighbourhood [37] before updating the velocity and position of a particle would be able to achieve this goal. The Voronoi neighbourhood used by Alizadeh et al. [37] is constructed by defining the site $V(s_i)$ which contains all of the points whose nearest site is s_i, that is,

$$V(s_i) = \{x : d(s_i, x) \le d(s_j, x), \forall j \ne i\}, \tag{3.53}$$

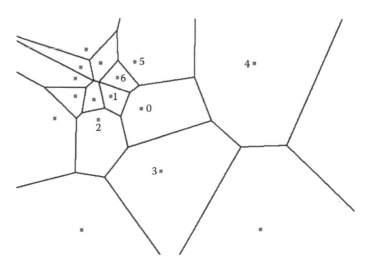

FIGURE 3.7   The Voronoi diagram in two-dimensional space.

where function $d(s_i, s_j)$ represents the Euclidean distance between s_i and s_j. The Voronoi tessellation [38] is defined as the union of the Voronoi regions of all the previous sites. Another definition used in the Voronoi neighbourhood is the adjacency of a site $adj(s_i)$, which consists of all the other sites that are adjacent to it in Voronoi tessellation. Figure 3.7 depicts the Voronoi neighbourhood using adjacency sites in which the points represent the sites, the edges represent Voronoi diagram, and the adjacency of s_0 is s_1, s_2, s_3, s_4, s_5 and s_6 [37].

This kind of neighbourhood is thus very suitable for optimum solutions outside the initialisation range. However, one shortcoming of the PSO with Voronoi neighbourhood is that the computing time of the Voronoi neighbours increases as the dimension of the problem increases.

The radius of the circle generates problems in the static Euclidean neighbourhood when it is being used to determine the neighbours in PSO. Figure 3.8 demonstrates such problem. In Figure 3.8a, the dots represent the particles which must coordinate themselves with their neighbours. To determine the neighbours of a particle, a circle is required to be drawn around the particle as shown in Figure 3.8b. Determining the radius is a tricky problem. On the one hand, some useful information of the particles outside the circle may be lost if the radius is too small. On the other hand, if the radius is too large, particles with useless information are included. The problem was first addressed by Alizadeh et al. [37] using a dynamic Euclidean neighbourhood. The method involves the use of a dynamic radius in each iteration. First, a fixed radius is assigned

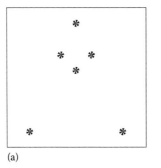 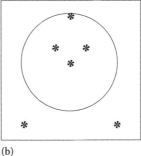

(a)   (b)

FIGURE 3.8   (a) A bird flock and (b) the neighbourhood of a particle.

at the start of the iterative process for all particles in the swarm. The particles that lie within the circle surrounding each particle are considered as its neighbours. Second, the radius R is dynamically calculated during the search based on the maximum distance between the particles, that is,

$$R = \frac{\max\left(d(X_{i,n}, X_{j,n})\right)}{\sqrt[N]{M}}, \quad i, j = 1, 2, \ldots, M, \ i \neq j, \tag{3.54}$$

where
  M is the swarm size
  N is the dimension of the search space
  Function $d(X_{i,n}, X_{j,n})$ represents the Euclidian distance between particles
    i and j at the nth iteration

Delaunay triangulation is also an efficient way of defining natural neighbours and produces useful spatial data structure for finding the nearest neighbours of a set of points. Similar concept was used by Lane et al. in a spatial sociometry for PSO [39]. Several other heuristic methods to leverage Delaunay neighbours for accomplishing diversity with local exploitation and global exploration were incorporated later [39]. Delaunay triangulation subdivides a set of points into triangles in two-dimensional (2D) cases (Figure 3.9) and tetrahedrons in three-dimensional (3D) cases [39].

Through the experiments performed on several benchmark functions Lane et al. found that their algorithm is comparatively successful with respect to the PSO with circle and von Neumann topologies.

Directed graphs were used by Mohais et al. to represent neighbourhoods for PSO [40]. A directed edge from a node u to another node v means that

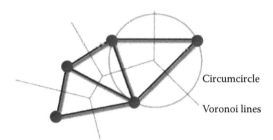

FIGURE 3.9 Delaunay triangulation of a set of points. (From Lane, J. et al., Particle swarm optimization with spatially meaningful neighbours, in *Proceedings of the 2008 IEEE Swarm Intelligence Symposium*, St. Louis, MO, 2008, pp. 1–8.)

u takes v as a neighbour, but not the other way. Two parameters were used in generating a random neighbourhood in the swarm. One is the size of the neighbourhood and the other is the out-degree of each node (the number of the outgoing edges). Two methods of modifying the structure of a neighbourhood were proposed in their work. The first method is called the 'random edge migration', which involves randomly selecting a node with the size of its neighbourhood larger than 1, detaching one of its neighbours from it and then re-attaching that neighbour to some other randomly selected node that does not have a full neighbourhood. The second method is referred to as the 'neighbourhood re-structuring'. In this approach, the neighbourhood structure of a particle is completely re-initialised, instead of being changed gradually, after it is kept fixed for a given amount of time. Only the parameters of the neighbourhood, that is, the size and the out-degree, are kept fixed and the connections are changed completely.

In graph theory a small world network is a graph in which most nodes are not neighbours of one another but can be reached from every other by a small number of hops or steps. This concept together with the selection of the best particle positions resulted in a dynamic neighbourhood topology as proposed by Liu et al. [41]. The velocity of the particle is updated according to the following equations:

$$V_{i,n+1}^{j} = wV_{i,n}^{j} + c_1 r_{i,n}^{j}\left(P_{ibest,n}^{j} - X_{i,n}^{j}\right) + c_2 R_{i,n}^{j}\left(G_n^{j} - X_{i,n}^{j}\right), \quad (3.55)$$

$$ibest = \{i \mid C_i = \max(C), L_i = \min(S_1, S_2)\}, \quad i \in S_1 \text{ or } i \in S_2, \quad (3.56)$$

$$C_i = \frac{3(K-2)}{4(K-1)}, \quad (3.57)$$

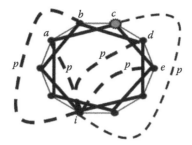

FIGURE 3.10 Learning exemplar of particle i in two small world networks. (From Liu, Y. et al., Particle swarm optimizer based on dynamic neighborhood topology, in *Proceedings of the Fifth International Conference on Intelligent Computing*, Changsha, China, 2009, pp. 794–803.)

where

S_1 and S_2 mean the entire swarms in two different small world networks

C_i is the biggest clustering coefficient for the ith particle

L is the average shortest distance from a node to another in the small world network

K is the degree of ith particle

ibest is the index of the best particle position

Note that *ibest* may or may not be the same as the index of the global best position. Figure 3.10 illustrates the update of the particle velocity using two small world networks as initial neighbourhood topologies (denoted as NT_1 and NT_2) [41]. Also, $C_i(1)$ and $C_i(2)$ represent the biggest clustering coefficient for particle i in NT_1 and NT_2, respectively; $L_i(1)$ and $L_i(2)$ are average shortest distance from particle i to another particle in NT_1 and NT_2, respectively. The learning exemplar of the particle from other personal best positions is chosen according to the following seven criteria:

1. If $C_i(1) = C_i(2)$ and $L_i(1) \neq L_i(2)$, NT_1 wins. If $L_i(1) < L_i(2)$, particle i in NT_1 is chosen as the exemplar.

2. If $C_i(1) = C_i(2)$ and $L_i(1) = L_i(2)$, NT_1 and NT_1 are chosen at will.

3. If $C_i(1) \neq C_i(2)$ and $L_i(1) = L_i(2)$, NT_1 wins. If $C_i(1) > C_i(2)$, particle i in NT_1 is chosen as the exemplar.

4. If $C_i(1) \neq C_i(2)$ and $L_i(1) < L_i(2)$, NT_1 wins, and particle i in NT_1 is chosen as the exemplar.

5. If $C_i(1) > C_i(2)$ or $L_i(1) < L_i(2)$, NT_1 wins, and particle i in NT_1 is chosen as the exemplar.

6. If $C_i(1) > C_i(2)$ or $L_i(1) < L_i(2)$, NT_1 wins, the particle i in NT_1 is chosen as the exemplar.

7. If $C_i(1) < C_i(2)$ or $L_i(1) < L_i(2)$ and the average degree $<K>$ in NT_2 is more than the $<K>$ in NT_1, NT_2 wins, and particle i in NT_2 is chosen as the exemplar. Otherwise, swarm NT_1 wins.

In Figure 3.10, particles a, b, c, d and e connect with particle i with probability p. Particle c is chosen as an exemplar based on the parameters C_i and L_i. After the learning exemplar is chosen, particle *ibest* is generated according to the personal best positions of the other particles. In this algorithm, the best neighbour is determined by the parameters C_i and L_i avoiding any blind exemplar choice that might occur as in the global best model. PSO with this method is known to be effective in solving multi-modal problems.

3.3.6 Species-Based PSO

The concept of forming species around different optima and using species seeds to provide the right guidance for particles in different species may assist locating multiple optima. The use of species around different optima is in essence determining neighbourhoods. Such concept has been used by Li in applying PSO to multi-modal problems [42]. The algorithm largely depends on how the species seeds are found [43]. First, the fitness values of all the individuals are calculated and sorted in descending order. Second, the Euclidean distance between two individuals is calculated and compared with the user-defined radius. If the distance is less than the radius, then the individual can be identified as species seeds. If an individual does not fall within the radius of all the seeds, then this individual becomes a new seed. After all the species seeds are determined, each species seed is assigned as the local best position to all individuals identified in the same species, and consequently the neighbourhood topology is determined. The next step of the algorithm is to update the velocity and position according to the local best model. After completing the iteration process, the species is able to locate multiple optima simultaneously.

3.4 PROBABILISTIC ALGORITHMS

Work in the area of simulating the particle trajectories by direct sampling, using a random number generator, from a distribution of practical interest leads to several probabilistic versions of PSO. QPSO algorithm [44–46], inspired by quantum mechanics and trajectory analysis of PSO [22], is also a probabilistic algorithm which samples the new position of the particle with a double exponential distribution. On the other hand, the update equation in the searching process uses an adaptive strategy and has fewer parameters to be adjusted. One would expect that such algorithm leads to a good performance of the algorithm as an overall result. Since the topic QPSO algorithm covers most of Chapters 4 through 6, here attention is paid to other probabilistic strategies for PSO.

3.4.1 Bare Bone PSO Family

It was pointed out by Kennedy et al. that the trajectory of a particle can be described as a cyclic path centred around a point which is computed as a random-weighted mean of the previous best individual points and the previous best neighbouring points in each dimension [4,22,47,48]. The global best position and the mean of the individual current positions and its best local neighbour form an important part in the definition of the trajectory. On the other hand, the difference between the individual and the previous best neighbourhood points is also an important parameter for scaling the amplitude of the trajectories of the particles [4]. Using these information and the particle positions sampled from Gaussian distribution the bare bone PSO (BBPSO) was examined by Kennedy [4,49] without the velocity vectors. Denote $G(mean, s_d)$ as the Gaussian random number generator. Here *mean* is defined as the mean of the individual personal best position and the global best (or local best) position in each dimension. Also s_d denotes the standard deviation defined as the difference between the individual personal best and the global best (or local best) positions on each dimension. The equation for updating the particle position is given by

$$X_{i,n+1}^{j} = G\left(\frac{P_{i,n}^{j} + G_{n}^{j}}{2}, | P_{i,n}^{j} - G_{n}^{j} | \right)$$ (3.58)

The values of skewness and kurtosis in probability theory and statistics are often used in measuring the symmetry and peakedness of the probability distribution of a real-valued random variable. Interesting phenomenon for the skewness and kurtosis of the distribution of sampling results was found in the work presented in [4,49]. The iterations for sampling were increased from 0 to 1,000,000. During the iterative procedure the skewness was always near zero in all cases and the kurtosis was high and increased as the number of iterations increases. Based on the observed results, it was pointed out that a probability distribution may represent the particle swarm trajectories with characteristics which are symmetric, high kurtosis, high centre, slumped shoulders and fat tails. Therefore, the double-exponential distribution was selected as a candidate for replacing particle swarm trajectories. The position updating equation is thus given as

$$X^j_{i,n+1} = mean + w\left(\ln(r^j_{i,n}) - \ln(R^j_{i,n})\right), \tag{3.59}$$

where

    mean is the centre

    w is used to control the standard deviation of the resulting distribution of points

    $r^j_{i,n}$ and $R^j_{i,n}$ are two different random numbers uniformly distributed on (0,1)

By comparing the experimental results between the double-exponential particle swarm and the canonical particle swarm, Kennedy observed that the double-exponential distribution does not change with the number of iterations. Kennedy has also studied a distorted Gaussian distribution by taking powers of the standard variants in particle swarm [49]. The position updating equation is modified to

$$X^j_{i,n+1} = mean + G(0,1)^r s_d w, \tag{3.60}$$

where s_d is the required standard deviation. The parameter r is used to control the kurtosis and w is used to control the resulting standard deviation. Experiments were also done and compared with the canonical particle swarm as shown in [49]. These results show that PSO with the distorted Gaussian distribution performs nearly as well as the canonical PSO on most of the cases except on the Rosenbrock function. Note that the kurtosis simply grows in the particle swarm with the increasing iterations, but not in the random number generators. On the other hand, the presence of outliers that are far from the mean makes the kurtosis grow. A Gaussian 'burst' particle swarm was meant to handle the outliers by dividing the algorithm into two modes which are the non-burst mode and the burst mode. The first step of the Gaussian 'burst' particle swarm is to determine the mode by the following methods:

$$Burst = rand(0, maxpower) \text{ if } Burst = 0 \quad \text{and} \quad rand1(0,1) < PBurstStart,$$

(3.61)

$$Burst = 0 \text{ if } Burst > 0 \quad \text{and} \quad rand2(0,1) < PBurstEnd \quad (3.62)$$

where

PBurstStart, PBurstEnd and maxpower are three user-defined parameters

Burst is an indication of the algorithm mode

rand(0, maxpower) is a random number uniformly distributed on (0, maxpower)

rand1(0,1) and rand2(0,1) are two different random numbers uniformly distributed on (0,1)

If the algorithm is in burst mode, that is, $Burst > 0$, the position is sampled as

$$X_{i,n+1}^{j} = Gaussian(0,1),$$

(3.63)

$$X_{i,n+1}^{j} = \left(X_{i,n+1}^{j} \right)^{Burst}.$$

(3.64)

If the algorithm is performed in a normal mode, that is, $Burst = 1$, then it is identical to the BBPSO. Note that good results were observed for the Rastrigin function. For the other tested functions, the Gaussian 'Burst' particle swarm does not show significant performance improvement compared to the canonical particle swarm.

3.4.2 Trimmed-Uniform PSO and Gaussian-Dynamic PSO

The PSO algorithms proposed in [4,49] have a common characteristic that the next position does not relate to the previous position. Kennedy proposed to call these algorithms the probabilistic algorithms [50]. The concept in these models is to sample the next position on the basis of the personal best positions based on a probability distribution for the generation of random numbers in order to produce a candidate solution. This approach is different from that in the canonical PSO in which the new position is determined by the two previous best positions. Based on the update equations of the canonical particle swarm, two forms of dynamic-probabilistic search methods were developed as presented in [4,49]. These two forms are known as the trimmed-uniform particle swarm (TUPS) and the Gaussian-dynamic particle swarm (GDPS).

For the TUPS algorithm, the position is updated by the equation

$$X^j_{i,n+1} = X^j_{i,n} + \chi\left(X^j_{i,n} - X^j_{i,n-1}\right) + \chi\frac{\varphi}{2}\left(\frac{r_c + (2s-1)r_w}{2}\right), \qquad (3.65)$$

where
  χ and φ have the same meaning as those in Equation 3.41
  s is a random number uniformly distributed on (0,1)

The parameters r_c and r_w are calculated by using these two expressions:

$$r_c = \frac{\left(P^j_{i,n} - X^j_{i,n}\right) + \left(G^j_{i,n} - X^j_{i,n}\right)}{2}, \qquad (3.66)$$

$$r_w = |P^j_{i,n} - X^j_{i,n}| + |G^j_{i,n} - X^j_{i,n}|. \qquad (3.67)$$

Note that the sampling results obtained by using the previous update formulae are not uniformly distributed over the entire range. This can be overcome, as proposed by Kennedy, when r_w in Equation 3.67 is divided by a number greater than 2 preferably near 2.5. The effect of this division is to trim off the tails of the distribution.

For the GDPS algorithm, the new position is determined by the equation

$$X^j_{i,n+1} = X^j_{i,n} + w_1\left(X^j_{i,n} - X^j_{i,n-1}\right) + w_2\left(\bar{P}^j_n - X^j_{i,n}\right) + s_N\left(\frac{f^j_n}{2}\right), \qquad (3.68)$$

where
  $\bar{P}^j_n$ is the mean of the personal best positions of all particles on the
  jth dimension
  s_N is a standard normal random number
  the value of w_1 is 0.729 and w_2 is 2.187

the burst mode. The first step of the Gaussian 'burst' particle swarm is to determine the mode by the following methods:

$$Burst = rand(0, maxpower) \text{ if } Burst = 0 \quad \text{and} \quad rand1(0,1) < PBurstStart,$$

(3.61)

$$Burst = 0 \text{ if } Burst > 0 \quad \text{and} \quad rand2(0,1) < PBurstEnd \quad (3.62)$$

where
PBurstStart, PBurstEnd and maxpower are three user-defined parameters
Burst is an indication of the algorithm mode
rand(0, maxpower) is a random number uniformly distributed on (0, maxpower)
rand1(0,1) and rand2(0,1) are two different random numbers uniformly distributed on (0,1)

If the algorithm is in burst mode, that is, $Burst > 0$, the position is sampled as

$$X_{i,n+1}^{j} = Gaussian(0,1),$$

(3.63)

$$X_{i,n+1}^{j} = \left(X_{i,n+1}^{j} \right)^{Burst}.$$

(3.64)

If the algorithm is performed in a normal mode, that is, $Burst = 1$, then it is identical to the BBPSO. Note that good results were observed for the Rastrigin function. For the other tested functions, the Gaussian 'Burst' particle swarm does not show significant performance improvement compared to the canonical particle swarm.

3.4.2 Trimmed-Uniform PSO and Gaussian-Dynamic PSO

The PSO algorithms proposed in [4,49] have a common characteristic that the next position does not relate to the previous position. Kennedy proposed to call these algorithms the probabilistic algorithms [50]. The concept in these models is to sample the next position on the basis of the personal best positions based on a probability distribution for the generation of random numbers in order to produce a candidate solution. This approach is different from that in the canonical PSO in which the new position is determined by the two previous best positions. Based on the update equations of the canonical particle swarm, two forms of dynamic-probabilistic search methods were developed as presented in [4,49]. These two forms are known as the trimmed-uniform particle swarm (TUPS) and the Gaussian-dynamic particle swarm (GDPS).

For the TUPS algorithm, the position is updated by the equation

$$X_{i,n+1}^{j} = X_{i,n}^{j} + \chi\left(X_{i,n}^{j} - X_{i,n-1}^{j}\right) + \chi\frac{\varphi}{2}\left(\frac{r_c + (2s-1)r_w}{2}\right), \quad (3.65)$$

where

χ and φ have the same meaning as those in Equation 3.41
s is a random number uniformly distributed on $(0,1)$

The parameters r_c and r_w are calculated by using these two expressions:

$$r_c = \frac{\left(P_{i,n}^{j} - X_{i,n}^{j}\right) + \left(G_{i,n}^{j} - X_{i,n}^{j}\right)}{2}, \quad (3.66)$$

$$r_w = |P_{i,n}^{j} - X_{i,n}^{j}| + |G_{i,n}^{j} - X_{i,n}^{j}|. \quad (3.67)$$

Note that the sampling results obtained by using the previous update formulae are not uniformly distributed over the entire range. This can be overcome, as proposed by Kennedy, when r_w in Equation 3.67 is divided by a number greater than 2 preferably near 2.5. The effect of this division is to trim off the tails of the distribution.

For the GDPS algorithm, the new position is determined by the equation

$$X_{i,n+1}^{j} = X_{i,n}^{j} + w_1\left(X_{i,n}^{j} - X_{i,n-1}^{j}\right) + w_2\left(\bar{P}_n^{j} - X_{i,n}^{j}\right) + s_N\left(\frac{f_n^{j}}{2}\right), \quad (3.68)$$

where

$\bar{P}_n^{j}$ is the mean of the personal best positions of all particles on the jth dimension
s_N is a standard normal random number
the value of w_1 is 0.729 and w_2 is 2.187

The parameter f_n^j is determined by summing the absolute values of the differences between the personal best position of a particle and its neighbours divided by the number of neighbours, that is,

$$f_n^j = \sum_{m=1}^{M_i} \frac{|P_{nbr,m,n}^j - P_{i,n.}^j|}{M_i}, \tag{3.69}$$

where

M_i is the number of neighbours of the ith particle

$P_{nbr,m,n}^j$ is the jth dimension of the personal best position of the mth neighbour of the ith particle

There were experiments designed to test TUPS, GDPS, canonical PSO and the FIPS on a set of benchmark functions [50]. Results show that TUPS has similar computational capability compared to the canonical PSO, but the former one is not as good as the FIPS algorithm. It was also observed that GDPS was able to generate better results than the canonical particle swarm and FIPS on most of the test cases.

3.4.3 Gaussian PSO and Jump Operators

The canonical PSO requires the use of several parameters, including the inertia weight, acceleration coefficient, upper limit of velocity, the maximum number of iterations and the size of the swarm. When using the canonical PSO for optimisation problems, suitable values of these parameters lead to good algorithmic performance. A natural idea of improving PSO is to reduce the number of the parameters of PSO. One such example developed by Krohling is to update the velocity with the Gaussian distribution in PSO without specifying the inertia weight, the upper limit of velocity and the acceleration coefficients [51]. The update equation is

$$V_{i,n+1}^j = |s_G|\left(P_{i,n}^j - X_{i,n}^j\right) + |m_G|\left(G_n^j - X_{i,n}^j\right), \tag{3.70}$$

where $|s_G|$ and $|m_G|$ are positive random numbers generated according to the absolute value of the Gaussian probability distribution. Experiences with the canonical PSO show that the expected values for $(P_{i,n}^j - X_{i,n}^j)$ and $(G_n^j - X_{i,n}^j)$ are 0.729 and 0.85, respectively [22,52]. Therefore, it is sensible to take the expected values on [0.729, 0.85] for the random numbers generated by using the Gaussian probability distribution, when using the

Gaussian PSO algorithm. This indicates an asymmetric probability distribution. Since there is no momentum term in the velocity update (Equation 3.70), it is not necessary to specify the upper limit of velocity for the particle. The experimental results as shown in [51] reflect that this algorithm is effective in finding the global optimum and exhibits considerably better convergence than the canonical PSO for multi-modal optimisation problems in low dimensional search space.

For multi-modal optimisation problems with high dimensionality, the algorithm may also be trapped in the local minima. To overcome this shortcoming, Krohling incorporated a *jump* operator into the Gaussian PSO in order to escape from the local minima [53]. Note that when the position of a particle is equal to its personal best position and the global best position, the fitness value does not change any more, indicating that the swarm may have been trapped in the local minima. A parameter can be used to monitor the fitness value of each particle for a pre-assigned number of iterations in such a way that the value of the parameter increases if there is no improvement of the fitness value. Once the parameter reaches a threshold, the jump operator is activated and a new position is produced. The jump operator is used as a mutation operator according to the following equations:

$$X_{i,n+1}^{j} = X_{i,n+1}^{j} + \eta s_N,$$ (3.71)

$$X_{i,n+1}^{j} = X_{i,n+1}^{j} + \eta s_C,$$ (3.72)

where

η is a scale parameter

$s_N \sim N(0,1)$ and $s_C \sim C(0,1)$ are two random numbers generated according to the standard normal distribution and Cauchy probability distribution, respectively, with mean 0 and standard deviation 1

Numerical tests on a set of benchmark functions comparing the Gaussian PSO algorithm with two different jump operators, the canonical PSO and PSO-Co can be found in [53]. Results show that the Gaussian PSO with Gaussian jump and with Cauchy jump performed considerably better than the canonical PSO and PSO-Co for most of the test cases. Krohling also compared the Gaussian PSO with the standard self-adaptive evolutionary programming (EP) with Gaussian probability distribution presented in [54]. Results also indicate that the Gaussian PSO with jumps outperformed self-adaptive EP. This leads to the conclusion that the Gaussian PSO algorithms with Gaussian jump and Cauchy jump indeed have strong ability to escape from local minima.

The same jump strategy was employed by Krohling with the BBPSO [4,49] algorithm to help particles escaping from the local minima for functions with many local minima in high dimensional search space [55]. The BBPSO with Gaussian and Cauchy jump uses the particle position update according to the following equations:

$$X_{i,n+1}^{j} = P_{i,n}^{j}(1+\eta s_N),$$

(3.73)

$$X_{i,n+1}^{j} = P_{i,n}^{j}(1+\eta s_C),$$

(3.74)

where

$P_{i,n}^{j}$ is the jth component of the position of the ith particle at nth iteration

η, s_N, and s_C share the same meanings with those in (3.71) and (3.72)

The parameter η is held constant. The BBPSO with Gaussian and Cauchy jumps with a growing scaling factor η was tested. The algorithm exhibits almost the same results when a constant scaling factor is used. Various numerical tests suggest that the jump strategy can indeed generate good points in the search [55].

3.4.4 Exponential PSO

It was pointed out by Krohling and Coelho that the PSO algorithm with Gaussian or Cauchy probability distribution may also get stuck in local minima in multi-modal problems of high dimensionality [56]. To improve the performance it is possible to use the exponential probability distribution in PSO [56]. The density function of the proposed exponential probability distribution in [56] is given by

$$f(x) = \frac{1}{2b}\exp\left(\frac{-|x-a|}{b}\right), \quad -\infty \le x < \infty, \quad \text{with } a,b > 0,$$

(3.75)

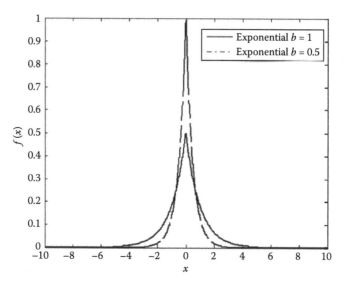

FIGURE 3.11 Exponential probability distribution function with location parameter $a = 0$ and scale parameter $b = 0.5$ and $b = 1$. (From Krohling, R.A. and dos Santos Coelho, L., PSO-E: Particle swarm with exponential distribution, in *Proceedings of the 2006 IEEE Congress on Evolutionary Computation*, Vancouver, BC, Canada, 2006, pp. 1428–1433.)

where location parameter a and scale parameter b are used to control the variance of the distribution. Figure 3.11 depicts the exponential probability distribution function with $a = 0$ and $b = 0.5$ and $b = 1$ [56].

The random number with the exponential probability distribution can be generated according to the equation

$$s_E = |a \pm b\log(s_u)|, \tag{3.76}$$

where

s_E denotes the exponential random number
s_u is a uniformly distributed random number over (0,1)

The sign $\pm$ is randomly chosen under the probability 50%. The weighting coefficients in the PSO algorithm according to the aforementioned exponential probability distribution lead to the update formula for the velocity

$$V_{i,n+1}^j = |s_E|\left(P_{i,n}^j - X_{i,n}^j\right) + |s_E'|\left(G_n^j - X_{i,n}^j\right), \tag{3.77}$$

where s_E and s_E' represent two different random numbers generated by the exponential probability distribution according to (3.70). Tests of the PSO algorithm with exponential distribution (PSO-E) on six benchmark functions show that the chance of escaping from local minima is better than the canonical PSO. Therefore, it improves the effectiveness of the PSO algorithm.

3.4.5 Alternative Gaussian PSO

Secrest and Lamont pointed out that there are two weaknesses in the distribution curve of PSO [57]. The first is the linearisation of the curve obtained in steady state and the second is the location of the median. With the median located between the global best and personal best positions, the area around the median is searched by the particle predominantly at the expense of the area around the global best and the personal best positions. To overcome the weaknesses, Secrest and Lamont proposed an alternative Gaussian PSO [57], in which the particle position is updated through three steps at each iteration. First, the magnitude of the new velocity is obtained by

$$|V_{i,n+1}| = (1-C_2)|P_{i,n}-G_n|s_N, \quad \text{when } s_u > C_1, \qquad (3.78)$$

or

$$|V_{i,n+1}| = C_2|P_{i,n}-G_n|s_N, \quad \text{when } s_u \le C_1, \qquad (3.79)$$

where
$|V_{i,n+1}|$ is the magnitude of the new velocity
$s_N \sim N(0,1)$ is a random number generated according to the standard normal distribution
$s_u \sim U(0,1)$ is a random number uniformly distributed on the interval $(0,1)$
C_1 and C_2 are two constants between 0 and 1
$|P_{i,n} - G_n|$ represents the distance between the global best and personal best positions

If both positions are the same, then set $|P_{i,n} - G_n|$ to one. Therefore, $(1-C_2)|P_{i,n}-G_n|$ or $C_2|P_{i,n}-G_n|$ provides the standard deviation in (3.78) or (3.79).

Second, the new velocity vector is determined by

$$V_{i,n+1} = |V_{i,n+1}|R, \qquad (3.80)$$

where R is a random vector with magnitude of one and angle uniformly distributed from 0 to 2π. It is essentially a unit vector indicating the direction of the new velocity vector. Thus, Equation 3.80 means that the direction of the new velocity vector is specified randomly after its magnitude is given.

Third, the new position vector of the particle is obtained by

$$X_{i,n+1} = P_{i,n} + V_{i,n+1}, \quad \text{when } s_u > C_1, \tag{3.81}$$

or

$$X_{i,n+1} = G_n + V_{i,n+1}, \quad \text{when } s_u \leq C_1, \tag{3.82}$$

where s_u has the same value as that in (3.78) or (3.79). Hence, it can be seen that C_1 is used to determine the distances from the particle to the global best position and personal best position. Larger C_1 means that more particles are placed around the global best position.

Numerical experiments were performed by using the Gaussian PSO and the canonical PSO; results show that the Gaussian PSO has performance significantly better than the canonical PSO [57].

3.4.6 Lévy PSO

Very often using Gaussian distribution in the swarm does not generate any new positions that are further away from the original position. Remedial action can be achieved by using the Lévy distribution which generates such new positions. This has advantages over Gaussian distribution in the optimisation of multi-modal function by evolutionary programming (EP) as shown in [54]. Similarly, attempts of using the Lévy distribution in the PSO algorithm in order to improve the performance of the algorithm by allowing escape from a local optimum can be found in [58]. The Lévy distribution has the following probability density:

$$L_{\alpha,\gamma}(z) = \frac{1}{\pi} \int_0^\infty \exp(-\gamma q^\alpha) \cos(qz) dq, \tag{3.83}$$

where
   the parameter α is used to control the sharpness of the graph which characterises the distribution
   the parameter γ is used to control the scale unit of the distribution

Figure 3.12 depicts the Lévy distribution with different values of α together with the Gaussian distribution as a comparison [58].

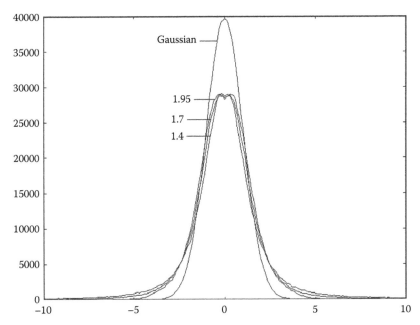

FIGURE 3.12   Gaussian and Lévy distributions. (From Richer, T.J. and Blackwell, T.M. *Proceedings of the 2006 Congress on Evolutionary Computation*, Vancouver, BC, Canada, 2006, pp. 808–815.)

It can be seen from Figure 3.12 that when α is smaller than 2, the Lévy distribution has shapes similar to the Gaussian distribution but with fatter tails which can produce more outliers in the search procedure. Instead of using the Gaussian-based PSO algorithms, such as Gaussian BBPSO [4,49], Gaussian PSO [57] and Gaussian Pivot, Richer and Blackwell replaced the Gaussian distribution by the Lévy distribution in these algorithms and tested them by using nine benchmark functions. Results with PSO-Co and the Gaussian-based PSO algorithms are also reported in [58]. It appears that the Lévy-based PSO algorithms performed as good as or better than PSO-Co and their Gaussian counterparts.

3.4.7 PSO with Various Mutation Operators

Mutation operations were also considered to improve the hybrid PSO with breeding and sub-population in [59]. As an example, Higashi and Iba have integrated a Gaussian mutation into the hybrid PSO [60] using the following equation:

$$mutation(x) = x(1 + G(0, \sigma)), \tag{3.84}$$

where

x denotes any component of the position of the particle where mutation
is applied

σ is set to be 0.1 times the length of the search space on each dimension

In the PSO with the aforementioned Gaussian mutation, the particles are selected at the predetermined probability and their positions are then calculated according to the previous equation governed by the Gaussian distribution. The algorithm was tested on a set of benchmark functions and a real-world problem about the inference of a gene network in order to show the effectiveness of the algorithm.

Recall that in the PSO algorithm, particle positions generally converge rapidly during the initial stages and slow considerably during the later stages and, very often, may be trapped in local minima. In an adaptation of the mutation by Stacey et al., a component of a selected particle is mutated as according to the mutation probability $1/N$, where N is the dimensionality of the problem [61]. The attempt was to speed up the convergence and to escape from, if the particle is being trapped in, local minima of the problem. Here, a Cauchy distribution is used instead of the Gaussian mutation operation [57,60] to produce random numbers as follows:

$$F(x) = \frac{0.2}{\pi} \cdot \frac{1}{x^2 + 0.2^2}. \tag{3.85}$$

The reason for using Cauchy distribution is that Cauchy distribution is similar to the normal distribution but is fatter in the tails. The effect of such choice is to increase the probability of generating larger values. Tests performed on four benchmark functions by Stacey et al. [61] using the Cauchy mutation operator in canonical PSO and PSO-Co show that the Cauchy mutation could provide a significant improvement in the performance of the algorithm on most of the test cases and convergence speed for the harder functions.

The aforementioned modifications involved random probability distributions. Certain situations may require less randomness, which could be achieved by the use of a quasi-random sequence. The systematic mutation (SM) operator, proposed by Pant et al., is based on such idea in order to perturb the swarm for the improvement of the algorithmic performance [62]. Here a quasi-random Sobol sequence [63] which covers the

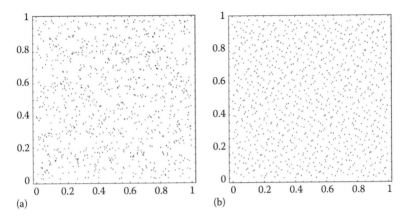

FIGURE 3.13   Sample points of pseudo-random sequence (a) and quasi-random sequence (b).

search space more evenly is used to generate new positions in the search space. Figure 3.13 illustrates the distributions using a random probability and the quasi-random sequence.

The Sobol mutation operator exerts on two kinds of individuals, the global best particle and the global worst particle, leading to two versions of PSO with SM mutation operator. In the PSO algorithm with Sobol mutation operator, the Sobol mutation is executed at every iteration after the personal best and global best positions have been updated according to the following equation:

$$temp_n^j = \frac{R_1 + R_2}{\ln R_1}, \tag{3.86}$$

where

   R_1 and R_2 are two random numbers in a Sobol sequence

   $temp_n^j$ is the jth component of a new position yielded by the Sobol mutation operator

After generating the new position, its fitness is evaluated and compared with the fitness value of the global best particle or the global worst particle. If the new position is better, then the new position replaces the global best particle or the global worst particle.

Adaptive mutation using beta distribution is an alternative method as discussed by Pant et al. [64]. The mutation applies on the personal best position of a particle or the global best particle, in order to improve the performance of PSO, using the following equations:

$$X_{i,n}^j = X_{i,n}^j + \sigma_i^{j'} Betarand^j, \tag{3.87}$$

$$\sigma_i^{j'} = \sigma_i^j \exp(\tau N(0,1) + \tau' N_j(0,1)), \tag{3.88}$$

$$\tau = \frac{1}{\sqrt{2M}}, \quad \tau' = \frac{1}{\sqrt{2\sqrt{M}}}, \tag{3.89}$$

where

$Betarand^j$ is a random number generated by beta distribution with parameters less than 1

M is the population size

$N(0,1)$ is a normally distributed random number with mean zero and deviation 1

$N_j(0,1)$ represents the random number generated for the jth component of the particle position drawn from a normal distribution

After each iterative updates of the velocity, position and personal best position of the particle and the global best position, the mutation operation will act on the personal best position of the particle or the global best particle according to certain rules described as follows. When $rand(0,1) < 1/N$, the personal best position or the global best particle is mutated. Here $rand(0,1)$ is a uniformly distributed random number on $(0,1)$, N is the dimensionality of the problem. The position before mutation is replaced by the mutated one if it is a better position.

Another concept of mutation is to diversify the swarm to avoid being trapped in local minima. The diversity can be obtained by means of a non-uniform mutation based on random numbers depending on the current iteration. This was first developed by Michalewicz for real-valued GA [65]. It was then used in the PSOs with global best model and local best model as proposed by Esquivel and Coello [66]. The non-uniform mutation operator proposed by Michalewicz is given by

$$mutation(X_{i,n}^j) = \begin{cases} X_{i,n}^j + \Delta(n, X_{max} - X_{i,n}^j), & r = 0, \\ X_{i,n}^j - \Delta(n, X_{i,n}^j - X_{min}), & r = 1, \end{cases} \tag{3.90}$$

where

$X_{i,n}^j$ is the jth component of the current position of particle i at the nth iteration

n is the current iteration number

X_{min} and X_{max} are the lower and upper bounds of each component of the particle's position

r is a binary integer whose value is randomly determined to be either 0 or 1

function $\Delta(n, y)$ returns a value on the interval $[0, y]$ governed by

$$\Delta(n, y) = y(1 - R^{(1-n/n_{max})^b}), \qquad (3.91)$$

where

R is a random number uniformly distributed on $(0,1)$

n_{max} is the maximum number of iterations

b is a system parameter determining the degree of dependency on iteration number and is set to 5 in [66] as suggested by Michalewicz [65]

It should note that $\Delta(n, y)$ decreases as the number of iterations increases. Four hard multi-modal functions were tested in [66] to demonstrate the convergence with the convergence-guaranteed PSO in [67] as one of the competitor algorithms. Results showed that PSO algorithms with the non-uniform mutation operator described earlier perform well.

Apart from mutating the positions it is also possible to mutate the velocities [68]. This concept has been used by Ratnaweera et al. in PSO. When the global best position does not change for a certain number of iterations, the mutation operator will be exerted on the particles. The particle to be mutated is randomly selected to surf a random perturbation on randomly selected components of its velocity vector under pre-defined probabilities. The mutation strategy is described as follows:

$$V_{k,n}^l = V_{k,n}^l + rand3(\cdot)\frac{V_{max}}{m} \quad \text{if } rand1(\cdot) < p_m \text{ and } rand2(\cdot) < 0.5, \quad (3.92)$$

$$V_{k,n}^l = V_{k,n}^l - rand4(\cdot)\cdot\frac{V_{max}}{m} \quad \text{if } rand1(\cdot) < p_m \text{ and } rand2(\cdot) \geq 0.5, \quad (3.93)$$

where

index k indicates the randomly selected particle

index l represents randomly selected component

rand1, rand2, rand3 rand4 are used to generate uniformly distributed
random numbers on (0,1) independently

p_m is the mutation probability

m is a constant

It was found that the performance of the PSO with the aforementioned mutation depends on the mutation step size. This leads to the use of a time-varying mutation step size which can overcome the difficulties of selecting a proper mutation step size.

Several different mutation operators [60,61,66,68] used in PSO were compared, finding optima of functions and solving real-world constrained optimisation problems, by Andrews to establish their effectiveness and performance [69]. The mutation operator used by Andrews in his studies is known as the random mutation operator which re-initialises each component of the particle position using a uniformly generated random value within the allowable range of each component. This version of mutation is often used in real-valued GA [65]. There are three possible ways to compute the mutation rate, namely constant mutations, linearly decreasing mutations, and mutations when stagnant. It was concluded that adding mutation operators to PSO would enhance the optimisation performance for PSOs with both the global best and local best models in general, but the improvement is often problem dependent.

A phenomenon known as stagnation [23] appears in the canonical PSO. This is due to the following drawback of PSO related to the velocity update, as pointed out by Esmin et al. [70]. If the current position of a particle coincides with the global best position, the particle will only move away from this point when the inertia weight and previous velocity are different from 0. All the particles stop moving once they catch up with the global best particle if their previous velocities are very close to zero. This leads to premature convergence of the algorithm—that is, the particles may stop exploring or exploiting any new space. In order to make the particle swarm escape from local optima and be able to search in different zones of the search space, Esmin et al. proposed a new model by incorporating the mutation operator,

$$mutate(X_{k,n}^j) = X_{k,n}^j + pert, \qquad (3.94)$$

where

$X^j_{k,n}$ is the *j*th component of the particle (with index *k*) being a randomly selected particle from the swarm at iteration *n*

pert is a random value between 0 and $0.1 \times (X_{max} - X_{min})$

The value of $(X_{max} - X_{min})$ denotes the length of the search range for each component. Experimental results performed by Esmin et al. on four benchmark functions and the loss power minimisation problem and compared with the canonical PSO confirm that the performance of this PSO algorithm exhibits a faster convergence speed.

It can be observed from (3.90) that the mutation space is kept unchanged throughout the search procedure and the space for the permutation of particles in the PSO is also fixed. It would be useful to consider varying the mutation space during the search [71,72]. Ling et al. implemented a dynamic mutating space which depends on the properties of a wavelet function. The wavelet is known to be a tool of modelling seismic signals by combing the dilations and the translation of a simple oscillatory function of finite duration [71]. There are two reasons why wavelet is used in mutation operations. One reason is that the admissibility criterion the mother wavelet must satisfy can improve the stability of the mutation operation. The other reason is that decreasing the amplitude of the wavelet function to constrain the search space, when the iteration number increases, can achieve a fine-tuning effect on the mutation operation. Therefore, solution quality can be improved by PSO with the wavelet mutation operation, which is given by

$$mut(X^j_{p,n}) = \begin{cases} X^j_{p,n} + \sigma(X^j_{max} - X^j_{p,n}) & \text{if } \sigma > 0, \\ X^j_{p,n} + \sigma(X^j_{p,n} - X^j_{min}) & \text{if } \sigma \le 0, \end{cases} \tag{3.95}$$

where

$$\sigma = \psi_{a,0}(\varphi) = \frac{1}{\sqrt{a}} \psi\left(\frac{\varphi}{a}\right), \tag{3.96}$$

$X^j_{p,n}$ is the *j*th component of the position of a randomly selected particle *p* to be mutated

$[X^j_{min}, X^j_{max}]$ is the search scope of the particle in the *j*th component

a is the dilation parameter varying according to n/n_{max} for fine-tuning purpose

φ is randomly generated between $-2.5a$ and $2.5a$

The meaning of n/n_{max} is the current number of iterations divided by the maximum number of iterations. Ling et al. suggested that the dilation parameter a may be determined by

$$a = e^{-\ln(g)\cdot\left(1-n/n_{max}\right)^{\zeta_{wm}}+\ln(g)},$$ (3.97)

where

ζ_{wm} is the shape parameter of the monotonic increasing function
g is the upper limit of the parameter a
$\psi(\varphi/a)$ in (3.96), known as the 'mother wavelet', is a continuous function of time

There are other mother wavelets as discussed in [71], the choice of the Morlet wavelet as the mother wavelet offers the best performance from experimental results. Therefore, the dilation parameter is given by

$$\sigma = \frac{1}{\sqrt{a}} e^{-(\varphi/a)^2/2} \cos\left(5\left(\frac{\varphi}{a}\right)\right).$$ (3.98)

Numerical solutions to the benchmark functions discussed in [73] show that the PSO with wavelet mutation operator again outperforms the PSO algorithm in terms of convergence speed, solution quality and solution stability.

Using the normal cloud model [74] in order to enhance the PSO algorithm is yet another idea [75]. Such a model is expressed by three digital characteristics, the expectations Ex, entropy En and hyperentropy He and was used by Wu et al. in the PSO algorithm with constrict factor to help the swarm escape the local minima [75]. The characteristics of the normal cloud model are determined by

$$Ex = f_{ave},$$ (3.99)

$$En = \frac{\left(f_{gbest} - f_{ave}\right)}{a_1},$$ (3.100)

$$He = \frac{En}{a_2},$$ (3.101)

where

f_{ave} is the average fitness of all the particles
f_{gbest} is the fitness of the global best particle

With these given parameters, one can create a normal random number S with the expected value En and standard deviation He by the equation

$$S = randn(En, He), \qquad (3.102)$$

where $randn(x, y)$ is the generator of a normal distributed random number with expected value and standard deviation being En and He, respectively.

Then, a normal random number $C_{i,n}^j$ with the expected value Ex and standard deviation S by the equation

$$C_{i,n}^j = randn(Ex, S), \qquad (3.103)$$

where $C_{i,n}^j$ represents a cloud drop in the universal space and is used by the following mutation operation.

At each iteration of the algorithm, after the position of each particle is updated, the components of the particle to be mutated are then randomly selected subject to the probability, p_m/N, where p_m is the user-defined mutation rate, and N is the dimension of a particle. The normal cloud mutation is then applied to the selected component of particle i, $X_{i,n}^j$ position via

$$mutate(X_{i,n}^j) = X_{i,n}^j + kC_{i,n}^j \qquad (3.104)$$

where k is a user-defined parameter for determining the step size of the mutation operation. Numerical experiments demonstrate the effectiveness of the PSO with normal cloud mutation operator as shown in [75].

The estimation of distribution algorithms (EDA) which build a probabilistic model by using selected individuals to capture the search space properties in order to generate new individuals was used by El-Abd in PSO in order to prevent the premature convergence of the PSO algorithm [76]. Two possible update methods for the position of the particle were used: one being the normal equation in canonical PSO and the other is to sample the position using the estimated distribution. A probability called the participation ratio is used to guide the particle in deciding which of the two methods is to be used to update its position. The participation ratio can be adaptively determined by

$$p^{n+1} = \frac{\displaystyle\sum_{k=1}^{n} \frac{sum\_PSO^k}{num\_PSO^k}}{\displaystyle\sum_{k=1}^{n} \frac{sum\_PSO^k}{num\_PSO^k} + \displaystyle\sum_{k=1}^{n} \frac{sum\_EDA^k}{num\_EDA^k}}, \qquad (3.105)$$

where

sum_PSO^k and num_PSO^k are the sum of improvements and the number of improvements done by PSO at the nth iteration, respectively

sum_EDA^k and num_EDA^k are the sum of improvements and the number of improvements done by EDA at the nth iteration respectively

The EDA described in [76] employs an independent univariate Gaussian distribution based on the best half of the swarm [77] with the mean and standard deviation calculated by

$$\mu_n^j = \frac{1}{M}\sum_{i=1}^{M} X_{i,n}^j, \tag{3.106}$$

$$\sigma_n^j = \sqrt{\frac{1}{M}\sum_{i=1}^{M}\left(X_{i,n}^j - \mu_n^j\right)^2}, \tag{3.107}$$

where M is the number of half of the swarm.

3.5 OTHER VARIANTS OF PSO

3.5.1 Discrete PSO Algorithms

In 1997, Kennedy and Eberhart proposed the first discrete binary version of PSO algorithm [78]. In the discrete binary PSO, vectors describing the positions and velocities are represented as a binary string of 0 or 1 and the personal best and the global best positions are updated just as in the continuous version. The trajectories are defined in terms of probabilities that a bit should be in one state or the other. Each component of the particle can only be restricted to move to either 0 or 1. Each component of the velocity represents the probability of the corresponding component of position taking the value 1 or 0. Kennedy and Eberhart proposed a sigmoid function for mapping the real-valued velocity to the range [0,1] as follows:

$$V_{i,n}'^j = sig\left(V_{i,n}^j\right) = \frac{1}{1+e^{-V_{i,n}^j}}, \tag{3.108}$$

where $V_{i,n}^j$ is calculated as that in the continuous PSO algorithm except that the value of the personal best position and global best position are integers of either 0 or 1. The position update equation is defined by

$$X_{i,n+1}^j = \begin{cases} 1 & r_{i,n}^j < sig\left(V_{i,n}^j\right), \\ 0 & otherwise, \end{cases} \tag{3.109}$$

where $r_{i,n}^j$ is a random number uniformly distributed on (0,1). Experimental results of the discrete binary PSO demonstrated by De Jong's five functions as shown in [78] show the flexibility and robustness of the algorithm.

There are shortcomings and problems exist in the aforementioned discrete binary PSO algorithm. First, the new position of the particle is updated randomly and independent of the previous position. This is different from the continuous PSO algorithms. Second, the different definitions of the parameters between the discrete binary PSO algorithm and the continuous PSO algorithm, such as the inertia weight and maximum velocity, affect the performance. Khanesar et al. proposed a different type of binary PSO algorithm with a different interpretation of velocity [79]. In their work two vectors, $V_{i,n+1,0}$ $V_{i,n+1,1}$, for each particle are used. The components of these two vectors are defined as

$$V_{i,n+1,1}^j = w \cdot V_{i,n,1}^j + d_{i,n,1}^{j,1} + d_{i,n,1}^{j,2}, \tag{3.110}$$

$$V_{i,n+1,0}^j = w \cdot V_{i,n,0}^j + d_{i,n,0}^{j,1} + d_{i,n,0}^{j,2}, \tag{3.111}$$

where
  w is the inertia weight and has similar meaning with that in the continuous PSO
  $d_{i,n,1}^j$ and $d_{i,n,0}^j$ are two temporary values which were determined according to the following four rules:

If $P_{i,n}^j = 1$ then $d_{i,n,1}^{j,1} = c_1 r_{i,n}^j$ and $d_{i,n,0}^{j,1} = -c_1 r_{i,n}^j$
If $P_{i,n}^j = 0$ then $d_{i,n,0}^{j,1} = c_1 r_{i,n}^j$ and $d_{i,n,1}^{j,1} = -c_1 r_{i,n}^j$
If $G_n^j = 1$ then $d_{i,n,1}^{j,2} = c_2 R_{i,n}^j$ and $d_{i,n,0}^{j,2} = -c_2 R_{i,n}^j$
If $G_n^j = 0$ then $d_{i,n,0}^{j,2} = c_2 R_{i,n}^j$ and $d_{i,n,1}^{j,2} = -c_2 R_{i,n}^j$

where
  $r_{i,n}^j$ and $R_{i,n}^j$ are two random numbers uniformly distributed on (0,1)
  c_1 and c_2 are two acceleration coefficients to be supplied by the user

By using the values of $V^j_{i,n+1,0}$ and $V^j_{i,n+1,1}$, the probability of velocity change in the jth bit of the ith particle is determined by

$$V^{j,c}_{i,n+1} = \begin{cases} V^j_{i,n+1,1} & \text{if } X^j_{i,n} = 0, \\ V^j_{i,n+1,0} & \text{if } X^j_{i,n} = 1. \end{cases} \tag{3.112}$$

The probability $V^{j,c}_{i,n+1}$ should be normalised by sigmoid function (3.124) as that used in [78], that is,

$$V'^{j,c}_{i,n+1} = sig(V^{j,c}_{i,n+1}) = \frac{1}{1+e^{-V^{j,c}_{i,n+1}}} \tag{3.113}$$

At last the new position of the particle can be updated according to

$$X^j_{i,n+1} = \begin{cases} \overline{X}^j_{i,n} & \text{if } rand < V'^{j,c}_{i,n+1} \\ X^j_{i,n} & \text{if } rand \geq V'^{j,c}_{i,n+1} \end{cases} \tag{3.114}$$

where

$rand$ is a random number uniformly distributed over $(0,1)$

$\overline{X}^j_{i,n}$ represents the complement of $X^j_{i,n}$

Other drawbacks of the discrete binary PSO were also discussed by Afshinmanesh et al. [80]. There the major concern is that the position update has a non-standard form and it is difficult to implement comparison with the continuous PSO. In addition, the non-monotonic shape of the changing probability function of a bit is in an unusual concave shape. Finally, the sigmoid function may also limit the values of velocity between 0 and 1 and lead to non-linear problems. The method proposed there is to use an artificial immune system [80]. The new position of the particle is generated by a set of operators between the previous position, the global best position and the personal best position using the binary form. These operators can be described as follows:

$$d_{1,i,n} = P_{i,n} \oplus X_{i,n}, \tag{3.115}$$

$$d_{2,i,n} = G_n \oplus X_{i,n}, \tag{3.116}$$

$$c_1 = rand(1, len), \quad c_2 = rand(1, len), \tag{3.117}$$

$$V_{i,n+1} = c_1 \otimes d_{2,i,n} + c_2 \otimes d_{2,i,n}, \tag{3.118}$$

$$X_{i,n+1} = X_{i,n} + V_{i,n+1}, \tag{3.119}$$

where
   len is the length of an individual string
   rand(1,len) is used to generate a vector in length of *len* with each element
      being a random number uniformly distributed over (0,1)

The operator $\oplus$ is used to compare two components using the exclusive disjunction operation, that is, the 'xor' operator, and the operator $\otimes$ is an 'add' operator.

Sadri and Suen incorporated the discrete binary proposed by Kennedy and Eberhart PSO in [78] with the concepts of birth and death operators in order to change the population dynamically [81]. These two operators are executed according to the birth and death rates. As a result of these two operators, the population size changes with every iterative step. Historical information such as the best solutions may also be added in the algorithm. The birth operator and death operator are executed after all the particles are updated at each iteration according to the discrete binary PSO algorithm in [78]. The birth operator is used to generate new particles and the death operator is used to remove the selected particles. The birth rate $b(t)$ and death rate $m(t)$ can be determined by

$$b(t) = \alpha_1 \cos(\omega t + \theta_1) + \beta_1, \quad \beta_1 \geq \alpha_1 \geq 0, \tag{3.120}$$

$$m(t) = \alpha_2 \cos(\omega t + \theta_2) + \beta_2, \quad \beta_2 \geq \alpha_2 \geq 0, \tag{3.121}$$

where
   $\alpha_1, \alpha_2, \beta_1, \beta_2$ and ω are set as very small positive values
   the values of θ_1 and θ_2 are selected in the range of $[0, 2\pi]$

Experimental results show that the algorithm could converge faster and better solutions can be found.

Finally, without providing the details, Lee et al. proposed an approach imitating the mechanism of genotype–phenotype concept in biology [82]. In this approach, the genotype is matched to the velocity and the phenotype to the position, which are in turn transformed by the sigmoid function [82]. The interpretation of velocity and position in the method is the same as that in the discrete binary PSO [78]. The genotype of an individual is the genetic information carried by the genes of each individual and the phenotype indicates all the observable characteristics of an individual. Details of the method can be found in [82]. Moreover, Marandi et al. proposed a kind of Boolean PSO algorithm based on the logical operator [83], which seems to be a promising binary version of PSO.

3.5.2 Hybrid PSO Algorithms

Hybridisation is a common strategy used to improve or enhance the performance of PSO. It aims at combining the desirable properties of different approaches to mitigate PSO weaknesses. Many hybrid PSO methods have been proposed and have been shown to have unique advantages. This section only covers some of those important ones in this area of research.

3.5.2.1 PSO with Selection Operator

In evolutionary algorithms [84,85], a selection operator is used to remove poor performing individuals and replace them with the other members. Therefore, it is useful for the good performing individuals to spread their information in the search region. One hybrid PSO is to incorporate the tournament selection mechanism similar to that used in the traditional evolutionary algorithms [86]. In the algorithm, the fitness values of all the particles are computed after the positions are updated at each iteration and the positions of these particles are sorted according to the fitness in descending order. The sorted particles are divided into two parts: the best half of the swarm and the worst half of the swarm. The velocities and the positions of the best half replace the velocities and the positions of the worst half. During the replacement procedure, all the personal best positions of the particles are kept unchanged in order to influence the update of their next positions. Through this kind of selection operation, the particles in the worst half of the swarm are moved to the relatively better positions than their previous positions. This provides better exploitative abilities for the particles and is helpful for finding better solutions more consistently than the original PSO algorithm. Numerical tests demonstrate that the algorithm performed poorly on the Griewank function.

3.5.2.2 *PSO with Differential Evolution*

Note that when the terms $V_{i,n}^j$, $|P_{i,n}^j - X_{i,n}^j|$, and $|G_n^j - X_{i,n}^j|$ in the velocity Equation 3.3 are all small, the particle may lose the exploration capacity during some of the iterative steps. A hybrid PSO with differential evolution (DE) operator (DEPSO) [87] and the bell-shaped mutation [88] may be used in the PSO to diversify the particle swarm and keep the swarm dynamical. The idea used in this approach is to calculate the differential change of the *j*th component of two personal best positions randomly chosen from the swarm at the *n*th iteration. Such differential change is added to the position of the particle when a random number generated is less than the crossover constant. It was suggested to perform the DE operator and the canonical PSO alternatively.

3.5.2.3 *PSO with Linkage Learning*

For some optimisation problems, decision variables (or the elements of the decision vector) may be closely related and should be evolved simultaneously as the problem is being solved. In these situations, the effects of the linkages among the elements when GA and other evolutionary algorithms are used should not be neglected. It is clear that the canonical PSO has no linkage operation. Devicharan and Mohan first proposed a hybrid PSO with linkage learning known as linkage-sensitive PSO (LPSO) [89]. The linkage learning could enable recognition and exploitation leading to effective search for high-quality candidate solutions. The linkage relations among components of the particle are represented by a linage matrix computed based on observation of the results of perturbations performed in some randomly generated particles. The values of elements in the linkage matrix are in the range of [0, 1] and indicate the strength of the linkage between two components of the particle position. Three different approaches may be used to determine the linkage matrix, namely adaptive linkage, hand-tailored linkage and fixed linkage. In adaptive linkage method, the linkage matrix is adaptively generated according to the linkage learning algorithm which samples the problem space with a large number of particles. In hand-tailored linkage approach, the linkage matrix is specified according to the domain-specific knowledge. In the fixed linkage approach, all the elements of the linkage matrix are set to 1. According to the linkage matrix, only the components of the position of the particle with strong linkage update their positions simultaneously. This search strategy is very different from that in the canonical PSO, since all the position components of the particle in the canonical PSO are attracted by the global best position. However, in the LPSO, the attractor for each particle is selected separately, depending on the subset of elements according to the linkage strengths. The attractor selection for the ith particle at the nth iteration using the subset of components $S_{i,n}$ can be realised by

$$g(S_{i,n}, y, X_{i,n}) = \frac{\left[f(P_{y,n}) - f(X_{i,n})\right]}{\left\|Proj(P_{y,n}, S_{i,n}) - Proj(X_{i,n}, S_{i,n})\right\|}, \qquad (3.122)$$

where
    y is the index of the particle acting as the attractor of the ith particle
    f is the fitness function (i.e. objective function)
    $P_{y,n}$ is the best position discovered by yth particle up to iteration n
      (i.e. the personal best position of particle y)
    $\|\cdot\|$ is the Euclidean norm
    $Proj(Z, S_{i,n})$ is the projection of Z (Z is replace by $P_{y,n}$ and $X_{i,n}$ in the
      equation) onto the positions in subset $S_{i,n}$

The function $g(S_{i,n}, y, X_{i,n})$ favours those particles where a small change in the strongly linked components results in a large improvement in the corresponding particle fitness. Using Equation 3.122 the personal best position with the most fitness improvement modulated by a distance constraint is selected as the attractor. The personal best positions with better fitness and greater proximity to X have a greater potential to attract particle i. Therefore, after the attractor has been found, the jth component of all the positions of the particles in the subset $S_{i,n}$ is updated according to

$$X^j_{i,n+1} = (1-c)X^j_{i,n} + cP^j_{y,n}, \qquad (3.123)$$

where c is a learning rate parameter valued in the range of $(0,1)$. It is employed to control the magnitude of particle position update. The optimal value of c can be determined by the users in the light of the empirical experiments or by the adaptation algorithms.

Experimental results demonstrated the robustness of the PSO with hand-tailored linkage. It should also be noted that the non-zero linkage values allow more elements to be updated simultaneously and the strongly linked components could preserve a bias towards preferentially updating.

Dynamic linkage is also a good way forward. Chen et al. proposed a dynamic linkage discovery technique to detect the building blocks of the objective function [90]. They also introduced a recombination operator for the building blocks in order to enhance the performance of PSO with the linkage. The method uses a randomly assigned linkage group in the initialisation according to the mechanism of natural selection. The current method ensures that the linkage configuration varies with the fitness landscape and the population distribution in different stages of the optimisation process. Several iterations of the procedure of the canonical PSO are executed first, followed by the search iterations using the recombination operator according to the linkage information. The recombination operation involves the selection of a set of particles with good fitness from the swarm in order to make up the building block pool. These particles can be regarded as the multi-parent of the new particles which are generated by choosing and recombining building blocks from the pool at random. Good results are demonstrated in [90].

3.5.2.4 Hybrids of PSO with GA

3.5.2.4.1 GA/PSO Algorithm   The interactions amongst particles in PSO enable the particle swarm with good ability to search for an optimal solution with high precision. However, this would weaken the global search ability at the later stage of the search. On the other hand, GAs have good global search ability but suffer from difficulties in finding good solutions. Settles and Soule have taken the advantages of both methods and proposed a hybrid PSO algorithm with GA (GA/PSO) to overcome the aforementioned issues [91].

There are three steps involved in the GA/PSO algorithm. First, a part of the best particles are selected and put into a temporary swarm considered to be the elitism. Second, a set of particles are selected to perform the standard velocity and position update rules of PSO. The number of particles in this set is determined by the formula

$$(M - M_{Elitism})Breed\_Ratio \tag{3.124}$$

where

M is the swarm size
$M_{Elitism}$ is the number of elitism selected in the first step
$Breed\_Ratio$ is the specified within [0,1] by users

The breeding ratio may be set static, stochastic or adaptive. Third, the crossover and mutation operators are exerted on the remaining particles.

The crossover operation is to update the position of the particle according to the velocity propelled averaged crossover (VPAC) method,

$$X_{p,n}'^j = \frac{\left(X_{p,n}^j + X_{q,n}^j\right)}{2 - \varphi_1 V_{q,n}^j}, \qquad (3.125)$$

$$X_{q,n}'^j = \frac{\left(X_{p,n}^j + X_{q,n}^j\right)}{2 - \varphi_2 V_{p,n}^j}, \qquad (3.126)$$

for $1 \le j \le N$, where $X_{p,n}'^j$ and $X_{q,n}'^j$ are two children generated by the parent particles p and q, whose current positions and velocities are $X_{p,n}^j$, $V_{p,n}^j$, $X_{q,n}^j$ and $V_{q,n}^j$; φ_1 and φ_2 are two random numbers uniformly distributed on (0,1).

Numerical tests indicate that the hybrid algorithm is competitive with both the GA and PSO and able to find the optimal or near optimal solution with a significantly faster speed than the GA and PSO.

3.5.2.4.2 Hybrid PSO with Breeding and Sub-Population   Løvbjerg et al. proposed a hybrid model incorporating breeding, that is, the reproduction and recombination operators used in standard GA, into PSO [59]. First, particles are randomly selected under the given breeding probability without considering their fitness values. These particles are treated as the parents and are replaced by their offspring particles through the breeding. The arithmetic crossover used to obtain the positions of the offsprings is given by

$$X_{p,n}'^j = rX_{p,n}^j + (1-r)X_{q,n}^j, \qquad (3.127)$$

$$X_{p,n}'^j = rX_{q,n}^j + (1-r)X_{p,n}^j, \qquad (3.128)$$

where r is a random number uniformly distributed over (0,1). The velocity vector of the offspring is computed as the sum of the values of the two parents normalised to the original length of each parent velocity

$$V_{p,n}' = \frac{V_{p,n} + V_{q,n}}{|V_{p,n} + V_{q,n}|} |V_{p,n}| \qquad (3.129)$$

$$V_{q,n}' = \frac{V_{p,n} + V_{q,n}}{|V_{p,n} + V_{q,n}|} |V_{q,n}| \qquad (3.130)$$

Through this kind of arithmetic crossover, the position of offspring is influenced by the parent particles and is located in a hypercube spanned by the parent particles. Experimental results show that the method has faster convergence speed than PSO and GA for uni-modal problems and performs better than the canonical PSO and GA with faster convergence speed for multi-modal problems.

3.5.3 Other Modified PSO

3.5.3.1 Attractive and Repulsive PSO

It is widely known that the canonical PSO algorithm is prone to suffer from premature convergence when solving strongly multi-modal problems. As Riget and Vesterstroem commented [92], this may be due to the low population diversity during the later stage of search that makes further global search impossible. To circumvent this shortcoming, they designed an attractive and repulsive PSO (ARPSO) in which the search of the swarm is guided by the population diversity measure. The diversity measure in ARPSO is defined as the mean of the distances of the particles from the average point of all the particles. The method also required to pre-assign two values, the upper and lower bounds, d_{high} and d_{low}, of the diversity. The particle swarm then alternates between two phases, that is, attraction and repulsion phases, during the search process. The particle swarm evolves in attraction phase after initialisation. If the diversity decreases to below d_{low}, the swarm shifts to the repulsion phase and all the particles are repelled from the global best position and their own personal best position, leading the diversity to increase until it reaches d_{high}. On the other hand, once the diversity is larger than d_{high}, the swarm returns to the attraction phase and the particles are attracted by the global best position and their own personal best positions resulting in decrease of the diversity. Guided by the diversity, the swarm switches its search mode between the attraction and repulsion phases so that it can maintain the diversity above d_{low} to prevent its search from stagnation.

The effectiveness of the ARPSO algorithm was demonstrated by testing it on some benchmark functions and comparing with GA and the canonical PSO [92]. The experimental results showed that ARPSO outperformed its competitors significantly for most cases except for the Griewank function. This implies that the ARPSO has a stronger ability to escape from the local minima than the canonical PSO when solving the multi-modal problems.

3.5.3.2 Comprehensive Learning PSO

In the PSO, each particle learns from the global best position and its own personal best position simultaneously during the iterative search. Restricting the social learning aspect to only the global best position certainly results in fast convergence of the canonical PSO. However, if the global best position is far from the global optimum, the particles may easily be attracted to the region around the global best position and be trapped in a local optimum if the search environment is complex with many local solutions. On the other hand, since the fitness value of a particle is determined by the values of all its position components, and a particle that has discovered the region corresponding to the global optimum in some dimensions may have low fitness value because of the poor solutions in other dimensions. Therefore, in order to make better use of beneficial information, Liang et al. employed a kind of learning strategy in the PSO and proposed an improved version of PSO referred to as the comprehensive learning PSO (CLPSO) [93]. In the CLPSO, a particle is attracted by the personal best position learning from all the other personal best positions instead of only its own and the global best position. For each component of a particle, the learning probability pc_i is assigned to decide the corresponding component of this personal best position. If the value of pc_i for this dimension is smaller than a random number generated uniformly over (0,1), its component in each dimension learns from the particle's own personal best position; otherwise, the component learns from another particle's personal best position through a tournament selection procedure. In order to ensure a particle learn from good exemplars and minimise the time waste on poor directions, Liang et al. suggested to reassign a learning exemplar for a particle whose performance stops improving for a given number of generations [94]. They analysed the algorithmic parameters in CLPSO, including the learning probability and the refreshing gap, and proposed a new method in order to avoid the particles moving out of the search domain. A suite of benchmark functions were used to test CLPSO and eight other PSO variants. These functions were unimodal, unrotated multi-modal, rotated multi-modal, or composed of several functions. The experimental results show that CLPSO could obtain better quality solutions than the other eight PSO variants, particularly for separable functions.

3.5.3.3 Cooperative PSO
The goal of developing the cooperative PSO (CPSO) is similar to that of the CLPSO algorithm. Its cooperative strategy was borrowed from the decomposition technique proposed by Potter for improving GA [94]. In CPSO [95], if the search space is N-dimensional, the original swarm splits into N swarms of one-dimensional (1D) particles and each swarm aims at optimising its corresponding dimension. When the fitness value of a particle is

being evaluated, it requires an N-dimensional position vector. Its position vector takes the components in the other $(N-1)$ dimensions from the global best particle and concatenates them with the components of the dimension being optimised. It was suggested that in CPSO, if some of the components are correlated, then they should be grouped into the same swarm and optimised together. In general, the relationship between components is unknown in advance. In order to tackle this problem, van den Bergh and Engelbrecht proposed a simple approximation method that splits the D dimensions into K parts blindly [95]. Knowing that the CPSO with K parts may also be trapped in a sub-optimal solution, they proposed a hybrid CPSO by combining the CPSO with the canonical PSO. A thorough comparison was made between the canonical PSO, CPSO, CPSO with K parts, hybrid CPSO, GA and CCGA as discussed in [94]; experimental results show that the CPSO algorithms provided significant improvement in solution quality and robustness over the canonical PSO and other competitors.

3.5.3.4 Self-Organised Criticality PSO

Lovbjerg and Krink proposed a variant of PSO extended with self-organised criticality (SOC PSO) [96]. The effect of self-organised criticality (SOC) is to diversify the swarm in order to enhance the global search ability of PSO. In the SOC PSO, a critical value is assigned to each particle subject to a globally set threshold. During the search process, if the critical value of a particle is smaller than the threshold, the particle disperses its criticality within a certain surrounding neighbourhood, and then relocates so that the population diversity could be increased. Lovbjerg and Krink proposed two relocation schemes, one is a random immigrant technique and the other is a pushing technique. The former relocating scheme re-initialises the particle and discards its personal best positions. The latter one pushes the particle a little further in the direction it is heading, with a magnitude of the criticality limit added to its critical value, but maintains its remembrance of its personal best position. The numerical results on several benchmark functions showed that the SOC PSO had better performance compared to the canonical PSO.

3.5.3.5 Dissipative PSO

As found by Xie et al. [97], the social model in the canonical PSO has some characteristics for self-organisation of the dissipative structure as the swarm keeps historical best positions (personal best and global best positions) through the evolution process. Once the algorithm has particles trapped in a local optimum, the particle swarm falls into the equilibrium state so that the evolution process stagnates without any fitness improvement. One idea brought by Xie et al. was to use the self-organisation of the dissipative structure with negative entropy introduced by additional chaos [97]. The resulting version of the algorithm is known as the dissipative PSO, in which the dissipative structure with negative entropy helps to maintain the particle swarm in a far-from-equilibrium state and drives the irreversible evolution process to obtain further better fitness. Experimental results for two benchmark functions demonstrated that introducing negative entropy into the dissipative structure for PSO was able to improve the algorithmic performance efficiently.

3.5.3.6 PSO with Spatial Particle Extension

To achieve a better algorithmic performance, one can also use the canonical PSO with the spatial extension technique proposed by Krink et al. [98]. This method is to bounce the particles away from the current clustered position once they gather too close. Each particle is assigned a radius to check the status of two particles to determine whether they collide. Once two particles are detected to be in collision, some measures should be taken to make them bounce off to avoid the collision and thus increase the diversity. Krink et al. proposed three methods of bouncing the particles away from the gathering place [98]. These methods are the random direction bouncing, the realistic physical bouncing and the random velocity bouncing. Testing the PSO with spatial particle extension on several benchmark functions, they found that the algorithm performed well on all the cases due to its ability to avoid stagnation of the search process.

REFERENCES

1. Y. Shi, R.C. Eberhart. A modified particle swarm optimizer. In *Proceedings of the 1998 IEEE International Conference on Evolutionary Computation*, Anchorage, AK, 1998, pp. 69–73.
2. M. Clerc. The swarm and the queen: Towards a deterministic and adaptive particle swarm optimization. In *Proceedings of the Congress on Evolutionary Computation*, Washington, DC, vol. 3, 1999, pp. 1951–1957.
3. R.C. Eberhart, J. Kennedy. A new optimizer using particle swarm theory. In *Proceedings of the Sixth International Symposium on Micro Machine and Human Science*, Nagoya, Japan, 1995, pp. 39–43.
4. J. Kennedy. Bare bones particle swarms. In *Proceedings of the 2003 IEEE Swarm Intelligence Symposium*, Indianapolis, IN, 2003, pp. 80–87.
5. J. Kennedy, R.C. Eberhart. Particle swarm optimization. In *Proceedings of the IEEE International Conference on Neural Networks*, Perth, WA, Australia, vol. IV, 1995, pp. 1942–1948.
6. Y. Shi, R.C. Eberhart. Parameter selection in particle swarm optimization. In *Proceedings of the Seventh International Conference on Evolutionary Programming*, San Diego, CA, vol. VII, 1998, pp. 591–600.

7. Y. Shi, R.C. Eberhart. Empirical study of particle swarm optimization. In *Proceedings of the 1999 Congress on Evolutionary Computation*, Washington, DC, vol. 3, 1999, pp. 1945–1950.

8. R.C. Eberhart, Y. Shi. Tracking and optimizing dynamic systems with particle swarms. In *Proceedings of the 2001 Congress on Evolutionary Computation*, Seoul, Korea, vol. 1, 2001, pp. 94–100.

9. M. Pant, T. Radha, V.P. Singh. Particle swarm optimization using Gaussian inertia weight. In *Proceedings of the International Conference on Computational Intelligence and Multimedia Applications*, Sivakasi, Tamil Nadu, India, 2007, pp. 97–105.

10. Y.-L. Zheng, L.-H. Ma, L.-Y. Zhang, J.-X. Qian. Empirical study of particle swarm optimizer with an increasing inertia weight. In *Proceedings of the Congress on Evolutionary Computation*, Canberra, ACT, Australia, vol. 1, 2003, pp. 221–226.

11. A. Chatterjee, P. Siarry. Nonlinear inertia weight variation for dynamic adaptation in particle swarm optimization. *Computers and Operations Research*, 2006, 33(3): 859–871.

12. G. Chen, X. Huang, J. Jia, Z. Min. Natural exponential inertia weight strategy in particle swarm optimization. In *Proceedings of the Sixth World Congress on Intelligent Control and Automation*, Dalian, Liaoning, China, 2006, pp. 3672–3675.

13. B. Jiao, Z. Lian, X. Gu. A dynamic inertia weight particle swarm optimization algorithm. *Chaos, Solitons & Fractals*, 2008, 37(3): 698–705.

14. A. Adriansyah, S.H.M. Amin. Analytical and empirical study of particle swarm optimization with a sigmoid decreasing inertia weight. In *Proceedings of the Regional Postgraduate Conference on Engineering and Science*, Johor, Bahru, Malaysia, 2006, pp. 247–252.

15. R.F. Malik, T.A. Rahman, S.Z.M. Hashim, R. Ngah. New particle swarm optimizer with sigmoid increasing inertia weight. *International Journal of Computer Science and Security*, 2007, 1(2): 43–52.

16. Y. Shi, R.C. Eberhart. Fuzzy adaptive particle swarm optimization. In *Proceedings of the 2001 Congress on Evolutionary Computation*, Seoul, Korea, vol. 1, 2001, pp. 101–106.

17. C.S. Feng, S. Cong, X.Y. Feng. A new adaptive inertia weight strategy in particle swarm optimization. In *Proceedings of the 2007 IEEE Congress on Evolutionary Computation*, Singapore, 2007, pp. 4186–4190.

18. K. Suresh, S. Ghosh, D. Kundu, A. Sen, S. Das, A. Abraham. Inertia-adaptive particle swarm optimizer for improved global search. In *Proceedings of the 2008 Eighth International Conference on Intelligent Systems Design and Applications*, Kaohsuing, Taiwan, vol. 2, 2008, pp. 253–258.

19. X. Zhang, Y. Du, G. Qin, Z. Qin. Adaptive particle swarm algorithm with dynamically changing inertia weight. *Journal of Xian Jiaotong University*, 2005, 39(10): 1039–1042.

20. X.-M. Yang, J.-S. Yuan, J.-Y. Yuan, H. Mao. A modified particle swarm optimizer with dynamic adaptation. *Applied Mathematics and Computation*, 2007, 189(2): 1205–1213.

21. A. Miao, X. Shi, J. Zhang, E. Wang, S. Peng. A modified particle swarm optimizer with dynamical inertia weight. In *Fuzzy Information and Engineering*, vol. 2, B, Springer, Berlin, Heidelberg, Germany, 2009, pp. 767–776.

22. M. Clerc, J. Kennedy. The particle swarm-explosion, stability and convergence in a multidimensional complex space. *IEEE Transactions on Evolutionary Computation*, 2002, 6(2): 58–73.

23. R.C. Eberhart, Y. Shi. Comparing inertia weights and constriction factors in particle swarm optimization. In *Proceedings of the 2000 Congress on Evolutionary Computation*, San Diego, CA, vol. 1, 2000, pp. 84–88.

24. L. Bui, O. Soliman, H.A. Abbass. A modified strategy for the constriction factor in particle swarm optimization. In *Proceedings of the Third Australian Conference on Progress in Artificial Life*, Gold Coast, QLD, Australia, 2007, pp. 333–344.

25. R.C. Eberhart, Y. Shi. Particle swarm optimization: Developments, applications and resources. In *Proceedings of the 2001 Congress on Evolutionary Computation*, Seoul, Korea, vol. 1, 2001, pp. 81–86.

26. J. Kennedy. Small worlds and mega-minds: Effects of neighborhood topology on particle swarm performance. In *Proceedings of the Congress on Evolutionary Computation*, Washington, DC, vol. 3, 1999, pp. 1931–1938.

27. D. Bratton, J. Kennedy. Defining a standard for particle swarm optimization. In *Proceedings of the IEEE Swarm Intelligence Symposium*, Honolulu, HI, 2007, pp. 120–127.

28. P.N. Suganthan. Particle swarm optimizer with neighborhood operator. In *Proceedings of the Congress on Evolutionary Computation*, Piscataway, NJ, 1999, pp. 1958–1961.

29. R. Mendes, J. Kennedy, J. Neves. The fully informed particle swarm: Simpler, maybe better. *IEEE Transactions on Evolutionary Computation*, 2004, 8(3): 204–210.

30. J. Kennedy, R. Mendes. Population structure and particle swarm performance. In *Proceedings of the 2002 Congress on Evolutionary Computation*, Honolulu, HI, vol. 2, 2002, pp. 1671–1676.

31. S.A. Hamdan. Hybrid particle swarm optimiser using multi-neighborhood topologies. *Journal of Computer Science*, 2008, 7(1): 36–43.

32. X. Hu, R.C. Eberhart. Multiobjective optimization using dynamic neighborhood particle swarm optimization. In *Proceedings of the 2002 Congress on Evolutionary Computation*, Honolulu, HI, vol. 2, 2002, pp. 1677–1681.

33. T. Peram, K. Veeramachaneni, C.K. Mohan. Fitness-distance-ratio based particle swarm optimization. In *Proceedings of the 2003 IEEE Swarm Intelligence Symposium*, Indianapolis, IN, 2003, pp. 174–181.

34. J.J. Liang, P.N. Suganthan. Dynamic multi-swarm particle swarm optimizer. In *Proceedings of the 2005 IEEE Swarm Intelligence Symposium*, Pasadena, CA, 2005, pp. 124–129.

35. H. Liu, B. Li, Y. Ji, T. Sun. Particle swarm optimisation from lbest to gbest. In *Applied Soft Computing Technologies: The Challenge of Complexity*, A. Abraham et al., Eds. Springer, Berlin, Germany, 2006, pp. 537–545.

36. B. Jiao, Z.G. Lian, Q.X. Chen. A dynamic global and local combined particle swarm optimization algorithm. *Chaos Solitons & Fractals*, 2009, 42(5): 2688–2695.

37. M. Alizadeh, E. Fotoohi, V. Roshanaei, E. Safavieh. Particle swarm optimization with voronoi neighborhood. In *Proceedings of the 14th International CSI Computer Conference*, Tehran, Iran, 2009, pp. 397–402.
38. F. Aurenhammer. Voronoi diagrams—A survey of a fundamental geometric data structure. *ACM Computing Surveys*, 1991, 23(3): 345–405.
39. J. Lane, A. Engelbrecht, J. Gain. Particle swarm optimization with spatially meaningful neighbours. In *Proceedings of the 2008 IEEE Swarm Intelligence Symposium*, St. Louis, MO, 2008, pp. 1–8.
40. A.S. Mohais, R. Mendes, C. Ward, C. Posthoff. Neighborhood re-structuring in particle swarm optimization. In *Proceedings of the 18th Australian Joint Conference on Artificial Intelligence*, Sydney, NSW, Australia, 2005, pp. 776–785.
41. Y. Liu, Q. Zhao, Z. Zhao, Z. Shang, C. Sui. Particle swarm optimizer based on dynamic neighborhood topology. In *Proceedings of the Fifth International Conference on Intelligent Computing*, Changsha, China, 2009, pp. 794–803.
42. X. Li. Adaptively choosing neighbourhood bests using species in a particle swarm optimizer for multimodal function optimization. In *Proceedings of the 2004 Genetic and Evolutionary Computation Conference*, Seattle, WA, 2004, pp. 105–116.
43. J.-P. Li, M.E. Balazs, G.T. Parks, P.J. Clarkson. A species conserving genetic algorithm for multimodal function optimization. *Evolutionary Computation*, 2002, 10(3): 207–234.
44. J. Sun, B. Feng, W.B. Xu. Particle swarm optimization with particles having quantum behavior. In *Proceedings of the Congress on Evolutionary Computation*, Portland, OR, vol. 1, 2004, pp. 326–331.
45. J. Sun, W.B. Xu, B. Feng. A global search strategy of quantum-behaved particle swarm optimization. In *Proceedings of the IEEE Conference on Cybernetics and Intelligent Systems*, Singapore, 2004, pp. 111–116.
46. J. Sun, W.B. Xu, B. Feng. Adaptive parameter control for quantum-behaved particle swarm optimization on individual level. In *Proceedings of the 2005 IEEE International Conference on Systems, Man and Cybernetics*, Waikoloa, HI, vol. 4, 2005, pp. 3049–3054.
47. J. Kennedy. The behavior of particles. In *Proceedings of the Seventh International Conference on Evolutionary Programming*, San Diego, CA, vol. VII, 1998, pp. 581–589.
48. J. Kennedy, R.C. Eberhart, Y. Shi. *Swarm Intelligence*. Morgan Kaufmann/ Academic Press, San Francisco, CA, 2001.
49. J. Kennedy. Probability and dynamics in the particle swarm. In *Proceedings of the 2004 Congress on Evolutionary Computation*, Portland, OR, vol. 1, 2004, pp. 340–347.
50. J. Kennedy. Dynamic-probabilistic particle swarms. In *Proceedings of the 2005 Conference on Genetic and Evolutionary Computation*, Washington, DC, 2005, pp. 201–207.
51. R.A. Krohling. Gaussian swarm: A novel particle swarm optimization algorithm. In *Proceedings of the 2004 IEEE Conference on Cybernetics and Intelligent Systems*, Singapore, 2004, pp. 372–376.
52. I.C. Trelea. The particle swarm optimization algorithm: Convergence analysis and parameter selection. *Information Processing Letters*, 2003, 85(6): 317–325.

53. R.A. Krohling. Gaussian particle swarm with jumps. In *Proceedings of the 2005 IEEE Congress on Evolutionary Computation*, Edinburgh, U.K., vol. 2, 2005, pp. 1226–1231.

54. C.-Y. Lee, X. Yao. Evolutionary programming using mutations based on the Levy probability distribution. *IEEE Transactions on Evolutionary Computation*, 2004, 8(1): 1–13.

55. R.A. Krohling, E. Mendel. Bare bones particle swarm optimization with Gaussian or cauchy jumps. In *Proceedings of the 11th Conference on Congress on Evolutionary Computation*, Montreal, Quebec, Canada, 2009, pp. 3285–3291.

56. R.A. Krohling, L. dos Santos Coelho. PSO-E: Particle swarm with exponential distribution. In *Proceedings of the 2006 IEEE Congress on Evolutionary Computation*, Vancouver, BC, Canada, 2006, pp. 1428–1433.

57. B.R. Secrest, G.B. Lamont. Visualizing particle swarm optimization— Gaussian particle swarm optimization. In *Proceedings of the 2003 IEEE Swarm Intelligence Symposium*, Indianapolis, IN, 2003, pp. 198–204.

58. T.J. Richer, T.M. Blackwell. The Levy particle swarm. In *Proceedings of the 2006 Congress on Evolutionary Computation*, Vancouver, BC, Canada, 2006, pp. 808–815.

59. M. Løvbjerg, T.K. Rasmussen, T. Krink. Hybrid particle swarm optimizer with breeding and subpopulation. In *Proceedings of the the Genetic and Evolutionary Computation Conference*, Portland, OR, 2001.

60. N. Higashi, H. Iba. Particle swarm optimization with Gaussian mutation. In *Proceedings of the 2003 IEEE Swarm Intelligence Symposium*, Indianapolis, IN, 2003, pp. 72–79.

61. A. Stacey, M. Jancic, I. Grundy. Particle swarm optimization with mutation. In *Proceedings of the 2003 Congress on Evolutionary Computation*, Canberra, ACT, Australia, vol. 2, 2003, pp. 1425–1430.

62. M. Pant, R. Thangaraj, V.P. Singh, A. Abraham. Particle swarm optimization using Sobol mutation. In *Proceedings of the First International Conference on Emerging Trends in Engineering and Technology*, Nagpur, India, 2008, pp. 367–372.

63. H.M. Chi, B. Peter, D.W. Evans, M. Mascagni. On the scrambled sobol sequence. In *Proceedings of the Workshop on Parallel Monte Carlo Algorithms for Diverse Applications in a Distributed Setting*, Atlanta, USA, 2005, pp. 775–782.

64. M. Pant, R. Thangaraj, A. Abraham. Particle swarm optimization using adaptive mutation. In *Proceedings of the 19th International Workshop on Database and Expert Systems Application*, Turin, Italy, 2008, pp. 519–523.

65. Z. Michalewicz. *Genetic Algorithms + Data Structures = Evolution Programs*. Springer-Verlag, Heidelberg, Germany, 1996.

66. S.C. Esquivel, C.A.C. Coello. On the use of particle swarm optimization with multimodal functions. In *Proceedings of the 2003 Congress on Evolutionary Computation*, Canberra, ACT, Australia, vol. 2, 2003, pp. 1130–1136.

67. E.S. Peer, F. van den Bergh, A.P. Engelbrecht. Using neighbourhoods with the guaranteed convergence PSO. In *Proceedings of the 2003 IEEE Swarm Intelligence Symposium*, Indianapolis, IN, 2003, pp. 235–242.

68. A. Ratnaweera, S.K. Halgamuge, H.C. Watson. Self-organizing hierarchical particle swarm optimizer with time-varying acceleration coefficients. *IEEE Transactions on Evolutionary Computation*, 2004, 8(3): 240–255.

69. P.S. Andrews. An investigation into mutation operators for particle swarm optimization. In *Proceedings of the 2006 IEEE Congress on Evolutionary Computation*, Vancouver, BC, Canada, 2006, pp. 1044–1051.

70. A.A.A. Esmin, G. Lambert-Torres, A.C. Zambroni de Souza. A hybrid particle swarm optimization applied to loss power minimization. *IEEE Transactions on Power Systems*, 2005, 20(2): 859–866.

71. S.H. Ling, H.H.C. Iu, K.Y. Chan, H.K. Lam, B.C.W. Yeung. Hybrid particle swarm optimization with wavelet mutation and its industrial applications. *IEEE Transactions on Systems, Man, and Cybernetics, Part B*, 2008, 38(3): 743–763.

72. S.H. Ling, C.W. Yeung, K.Y. Chan, H.H.C. Iu, F.H.F. Leung. A new hybrid particle swarm optimization with wavelet theory based mutation operation. In *Proceedings of the 2007 IEEE Congress on Evolutionary Computation*, Singapore, 2007, pp. 1977–1984.

73. J.J. Liang, B. Baskar, P.N. Suganthan, A.K. Qin. Performance evaluation of multiagent genetic algorithm. *Natural Computing*, 2006, 5(1): 83–96.

74. D. Li, C. Liu, L. Liu. Study on the universality of the normal cloud model. *Engineering Science*, 2004, 6(8): 28–34.

75. X. Wu, B. Cheng, J. Cao, B. Cao. Particle swarm optimization with normal cloud mutation. In *Proceedings of the Seventh World Congress on Intelligent Control and Automation*, Chongqing, China, 2008, pp. 2828–2832.

76. M. El-Abd. Preventing premature convergence in a PSO and EDA hybrid. In *Proceedings of the 2009 IEEE Congress on Evolutionary Computation*, Trondheim, Norway, 2009, pp. 3060–3066.

77. Y. Zhou, J. Jin. EDA-PSO—A new hybrid intelligent optimization algorithm. In *Proceedings of the Michigan University Graduate Student Symposium*, Singapore, 2006.

78. J. Kennedy, R.C. Eberhart. A discrete binary version of the particle swarm algorithm. In *Proceedings of the 1997 IEEE International Conference on Systems, Man, and Cybernetics*, Orlando, FL, 1997, pp. 4104–4108.

79. M.A. Khanesar, M. Teshnehlab, M.A. Shoorehdeli. A novel binary particle swarm optimization. In *Proceedings of the 2007 Mediterranean Conference on Control & Automation*, Athens, Greece, 2007, pp. 1–6.

80. F. Afshinmesh, A. Marandi, A. Rahimi-Kian. A novel binary particle swarm optimization method using artificial immune system. In *Proceedings of the 2005 International Conference on Computer as a Tool*, Belgrade, Serbia, 2005, pp. 217–220.

81. J. Sadri, C.Y. Suen. A genetic binary particle swarm optimization model. In *Proceedings of the 2006 IEEE Congress on Evolutionary Computation*, Vancouver, BC, Canada, 2006, pp. 656–663.

82. S. Lee, S. Soak, S. Oh, W. Pedrycz, M. Jeon. Modified binary particle swarm optimization. *Progress in Natural Science*, 2008, 18(9): 1161–1166.

83. A. Marandi, F. Afshinmesh, M. Shahabadi, F. Bahrami. Boolean particle swarm optimization and its application to the design of a dual-band dual-polarized planar antenna. In *Proceedings of the 2006 IEEE Congress on Evolutionary Computation*, Vancouver, BC, Canada, 2006, pp. 3212–3218.

84. D. Fogel. *Evolutionary Computation: Towards a New Philosophy of Machine Intelligence.* IEEE Press, Piscataway, NJ, 1996.

85. T. Bäck. *Evolutionary Algorithms in Theory and Practice.* Oxford University Press, New York, 1996.

86. P.J. Angeline. Using selection to improve particle swarm optimization. In *Proceedings of the 1998 IEEE International Conference on Evolutionary Computation,* Anchorage, AK, 1998, pp. 84–89.

87. R. Storn, K. Price. Differential evolution—A simple and efficient adaptive scheme for global optimization over continuous spaces. In *Technical Report of International Computer Science Institute,* Berkeley, CA, 1995.

88. W.-J. Zhang, X.-F. Xie. DEPSO: Hybrid particle swarm with differential evolution operator. In *Proceedings of the 2003 IEEE International Conference on Systems, Man and Cybernetics,* Washington, DC, vol. 4, 2003, pp. 3816–3821.

89. D. Devicharan, C.K. Mohan. Particle swarm optimization with adaptive linkage learning. In *Proceedings of the 2004 Congress on Evolutionary Computation,* Portland, OR, vol. 1, 2004, pp. 530–535.

90. Y.-P. Chen, W.-C. Peng, M.-C. Jian. Particle swarm optimization with recombination and dynamic linkage discovery. *IEEE Transactions on Systems, Man, and Cybernetics, Part B,* 2007, 37(6): 1460–1470.

91. M. Settles, T. Soule. Breeding swarms: A GA/PSO hybrid. In *Proceedings of the 2005 Conference on Genetic and Evolutionary Computation,* Washington, DC, 2005, pp. 161–168.

92. J. Riget, J. Vesterstroem. A diversity-guided particle swarm optimizer—The ARPSO. Department of Computer Science, University of Aarhus, Aarhus, Denmark, 2002.

93. J.J. Liang, A.K. Qin, P.N. Suganthan, S. Baskar. Comprehensive learning particle swarm optimizer for global optimization of multimodal functions. *IEEE Transactions on Evolutionary Computation,* 2006, 10(3): 281–295.

94. M. Potter, K. De Jong. A cooperative coevolutionary approach to function optimization. In *Proceedings of the 1994 International Conference on Parallel Problem Solving from Nature,* Jerusalem, Israel, 1994, pp. 249–257.

95. F. van den Bergh, A.P. Engelbrecht. A cooperative approach to particle swarm optimization. *IEEE Transactions on Evolutionary Computation,* 2004, 8(3): 225–239.

96. M. Lovbjerg, T. Krink. Extending particle swarm optimisers with self-organized criticality. In *Proceedings of the 2002 Congress on Evolutionary Computation,* Honolulu, HI, vol. 2, 2002, pp. 1588–1593.

97. X.-F. Xie, W.-J. Zhang, Z.-L. Yang. Dissipative particle swarm optimization. In *Proceedings of the 2002 Congress on Evolutionary Computation,* Honolulu, HI, 2002, pp. 1456–1461.

98. T. Krink, J.S. Vesterstrom, J. Riget. Particle swarm optimisation with spatial particle extension. In *Proceedings of the 2002 Congress on Evolutionary Computation,* Honolulu, HI, 2002, pp. 1474–1479.

Quantum-Behaved Particle Swarm Optimisation

THIS CHAPTER BEGINS WITH a thorough overview of the literature on quantum-behaved particle swarm optimisation (QPSO) followed by the formulation based on the motivations of the QPSO algorithm. A discussion is given of the fundamental model—a quantum δ potential well model—for QPSO algorithm. Details of applying the algorithm to solve typical optimisation problems are provided. Several variations of the QPSO method are also included towards the end of this chapter.

4.1 OVERVIEW

4.1.1 Introduction

QPSO, a relatively new version of PSO, was motivated from quantum mechanics and dynamical analysis of PSO [1]. The algorithm uses a strategy based on a quantum delta potential well model to sample around the previous best points [2] and receives help from the mean best position in order to enhance the global search ability of the particle [3]. QPSO is a kind of probabilistic algorithm [4], and the iterative equation of QPSO is very different from that of PSO. It does not require velocity vectors for particles and has fewer parameters to adjust, making it easier to implement. Numerical experiments and analysis show that the QPSO algorithm offers good performance in solving a wide range of continuous optimisation problems, and many efficient strategies have been proposed to improve the algorithm. In order to make a clear vision of the improvements and applications, this section attempts to give a *compendious and timely* review on QPSO by categorising most publications on improvements and applications of QPSO.

4.1.2 Various Improvements of QPSO

The performance of a population-based random search technique lies in its global search ability, convergence speed, solution precision, robustness, etc. In order to improve the performance of QPSO on the multimodal problems and avoid the premature convergence of QPSO, many strategies have been proposed from different aspects. It should be noted that mutation and selection are operators in evolutionary algorithms (EAs) and are often used to improve QPSO by increasing the population diversity in a search process in order to explore any undiscovered solution space. In this book, the authors classify these improvements into six categories. The classification here is by no means the best, but it covers almost all kinds of improvements currently adapted for QPSO.

4.1.2.1 Selecting and Controlling the Parameters Effectively

The most important and the simplest way of improving QPSO without increasing the complexity of the algorithm and its computational cost is to select and control the algorithmic parameters effectively. Apart from the swarm size, problem dimension and the number of maximum iterations, the most important parameter involved in QPSO is the contraction–expansion (CE) coefficient α, which is vital to the dynamical behaviour of individual particle and the convergence of the algorithm. Much of the research work involved in this area concentrates on finding effective control and selection strategies for the CE coefficient in order to provide good algorithmic performance. Such attempt was first made by Sun, the first author of this book, who implemented a stochastic simulation of obtaining values of the CE coefficient in order to prevent the particle from explosion [5]. Results from stochastic simulation, obtained by Sun, show that the particle position either converges to the point of its local focus or is bounded when $\alpha < 1.78$ and the particle position diverges when $\alpha > 1.8$. As a result, there must be a threshold $\alpha_0 \in [1.7, 1.8]$ such that the particle position converges or is bounded if $\alpha \leq \alpha_0$, otherwise it diverges. A comprehensive theoretical analysis, as given in Chapter 5, of the influence of

α on the behaviour of the particle shows that the value of α_0 is $e^\gamma \approx 1.781$, where $\gamma \approx 0.577215665$ is known as the Euler constant [6].

It should be noted that $\alpha < \alpha_0$ does not always mean a good algorithmic performance would be achieved. There are various ways of selecting and controlling the CE coefficient. Effectiveness and performance certainly depend on these choices. Two simple and effective control methods for this parameter were investigated in depth [6]. One is fixing α at a value within (0, 1.781) over the course of the search process and the other is decreasing the value of α linearly with respect to the number of iterations. A detailed description of the two parameter control methods is also presented in Chapter 5. It is also possible to use adaptive parameter selection methods. One such technique can be found in [7], in which α is selected dynamically according to the distance between the local focus and the current position of the particle, or according to the difference between the fitness value of the personal best position of the particle and the global best position. These adaptive methods indeed show better performance on some benchmark functions as shown in [7].

4.1.2.2 Using Different Probability Distribution

Another method of achieving improvement of QPSO without complicating the algorithm significantly is by using different probability distributions to sample the positions of the particles other than the double exponential distribution used in the original QPSO. A hybrid of Gaussian distribution and double exponential distribution was used to replace the double exponential distribution used in QPSO [8]. With the two different distributions, there are respectively two random number sequences. In the hybrid method, a new random number sequence is generated by adding the two random number sequences to form a new random number sequence which is then employed to yield the sequence of the position of the particle. It was shown that the QPSO with this hybrid probability distribution (HPD) outperforms the original QPSO for some well-known benchmark functions.

Moreover, there are several reports describing the use of Gaussian probability distribution to sample the positions of the particles in QPSO [9–12]. For example, the logarithmic value of a random number with Gaussian distribution replaces the random number with double exponential distribution in order to generate the new position of the particle [9]. This was shown to have added effectiveness in finding the solutions of several benchmark functions.

4.1.2.3 Controlling the Diversity Measures

It is likely that by finding proper parameter control and selection methods or probability distributions for sampling the particle positions may only lead to limited performance improvement for QPSO. For further enhancement of the search ability of the algorithm, particularly when harder problems are being solved, diversity control strategies may be incorporated into QPSO to guide the search process.

As pointed out by Ursem, loss of diversity in EAs is often blamed for being the main reason for premature convergence [13]. The work in the reference provides an insight into some of the diversity control methods in order to improve the ability of QPSO to escape from the local optima. The diversity control method requires to set a lower bound d_{low} as suggested in [14] for the diversity of the swarm. If the diversity measure of the swarm declines to below d_{low}, the swarm needs to explode in order to increase the diversity until it is larger than d_{low}. Two methods [14] were proposed to allow the explosion of the particles. One method sets the value of the CE coefficient to be large enough to make the particles diverge. The other reinitialises the mean best position so that the distance from the particle to the mean best position can be increased to make the particles volatile. A further method capable of diversifying the swarm proposed in [15] is to exert the mutation on the particle with the global best position (i.e. the global best particle) in order to activate the particles. When the mutation operation is exerted, the displacement of the global best particle takes place and thus increases the average distance from personal best positions of the particles to the mean best position. As a result, there is an increase in the swarm diversity. Instead of measuring the diversity of the swarm using the current positions of all particles, the distance-to-average-point of the personal best positions of all particles is used as the diversity measure to guide the search process [14,16]. In order to prevent the swarm from clustering, in which case the convergence will be affected, a diversity-controlled QPSO relying on a lower bound value d_{low} was proposed by Sun et al. [16]. The method relies on a mutation operation being exerted on the personal best position of a randomly selected particle until the diversity returns to above d_{low}.

4.1.2.4 Hybridising with Other Search Techniques

No free lunch theorem implies that each random search technique has its own advantages on a certain set of problems over other methods. Hybridising a random optimisation algorithm with others thus becomes an important way of bringing advantages together to improve the algorithm. The aim to combine the desirable properties of different approaches with QPSO is to mitigate its weakness.

An interesting hybridisation for QPSO was proposed in [17] in which the main idea is to incorporate a local search operator while the main QPSO performs global exploration search. This technique relies on a generalised local search operator while the local search operator exploits the neighbourhood of the current solution provided by the main QPSO method. Another possibility is to use the public history for searching variant particles in order to improve convergence speed and enhance the global search ability [18]. On the other hand, a chaotic search method may also be used to diversify the swarm at a later stage of the search process so as to help the system escape from local optima [19]. However, most of the hybridisations involve combining QPSO with other meta-heuristic methods. One typical example as discussed in [20] is to combine the search mechanism of simulated annealing (SA) to jump out of the local optima and enhance the global search ability of the QPSO algorithm. The methodology of using artificial immune system has also been considered to improve QPSO. For examples, the immune operation based on the regulation of the antibody thickness and vaccination operation in immune system was introduced into QPSO [21]. It is also possible to calculate the density of an antibody based on the vector distance as an immune operator [22] such that it is able to restrain the degeneration in the process of optimisation effectively. It is also interested to note that a hybrid of QPSO and the differential evolution (DE) operation is able to strengthen the search ability of the particles [23].

Mutation operation is generally used by EAs to diversify the populations in order to enhance the global search abilities of the algorithms. QPSO and PSO have no mutation operation in their evolution process. Thus, mutation is often being used to assist the particles in QPSO to escape from the local optima when the algorithm is used to solve harder problems. The global best and the mean best positions in QPSO can be mutated by using Cauchy distribution in such a way that the scaling parameter of the resulting mutation operator can be annealed to make the algorithm adaptive [24]. It is also possible to have mutation based on chaotic sequences [25] which significantly diversify the swarm and improve the QPSO performance by preventing the algorithm from premature convergence to local optima. Readers who are interested in the performance of these algorithms should consult [26].

Selection operation in EAs is used to enhance the local search ability of the algorithms by discarding the individuals with poor fitness values. Selection may also be used in QPSO to exert on the global best solution [27]. However, this selection method is for the purpose of enhancing the global search ability, instead of the local search ability, of the QPSO algorithm. On the other hand, elitist selection is to speed up the convergence of the QPSO by assigning different weights to the personal best positions of the particles in computing the mean best positions according to the fitness values of their personal best positions. The particles with better fitness values are assigned larger weights so that the particles could gather rapidly towards the better particles. By considering elitist selection, which incorporates the thinking model of humans as accurate as possible, on the mean best particles, it would balance the global and local search abilities [28]. In addition, it is possible to rank all the particles according to their fitness values so that suitable weight coefficients can be assigned accordingly [29].

4.1.2.5 Cooperative Mechanism

Cooperative method is a sophisticated strategy that shows to be effective in solving separable functions which can be optimised separately for each dimension. In [30], the particle swarm is partitioned into N subswarms where N is the dimensionality of the search space. Each subswarm contains M particles whose position is one dimensional (1D) and represents a component of the solution vector. The particles in a certain subswarm search on a certain dimension of the solution space and after the positions of all the particles in the whole swarm are updated, N particles with each from different subswarm cooperate to build up a new potential solution, with each particle's position being its component on the corresponding dimension. The QPSO with this mechanism is known as the cooperative QPSO (CQPSO).

A hybrid CQPSO which combines the CQPSO and the original QPSO is described in [31]. In this variant, the CQPSO is executed followed by the QPSO algorithm at each iteration. Under the protection of the global best position of any of the subswarms, CQPSO and QPSO exchange information and in turn enhance the search ability of the whole algorithm. A competitive scheme [31] has been incorporated into the hybrid CQPSO [32]. The new scheme further diversifies the swarm to help to escape from local optima. Finally the multiswarms may be cooperating in a two-layer framework [33] where a single QPSO swarm is employed to track with the global best solution in the top layer and several other QPSO swarms are employed to search in the bottom layer.

4.1.3 Applications of the QPSO

Since QPSO was first introduced, there are numerous papers reporting successful applications in various aspects of the method. In this section, an overview is given of these applications. In providing such an overview, the authors have adopted the categories in line with those used in [34]. Note that the discussion in the following does not represent an exhaustive list of all applications involving QPSO, but provides with the readers a summary of important areas that have been examined by various researchers. The topics are grouped in alphabetical order.

4.1.3.1 Antenna Design

The problem of finding a set of infinitesimal dipoles to represent an arbitrary antenna with known near-field distribution is a challenging problem. It involves the formulation of the task as an optimisation problem in which the function to be optimised is a cost function defined as the average difference between the near-field distribution and the field generated by the dipoles over a series of spatial positions. Mikki et al. used the QPSO technique for the problem of dielectric resonator antennas [35].

4.1.3.2 Biomedicine

The codon usage of synthetic gene is an important area in computational biomedical science. Computational methods such as genetic algorithm (GA) have been developed for such applications. Cai et al. explored the technique of QPSO for optimising the codon usage of synthetic gene and have shown to generate better results when the DNA/RNA sequence length is less than 6 kb [36]. Liu et al. coupled RBF neural network to the QPSO technique for the culture conditions of hyaluronic acid production by *Streptococcus zooepidemicus* [37]. In [38], Lu and Wang employed QPSO to estimate parameters from kinetic model of batch fermentation. An intelligent type 2 diabetes diagnosis based on QPSO and weighted least squares support vector machines (WLS-SVM) was presented in [39] to overcome the disadvantage of low diagnostic accuracy.

4.1.3.3 Mathematical Programming

Mathematical programming problems that have been solved by using QPSO include integer programming, constrained non-linear programming and layout optimisation. Constrained optimisation problems have been tackled as reported in [40] and combinatorial optimisation problems have also been considered by using a novel discrete QPSO based on the estimation of distribution algorithm [41]. The first attempt of using QPSO for integer programming was examined by Liu et al. [42] and for constrained and unconstrained non-linear programming problems was discussed by Sun et al. [43]. An improved QPSO was applied to solve a non-linear substation locating and sizing planning model [44]. It was also reported that a new generic method/model for multiobjective design optimisation of laminated composite components has been studied by using a novel multiobjective optimisation algorithm based on QPSO [45].

4.1.3.4 Communication Networks

QoS multicast routing, network anomaly detection, mobile IP routing and channel assignment in communication networks are important topics in computer science and have been studied in many literatures. In particular, QoS multicast routing is an NP-hard problem. This problem may be converted into an integer programming problem and has been solved by using a QPSO-based idea developed by Sun et al. [46]. It was also reported that the QPSO was used to solve the selection problem of the mobile IP routing [47] and the network anomaly detection problem. The hybrid of QPSO with gradient descent algorithm has been employed to train RBFNN [48] and used for network anomaly detection. On the other hand, a combination of QPSO with conjugate gradient algorithm has been used to train wavelet neural network and for network anomaly detection [49]. It was also reported that the QPSO method was used to train the WLS-SVM which was subsequently applied to the network anomaly detection problem [50]. Finally, the mathematic model of channel assignment problem was also tackled as discussed in [51].

4.1.3.5 Control Engineering

Applications of the QPSO method in control engineering have been reported in areas such as the design of controllers, PID controllers, H_∞ controllers, thruster fault-tolerant control and chaotic control. The selection of the output feedback gains for the unified power flow controllers can be converted to an optimisation problem with the time domain-based objective function solved by using QPSO as described in [52]. Gaussian distribution can be used in QPSO in order to tune the design parameters of a fuzzy logic control with PID conception [10]. In the design of H_∞ controller and

PID parameter optimisation, Xi et al. have employed the QPSO technique [53,54]. In the control reallocation problem, where the control energy cost function is required to be minimised, the thruster fault-tolerant approach for unmanned underwater vehicles has been reported successfully solved [55]. Finally, experience in estimating the unknown parameters of chaos systems in chaotic control and synchronisation was reported in [56].

4.1.3.6 Clustering and Classification

A QPSO-based data clustering technique was first proposed in [57] where a particle represents the cluster centroid vector. The results as shown in the paper suggest that QPSO could be used to generate good results in clustering data vectors with tolerable time consumption. A similar data clustering algorithm has been applied in gene expression data clustering by partitioning the N patterns of the gene expression dataset into K categories in order to minimise the fitness function of total within-cluster variation [58], where K is the number of clusters prespecified by the user. By combining QPSO and K-harmonic means clustering algorithm, Lu et al. [59] proposed a hybrid clustering method.

In another application related to alert correlation methods used for attack detection, the QPSO method can be used to optimise the weights and similarity value of the alerts in the cluster [60], which can aggregate the relational alerts by computing the similarity between alert attributes and discover new, simple high-level attacks.

Spatial clustering is an active research area in the data mining community. A novel clustering algorithm based on QPSO and K-Medoids to cluster spatial data with obstacles was proposed in [61]. In addition, combining the fuzzy C-means algorithm and QPSO was used to improve the performance of the fuzzy C-means algorithm with the gradient descent [62].

Text classification is an important research area of data mining. It is possible to improve the classification precision and accuracy for the text topic mining and classification and acquisition of classification rule [63,64]. Furthermore, in the area of text categorisation, an adaptive QPSO has been used to train the central position and the width of the basis function adopted in the RBFNN which was then used for the feature selection problem [65].

QPSO may also be transformed into a binary QPSO and then combined with the attribute reduction algorithm [66] to solve the attribute reduction problem. On the other hand, QPSO was also used to optimise the parameter selection in the minimal attribute reduction problem [67].

4.1.3.7 Electronics and Electromagnetics

The first paper reported on the application of QPSO to electromagnetic appeared in [68] for linear array antenna synthesis problems. Several different versions of QPSO have also been developed for linear array antenna synthesis problems and for finding an equivalent circuit model [69]. Other applications include finding infinitesimal dipole models for antennas with known near fields [70] and electromagnetic design optimisation using mutation operator based on Gaussian and exponential probability distributions [9,71].

Finally the QPSO has also found applications in optimal design of CMOS operational amplifier circuit performance [72] and solution to large systems of matrix equations in transistor simulation [73].

4.1.3.8 Finance

Work in the area of financial engineering concerns investment decision making, including multistage portfolio [74], and parameter optimisation of the GARCH model for stock market prediction [75].

4.1.3.9 Fuzzy Systems

In the area of fuzzy systems, QPSO was demonstrated to be a promising tool for the design of these systems. The design of fuzzy system can be treated as a space search problem. Each point in the space represents a fuzzy system with assigned parameters in each component. Optimisation algorithms, such as GA and PSO, can be used to search for the near-optimal point in the space composed of all the possible fuzzy systems. QPSO was shown to be applicable to design components of fuzzy system and was shown to be able to evolve both the fuzzy rules and the membership efficiently [76].

4.1.3.10 Graphics

Applications in the areas of computer graphics fall into three main areas, including rectangle-packing problems [77], polygonal approximation of curves [78] and irregular polygons layout [79]. For rectangle-packing problems, two-dimensional (2D) irregular objects are replaced by the rectangle packing and QPSO can be applied to find the optimal peaks of the rectangle. On the other hand, a binary QPSO was used for the polygonal approximation of curves [80].

4.1.3.11 Image Processing

QPSO has been applied to image processing, including face detection, image segmentation, image registration and image interpolation.

The QPSO method was used, instead of the gradient descent method, as a learning algorithm for backpropagation neural network which can then be applied to face detection as presented in [81].

QPSO can be combined with three kinds of image threshold selection methods, including 2D maximum entropy [82], 2D OTSU [83] and multilevel minimum cross-entropy threshold (MCET) [84], in image segmentation. The optimum solution is searched for after retrieving the 2D histogram of an image [82] in the case of 2D maximum entropy. In the case of 2D OTSU, the best 2D threshold vector representing the particle position is to be found. Results as shown in [83] demonstrate the reduction in computational costs and ideal segmentation results. Other research which includes the near optimal MCET thresholds can be found in [84].

QPSO can also be used as an optimiser in order to find the best rigid parameters based on a coarse-to-fine registration framework in a coarse registration step [85] involved in image registration of remote sensing images.

It is also possible to improve the quality of the interpolated image and enhance the resolution of images supplementing the conventional algorithm for image interpolation in order to avoid artefacts and to provide high-resolution images [86].

4.1.3.12 Power Systems

The objective of solving an economic dispatch (ED) problem is to simultaneously minimise the generation cost rate while satisfying various equality and inequality constraints. By using the harmonic oscillator potential well, QPSO was employed to solve ED problems in power systems [87]. Sun et al. used the QPSO algorithm with differential mutation operation to solve the ED problems of three different power systems [88]. Results observed there show high-quality stable and efficient solutions.

4.1.3.13 Modelling

It is well known that the regularisation parameter and the kernel parameter play an important role in the performance of the standard SVM model and LS-SVM model. Traditionally analytical, algebraic and heuristic techniques have been used for tuning the regularisation parameter and the kernel parameter. The QPSO method has found applications in selecting LS-SVM hyper-parameters [89,90], and a diversity-based adaptive QPSO has also been employed to determine automatically the free parameters of the SVM model [91]. In addition, the method can be used to train the

SVM model for solving quadratic programming problems [92]. Selection of optimal parameters for SVM model and using these parameters to establish the soft-sensor model of desorption-process in an organic solvent recovery was also reported in [93].

Finally modelling for a gallium arsenide-based metal semiconductor field effect transistor device [94], non-linear system identification of Hammerstien and Wiener model [95], and the detection of unstable periodic orbits in a non-Lyapunov way [96], all were found to be fruitful using the QPSO method.

4.1.3.14 Other Application Areas
The QPSO method is suitable for training the radial basis function neural networks [97], which can also be combined with QPSO and could handle many applications [65,98–100], including feature selection [65], short-term load forecasting in power systems [98], sidelobe suppression for chaotic frequency modulation signal [99] and identification of power quality disturbance [100].

Multiprocessor scheduling problems were considered in [101], where the QPSO search technique is combined with list scheduling to improve the solution quality in short time.

Finally the design of FIR filters [102,103], design of IIR filters [104,105] and design of 2D filters [106,107], which are essentially multiparameters optimisation problems, have been solved by using the QPSO and other adaptive versions.

4.2 MOTIVATION: FROM CLASSICAL DYNAMICS TO QUANTUM MECHANICS

QPSO was motivated by considering quantum mechanics and the analysis of the particle behaviour in PSO. Analysis of the dynamical behaviour of an individual particle in PSO shows that properly selected values of constriction factor and acceleration coefficients can prevent the particle position from exploding, that is, ensure the boundedness of the particle. The boundedness of a particle in PSO reflects the collective property of the particle swarm. This means that the particle either has a cyclic trajectory or converges to a focus. Such property guarantees the convergence of the particle swarm in finding optima or suboptima of the problem. On the other hand, explosion of the particle position makes the particle fly away from the search region where global optimal solution is located resulting in a poor solution of a problem. In order to ensure the particle swarm works properly, each particle must be in a bound state during the search process.

In this section, a dynamical analysis of the behaviour of individual particle is used to show the trajectories for different parameter selections and is then extended to the concept of boundedness, that is, the bound state, in classical dynamics to quantum mechanics. This leads to the motivation of developing a PSO algorithm which is able to ensure the bound state.

4.2.1 Dynamical Analysis of Particle Behaviour

Clerc proposed a dynamical analysis of particles in the original PSO with or without the constriction factor (PSO-Co) [1]. Here, the analysis proposed by Clerc is used to treat the PSO with inertia weight (PSO-In) as it is one of the two most widely used variants of the original PSO algorithm. The evolution equations of the particle in PSO-In are given by

$$V_{i,n+1}^{j} = wV_{i,n}^{j} + c_1 r_{i,n}^{j}(P_{i,n}^{j} - X_{i,n}^{j}) + c_2 R_{i,n}^{j}(G_n^{j} - X_{i,n}^{j}), \qquad (4.1)$$

$$X_{i,n+1}^{j} = X_{i,n}^{j} + V_{i,n+1}^{j}. \qquad (4.2)$$

Since the components of the velocity and position of the particle in each dimension update independently with the same form of the evolution equation, component-wise analysis of the asymptotic behaviour of the velocity and position vectors is sufficient. Without loss of generality, the case of a single particle in 1D space is considered and the evolution equations are simplified to

$$v_{n+1} = wv_n + c_1 r_n(P_n - x_n) + c_2 R_n(G_n - x_n), \qquad (4.3)$$

$$x_{n+1} = x_n + v_{n+1}, \qquad (4.4)$$

where v_n and x_n denote the velocity and position of the particle, respectively, at the nth iteration. Let $\varphi_1 = c_1 r_n$ and $\varphi_2 = c_2 R_n$. r_n, $R_n \sim U(0,1)$, $\varphi_1 \sim U(0,c_1)$ and $\varphi_2 \sim U(0,c_2)$ are two random variables uniformly distributed on $(0,c_1)$ and $(0,c_2)$ respectively. Consequently, Equation 4.3 can be written as

$$v_{n+1} = wv_n + \varphi_1(P_n - x_n) + \varphi_2(G_n - x_n). \qquad (4.5)$$

Writing the weighted mean of P_n and G_n as

$$p = \frac{\varphi_1 P_n + \varphi_2 G_n}{\varphi_1 + \varphi_2}, \qquad (4.6)$$

where p is a new random variable. Here, for simplification, φ_1, φ_2 and p are treated as constants as in [1]. Replacing the iteration number n by time t, the particle state can be formulated by using the following discrete dynamical system

$$\begin{cases} v_{t+1} = wv_t + \varphi(p - x_t), \\ x_{t+1} = x_t + v_{t+1}, \end{cases} \tag{4.7}$$

where $\varphi = \varphi_1 + \varphi_2$, and $w > 0$ and $\varphi > 0$. Let $y_t = p - x_t$, the system can be rewritten as

$$\begin{cases} v_{t+1} = wv_t + \varphi y_t, \\ y_{t+1} = -wv_t + (1 - \varphi)y_t. \end{cases} \tag{4.8}$$

Let the current evolution state in R^2 be

$$S_t = \begin{bmatrix} v_t \\ y_t \end{bmatrix}, \tag{4.9}$$

and the matrix of the system be

$$M = \begin{bmatrix} w & \varphi \\ -w & 1-\varphi \end{bmatrix}. \tag{4.10}$$

The dynamical equation of the system can be rewritten as

$$S_{t+1} = MS_t, \tag{4.11}$$

or in terms of S_0

$$S_t = M^t S_0. \tag{4.12}$$

Here S_{t+1} is the evolution state at time $t + 1$. The system is now defined completely by M where the eigenvalues are given by

$$\begin{cases} e_1 = \dfrac{1 + w - \varphi + \sqrt{(\varphi - w - 1)^2 - 4w}}{2}, \\ e_2 = \dfrac{1 + w - \varphi - \sqrt{(\varphi - w - 1)^2 - 4w}}{2}. \end{cases} \tag{4.13}$$

When $(\varphi - w - 1)^2 - 4w \neq 0$, $e_1 \neq e_2$ there exists an invertible matrix

$$A = \begin{bmatrix} \varphi + w - 1 + \sqrt{(\varphi - w - 1)^2 - 4w} & 2\varphi \\ \varphi + w - 1 - \sqrt{(\varphi - w - 1)^2 - 4w} & 2\varphi \end{bmatrix}, \quad (4.14)$$

such that

$$AMA^{-1} = L = \begin{bmatrix} e_1 & 0 \\ 0 & e_2 \end{bmatrix}. \quad (4.15)$$

It is possible to define $Q_t = AS_t$, that is,

$$Q_{t+1} = LQ_t, \quad (4.16)$$

or

$$Q_{t+1} = L^t Q_0 = \begin{bmatrix} e_1^t & 0 \\ 0 & e_2^t \end{bmatrix} Q_0. \quad (4.17)$$

The system converges, that is, $v_t \to 0$ and $y_t \to 0$ if and only if $|e_1| < 1$ and $|e_2| < 1$. In the remaining part of this section, an analysis of the trajectory of the particle for three different cases of e_1 and e_2 is given. The simulation results of the trajectory of the particle are both traced on x–t plane and v–x plane. In each simulation, the data for the initial position were chosen as $x_0 = 2$, the initial velocity as $v_0 = -0.1$ and $p = 0$.

4.2.1.1 Case When $(\varphi - w - 1)^2 - 4w = 0$

The system has two identical real eigenvalues, that is, $e_1 = e_2 = (1 + w - \varphi)/2$. The necessary and sufficient condition for the system to converge is given by

$$\begin{cases} \left| \dfrac{1 + w - \varphi}{2} \right| < 1, \\ (\varphi - w - 1)^2 - 4w = 0. \end{cases} \quad (4.18)$$

For example, if $w = 0.64$, $\varphi = 0.04$ and $w = 0.64$, $\varphi = 3.24$, the previous conditions are satisfied and the particle converges to its focus $p = 0$, as shown in Figures 4.1 and 4.2.

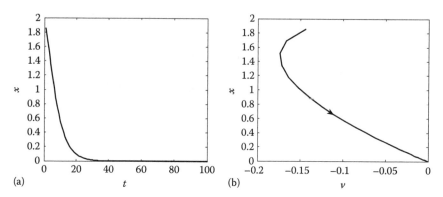

FIGURE 4.1   The trajectories of the particle on (a) x–t plane and on (b) v–x plane when $w = 0.64$ and $\varphi = 0.04$.

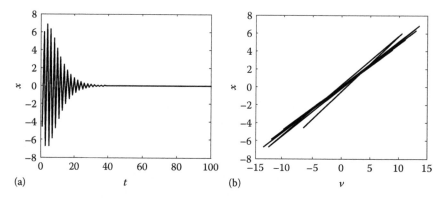

FIGURE 4.2   The trajectories of the particle on (a) x–t plane and on (b) v–x plane when $w = 0.64$ and $\varphi = 3.24$.

When $|(1 + w - \varphi)/2| = 1$, either $e_1 = e_2 = (1 + w - \varphi)/2 = 1$ or $e_1 = e_2 = (1 + w - \varphi)/2 = -1$, and the system diverges. If $(1 + w - \varphi)/2 = 1$, one has $(\varphi - w - 1)^2 - 4w = 0$, $w = 0$ and $\varphi = -1$, the conditions $w > 0$ and $\varphi > 0$ are not satisfied. If $(1 + w - \varphi)/2 = -1$, the condition $(\varphi - w - 1)^2 - 4w = 0$ leads to $w = 1$ and $\varphi = 4$, and the results of this diverging system are illustrated in Figure 4.3. In this case, the particle position vibrates around $p = 0$, but its amplitude increases gradually and goes to infinity.

When $|(1 + w - \varphi)/2| > 1$, the system also diverges. Figure 4.4 shows an example for when $w = 1.44$ and $\varphi = 0.04$, such that the inequality $|(1 + w - \varphi)/2| > 1$ and the equality $(\varphi - w - 1)^2 - 4w = 0$ are satisfied. The position of the particle tends to infinity without vibration.

Figure 4.5 illustrates another example for which $w = 1.44$ and $\varphi = 4.84$, such that the conditions $|(1 + w - \varphi)/2| > 1$ and $(\varphi - w - 1)^2 - 4w = 0$

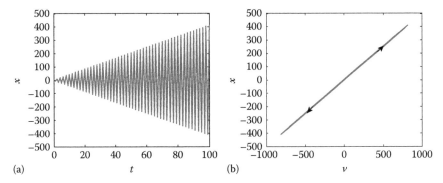

FIGURE 4.3 The trajectories of the particle on (a) x–t plane and on (b) v–x plane when $w = 1$ and $\varphi = 4$.

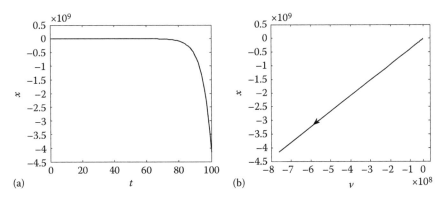

FIGURE 4.4 The trajectories of the particle on (a) x–t plane and on (b) v–x plane when $w = 1.44$ and $\varphi = 0.04$.

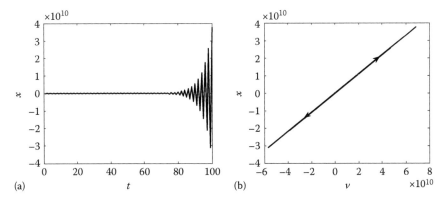

FIGURE 4.5 The trajectories of the particle on (a) x–t plane and on (b) v–x plane when $w = 1.44$ and $\varphi = 4.84$.

are satisfied. In this example, the particle vibrates with the amplitude diverges, as can be seen from the trajectory on the x–t plane as shown in Figure 4.5a.

4.2.1.2 Case When $(\varphi - w - 1)^2 - 4w > 0$

In this case, e_1 and e_2 are two distinct real eigenvalues. The system converges if and only if

$$
\begin{cases}
-2 < 1 + w - \varphi + \sqrt{(\varphi - w - 1)^2 - 4w} < 2, \\
-2 < 1 + w - \varphi - \sqrt{(\varphi - w - 1)^2 - 4w} < 2, \\
(\varphi - w - 1)^2 - 4w > 0,
\end{cases}
\tag{4.19}
$$

that is,

$$
\begin{cases}
-2 < 1 + w - \varphi < 2, \\
(\varphi - w - 1)^2 - 4w > 0.
\end{cases}
\tag{4.20}
$$

Figure 4.6 shows the convergence history when $w = 0.4$ and $\varphi = 2.7$, which lie within the convergence condition. Note that the particle converges to its focus $p = 0$ as the amplitude of the oscillation decreases. Figure 4.7 depicts another example when $w = 0.4$ and $\varphi = 0.1$. The trajectory in this example shows that the particle converges to its focus $p = 0$ monotonically without any oscillation.

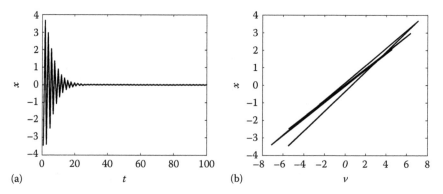

(a)

(b)

FIGURE 4.6   The trajectories of the particle on (a) x–t plane and on (b) v–x plane when $w = 0.4$ and $\varphi = 2.7$.

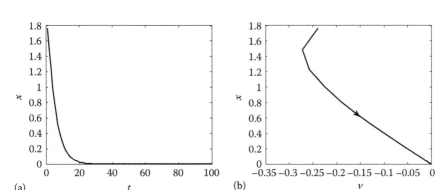

(a)

(b)

FIGURE 4.7 The trajectories of the particle on (a) x–t plane and on (b) v–x plane when $w = 0.4$ and $\varphi = 0.1$.

When $|1 + w - \varphi| = 2$, that is, either $1 + w - \varphi = 2$ or $1 + w - \varphi = -2$, the condition $(\varphi - w - 1)^2 - 4w > 0$ leads to $w < 1$. The equality $1 + w - \varphi = 2$ leads to $\varphi < 0$ which does not satisfy the condition $\varphi > 0$. From $1 + w - \varphi = -2$, one obtains $\varphi = 3 + w$ and $w < 1$, which leads to a diverging system. Figure 4.8 demonstrates one typical diverging example satisfying the conditions $|1 + w - \varphi| = 2$ and $(\varphi - w - 1)^2 - 4w > 0$ by taking $w = 0.9$ and $\varphi = 3.9$.

When $|1 + w - \varphi| > 2$, the system diverges as shown in Figure 4.9 by choosing $w = 1.2$ and $\varphi = 5$. The trajectory of this example in x–t plane has the amplitude oscillating towards infinity which means the particle position diverges.

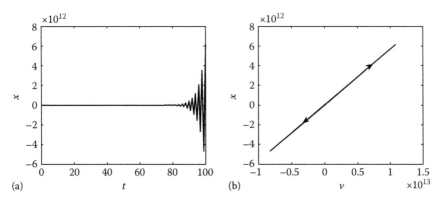

(a)

(b)

FIGURE 4.8 The trajectories of the particle on (a) x–t plane and on (b) v–x plane when $w = 0.9$ and $\varphi = 3.9$.

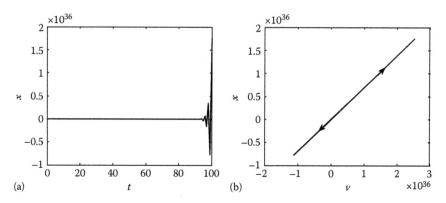

FIGURE 4.9   The trajectories of the particle on (a) x–t plane and on (b) v–x plane when $w = 1.2$ and $\varphi = 5$.

4.2.1.3 Case When $(\varphi - w - 1)^2 - 4w < 0$

The system has a pair of complex conjugate eigenvalues. The absolute value of this pair of complex conjugate eigenvalues is given by

$$|e_1| = |e_2| = \sqrt{\frac{(1+w-\varphi)^2}{4} + \frac{4w-(\varphi-w-1)^2}{4}} = \sqrt{w}. \qquad (4.21)$$

The system converges if and only if

$$\begin{cases} w < 1, \\ (\varphi - w - 1)^2 - 4w < 0. \end{cases} \qquad (4.22)$$

Two examples are used to illustrate the convergence, with w and φ set as 0.7 and 0.5, and 0.7 and 2.0, respectively. Figures 4.10 and 4.11 depict the trajectories of these two examples.

Choosing $w = 1$, the condition $(\varphi - w - 1)^2 - 4w < 0$ leads to $\varphi < 4$. The absolute values of the eigenvalues are 1, and the system has cyclic or quasi-cyclic solutions. Figure 4.12 shows an example of cyclic solution with cyclic period 4 by choosing $w = 1$ and $\varphi = 2$. Figure 4.13 shows a quasi-cyclic example by choosing $w = 1$ and $\varphi = 1.5$.

When $w > 1$ and $(\varphi - w - 1)^2 - 4w < 0$, the system diverges as illustrated in the trajectories traced in Figure 4.14 for the case of $w = 1.5$ and $\varphi = 2$.

Putting the previous three cases together reveals that φ and w should possess values within the shaded region in the $\varphi - w$ plane as shown in Figure 4.15. Values falling within the shaded region guarantee the convergence of the system.

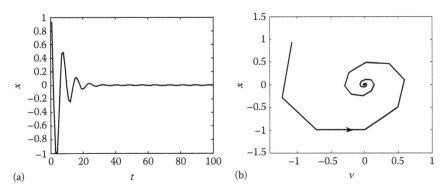

FIGURE 4.10 The trajectories of the particle on (a) x–t plane and on (b) v–x plane when $w = 0.7$ and $\varphi = 0.5$.

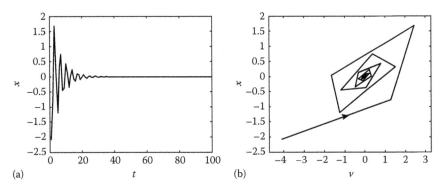

FIGURE 4.11 The trajectories of the particle on (a) x–t plane and on (b) v–x plane when $w = 0.64$ and $\varphi = 3.24$.

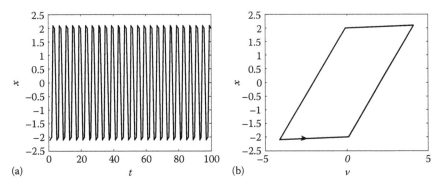

FIGURE 4.12 The trajectories of the particle on (a) x–t plane and on (b) v–x plane when $w = 1$ and $\varphi = 2$.

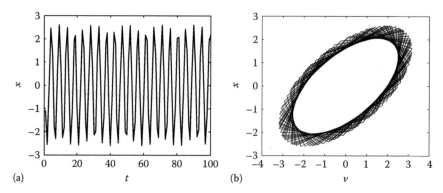

FIGURE 4.13 The trajectories of the particle on (a) x–t plane and on (b) v–x plane when $w = 0.64$ and $\varphi = 3.24$.

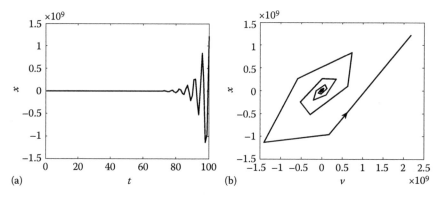

FIGURE 4.14 The trajectories of the particle on (a) x–t plane and on (b) v–x plane when $w = 0.64$ and $\varphi = 3.24$.

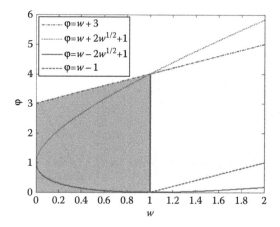

FIGURE 4.15 Values of φ and w within the shaded region on the $\varphi - w$ plane guarantee the convergence of the particle position to its focus.

The earlier analysis indicates that properly selected φ and w can lead the particle position to converge to its focus p. When PSO is used for an N-dimensional optimisation problem, the coordinates $p_{i,n}$ of the local focus of particle i are defined as

$$p_{i,n}^j = \frac{c_1 r_{i,n}^j P_{i,n}^j + c_1 R_{i,n}^j G_n^j}{c_1 r_{i,n}^j + c_1 R_{i,n}^j}, \tag{4.23}$$

or

$$p_{i,n}^j = \varphi_{i,n}^j P_{i,n}^j + (1 - \varphi_{i,n}^j) G_n^j, \tag{4.24}$$

where $\varphi_{i,n}^j = c_1 r_{i,n}^j \big/ (c_1 r_{i,n}^j + c_2 R_{i,n}^j)$ with random numbers $r_{i,n}^j$ and $R_{i,n}^j$ as defined in (4.1) and (4.2). In PSO, the acceleration coefficients c_1 and c_2 are generally set to be equal, and thus $\varphi_{i,n}^j$ is a sequence of uniformly distributed random numbers on $(0,1)$. As a result, Equation 4.20 can be rearranged as

$$p_{i,n}^j = \varphi_{i,n}^j P_{i,n}^j + (1 - \varphi_{i,n}^j) G_n^j, \quad \varphi_{i,n}^j \sim U(0,1). \tag{4.25}$$

This implies that $p_{i,n}$ is a random point which is located within the hyperplane constructed in between $P_{i,n}$ and G_n in the search space. As the particles are converging to their own local focuses, their current positions, personal best positions, local focuses and the global best positions are all converging to the same point. This way the PSO algorithm is said to be convergent. From the point of view of classical dynamics, the convergence history of a particle would be to move around and careen towards the point $p_{i,n}$ with its kinetic energy (or velocity) declining to zero, like a returning satellite orbiting the earth. As such, the particle in PSO can be considered as an object flying in an attraction potential field centred at the point $p_{i,n}$ in the Newtonian space. To avoid explosion and ensure convergence, one requires the distance from $p_{i,n}^j$ to each component of the particle position to be bounded which is known as the bound state, that is,

$$\sup_{n \geq 0} |X_{i,n}^j - p_{i,n}^j| < \infty, \quad 1 \leq i \leq M, \quad 1 \leq j \leq N. \tag{4.26}$$

In the case that the particle in PSO has quantum behaviour moving in the N-dimensional Hilbert space, the particle needs to move in a quantum potential field to ensure the bound state. Since the bound state in the quantum space is different from that in the Newtonian space, they lead to very different forms of PSO. The particle bound state from the perspectives of quantum mechanics is discussed in the next section.

4.2.2 Particle Behaviour from the Perspectives of Quantum Mechanics

In Section 4.2.1, using dynamical analysis of the particle behaviour, it is shown that each particle in PSO should be in a bound state to guarantee the convergence of the swarm during the search process. However, a social organism, such as the human society, is a system far more complex than that formulated by the evolution equation of PSO. Simulating the behaviour of human intelligence, instead of that of a bird flock or fish schooling, requires the thinking mode of an individual of the social organism which is not sufficient to be described by using a linear evolution equation. It is believed that human thinking is uncertain as if a particle having quantum behaviour.

In the quantum time-space framework, the state of a particle is described by the wave function $\Psi(X, t)$. In a three-dimensional (3D) space, the wave function $\Psi(X, t)$ of a particle satisfies the relation

$$|\Psi|^2 \, dxdydz = Qdxdydz, \qquad (4.27)$$

where $Qdxdydz$ is the probability for the particle to appear in an infinitesimal volume about the point (x, y, z). In other words, $|\Psi|^2 = Q$ represents the probability density function satisfying

$$\int_{-\infty}^{+\infty} |\Psi|^2 \, dxdydz = \int_{-\infty}^{+\infty} Qdxdydz = 1. \qquad (4.28)$$

Equation 4.27 or 4.28 gives the statistical interpretation of the wave function. Generally, $\Psi(X, t)$ varies in time according to the equation

$$i\hbar \frac{\partial}{\partial t} \Psi(X,t) = \hat{H}\Psi(X,t), \qquad (4.29)$$

where
$\hbar$ is the Planck constant
$\hat{H}$ is the Hamiltonian operator

$$\hat{H} = -\frac{\hbar^2}{2m}\nabla^2 + V(X), \qquad (4.30)$$

for a single particle of mass m in a potential field $V(X)$. Equation 4.29 is known as the time-dependent Schrödinger equation a system evolving with time. The equation, formulated by the Austrian physicist Erwin Schrödinger, is used to describe how the quantum state of a physical system changes in time. It is as central to quantum mechanics as Newton's laws to classical mechanics. Solutions to Schrödinger's equation describe not only molecular, atomic and subatomic systems, but also macroscopic systems. For a 1D system in its stationary state, removing the time-dependent term of the Schrödinger equation in Equation 4.29, it is sufficient to describe the state of the particle with a given energy level as follows:

$$\frac{d^2\psi}{dX^2} + \frac{2m}{\hbar^2}[E + V(X)]\psi = 0, \qquad (4.31)$$

where
   E is the energy of the particle
   ψ is the wave function of the particle which only depends on the position

In quantum mechanics, if the particle is in a bound state, its wave function should satisfy the condition

$$\psi \xrightarrow[|X|\to\infty]{} 0, \qquad (4.32)$$

which can be interpreted in terms of the probability measure P as

$$P\{\sup|X| < \infty\} = 1, \qquad (4.33)$$

that is, the probability of finding the particle at infinity is zero.

4.3 QUANTUM MODEL: FUNDAMENTALS OF QPSO

4.3.1 δ Potential Well Model

Assuming each single particle in QPSO is treated as spinless and being driven in the 1D Hilbert space with a given energy. Its state is characterised by the wave function which only depends on the position. Inspired by the convergence analysis of the particle in PSO, it is assumed that the jth component of particle i is being driven in the N-dimensional Hilbert space with a δ potential well centred at $p^j_{i,n}$, $1 \leq j \leq N$, at the nth iteration. For simplicity, the position of the particle is denoted as X and $p_{i,n}$ is written as p. With point p the centre of the potential field, the potential energy of the particle in the 1D δ potential well is represented as

$$V(X) = -\gamma\delta(X - p) = -\gamma\delta(Y), \tag{4.34}$$

where
$Y = X - p$
γ is the intensity of the potential well (see Figure 4.16)

Hence, for this bound state problem, the Hamiltonian operator is

$$\hat{H} = -\frac{\hbar^2}{2m}\frac{d^2}{dY^2} - \gamma\delta(Y). \tag{4.35}$$

The state of a particle satisfies the stationary Schrödinger equation

$$\frac{d^2\psi}{dY^2} + \frac{2m}{\hbar^2}[E + \gamma\delta(Y)]\psi = 0, \tag{4.36}$$

where
E is the energy of the particle
ψ is the wave function of the particle which depends on the position only

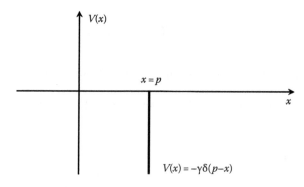

FIGURE 4.16   Potential energy of the δ potential well.

Theorem 4.1 presented in the following provides a method to obtain the normalised wave function of the particle in the bound state by solving the previous Schrödinger equation with the bound condition $\psi \to 0$ when $|Y| \to \infty$.

Theorem 4.1

For the particle moves in the 1D δ potential well formulated by Equation 4.36, its normalised wave function in bound state is given by

$$\psi(Y) = \frac{1}{\sqrt{L}} e^{-|Y|/L}, \tag{4.37}$$

where $L = 1/\beta = \hbar^2/m\gamma$.

Proof: Integrating Equation 4.36 with respect to Y from $-\varepsilon$ to ε and take $\varepsilon \to 0^+$ leads to

$$\psi'(0^+) - \psi'(0^-) = -\frac{2m\gamma}{\hbar^2} \psi(0). \tag{4.38}$$

For $Y \neq 0$, Equation 4.36 can be written as

$$\frac{d^2\psi}{dY^2} - \beta^2 \psi = 0, \tag{4.39}$$

where $\beta = \sqrt{-2mE}/\hbar$ $(E < 0)$. To satisfy the bound condition

$$\psi \to 0 \quad \text{as} \, |Y| \to \infty, \tag{4.40}$$

the solution of Equation 4.39 must be

$$\psi(Y) \propto e^{-\beta|Y|} \quad (Y \neq 0). \tag{4.41}$$

Note that it can be proved that only even wave function satisfies the bound condition (4.40). Introducing a normalised constant A, the solution in (4.41) can be written as

$$\psi(Y) = \begin{cases} Ae^{-\beta Y} & Y > 0, \\ Ae^{\beta Y} & Y < 0. \end{cases} \tag{4.42}$$

Using Equation 4.38 one obtains

$$-2A\beta = -\frac{2m\gamma}{\hbar^2}A.$$

Thus

$$\beta = \frac{m\gamma}{\hbar^2},$$
(4.43)

and

$$E = E_0 = -\frac{\hbar^2\beta^2}{2m} = -\frac{m\gamma^2}{2\hbar^2}.$$
(4.44)

The function $\psi(Y)$ satisfies the normalisation condition

$$\int_{-\infty}^{+\infty} |\psi(Y)|^2 dY = \frac{|A|^2}{\beta} = 1,$$
(4.45)

leading to $|A| = \sqrt{\beta}$. Let $L = 1/\beta = \hbar^2/m\gamma$ be the characteristic length of the δ potential well. Substituting $|A| = \sqrt{\beta} = 1/\sqrt{L}$ and $\beta = 1/L$ into Equation 4.42, the normalised wave function can then be written as

$$\psi(Y) = \frac{1}{\sqrt{L}} e^{-|Y|/L}.$$
(4.46)

This completes the proof. ■

In terms of the statistical interpretation of the wave function, the probability density function of Y is given by

$$Q(Y) = |\psi(Y)|^2 = \frac{1}{L} e^{-2|Y|/L},$$
(4.47)

and the corresponding probability distribution function is

$$F(Y) = 1 - e^{-2|Y|/L}.$$
(4.48)

4.3.2 Evolution Equation of QPSO

Given a probability distribution function, the position of a particle can be measured by using Monte Carlo inverse transformation. This technique is being described in the proof of Theorem 4.2. Such process of measuring particle positions in quantum mechanics is essentially achieved by collapsing the quantum state to the classical state.

Theorem 4.2

If a particle moves in a bound state in the 1D δ potential well as described by Equation 4.34, its position can be determined by using the stochastic equation

$$X = p \pm \frac{L}{2} \ln\left(\frac{1}{u}\right), \tag{4.49}$$

where u is a random number uniformly distributed on (0,1), that is, $u \sim U(0,1)$.

Proof: Let v be a random number uniformly distributed on (0,1), that is, $v \sim U(0,1)$. Substituting v for $F(Y)$ in (4.48) and following Monte Carlo simulation one obtains

$$1 - v = e^{-2|Y|/L}. \tag{4.50}$$

Since $1 - v \sim U(0,1)$, putting $u = 1 - v$ leads to $u \sim U(0,1)$. Thus Equation 4.50 can be written as

$$u = e^{-2|Y|/L}. \tag{4.51}$$

One obtains immediately

$$Y = \pm \frac{L}{2} \ln\left(\frac{1}{u}\right).$$

Putting $Y = X - p$ one obtains

$$X = p \pm \frac{L}{2} \ln\left(\frac{1}{u}\right), \quad u \sim U(0,1). \tag{4.52}$$

This completes the proof of the theorem. ∎

Since u is a random variable, the particle's position X in Equation 4.49 is also a random variable. If the particle positions are sampled successively, one can obtain a sequence of random variables given by

$$X_n = p \pm \frac{L}{2} \ln\left(\frac{1}{u_n}\right), \quad u_n \sim U(0,1), \tag{4.53}$$

where p is a random variable. Assuming the intensity of the potential well γ develops with the time steps, the characteristic length L, denoted as L_n at the nth iteration, of the δ potential well also varies with the time steps. The previous equation can be written as

$$X_{n+1} = p \pm \frac{L_n}{2} \ln\left(\frac{1}{u_{n+1}}\right), \quad u_{n+1} \sim U(0,1). \tag{4.54}$$

Equation 4.54 can be generalised to the case in the N-dimensional Hilbert space where each dimension of the particle position is bounded in a δ potential well and updated independently. It is assumed that the particle position, the local focus, the characteristic length of δ and the random variable u evolve with the number of iterations n. It is possible to use the following fundamental evolution equation of the QPSO to measure the jth component of the position of particle i at the $(n + 1)$th iteration.

$$X_{i,n+1}^j = p_{i,n}^j \pm \frac{L_{i,n}^j}{2} \ln\left(\frac{1}{u_{i,n+1}^j}\right), \quad u_{i,n+1}^j \sim U(0,1), \tag{4.55}$$

where
  $p_{i,n}^j$ is the jth component of the local focus $p_{i,n}$ of particle i at the nth iteration
  $u_{i,n+1}^j$ is a sequence of random numbers uniformly distributed on $(0,1)$

4.3.3 Conditions of Convergence

As far as the canonical PSO is concerned, the parameters w, c_1 and c_2 must be suitably chosen so that the particle position converges to its local focus leading to a convergent algorithm. Therefore, the particle in QPSO should also approach its local focus when the algorithm is used for practical applications. However, Equation 4.55 is of stochastic nature such that $X_{i,n}$ is a random variable irrespective of the randomness of the local focus $p_{i,n}$. Hence, the convergence of the particle position must be investigated using the framework of probability theory.

In probability theory, there are several kinds of convergence theory of random variables. In this chapter, one definition of convergence in probability is presented. Several other possible definitions are presented in Chapter 5.

Let $\xi_1, \xi_2, \ldots$ be random variables defined on a probability space $(\Omega, \mathcal{F}, P)$, where Ω is the sample space, $\mathcal{F}$ is a σ-algebra of subsets of Ω and P is a probability measure on $\mathcal{F}$. Convergence in probability of the sequence $\{\xi_n\}$ is defined as follows:

Definition 4.1

The sequence $\xi_1, \xi_2, \ldots$ of random variables converges in probability to the limit random variable ξ, that is, $\xi_n \xrightarrow{P} \xi$, if for every $\varepsilon > 0$

$$P\{|\xi_n - \xi| > \varepsilon\} \to 0, \quad n \to \infty, \tag{4.56}$$

or

$$P\{|\xi_n - \xi| < \varepsilon\} \to 1, \quad n \to \infty, \tag{4.57}$$

where $P\{A\}$ is the probability measure of the measurable set A in $\mathcal{F}$.

Since the convergence of the position of the particle can be reduced to the convergence of each dimension, for simplicity only 1D case of the evolution equation for the particle given by Equation 4.53 is considered. For the random variable defined in Equation 4.53, the probability density function is given by

$$Q(Y_{n+1}) = \frac{1}{L_n} e^{-2|Y_{n+1}|/L_n}, \tag{4.58}$$

where $Y_n = X_n - p$ for every $n > 0$. The following main theorem is needed.

Theorem 4.3

A necessary and sufficient condition for $X_n \xrightarrow{P} p$ is that

$$\lim_{n \to +\infty} I_m = 0. \tag{4.59}$$

Proof: There are two steps in proving the theorem, (i) the sufficient condition and (ii) the necessary condition.

(i) Proof of the sufficiency

Suppose the probability density function in Equation 4.58 is a function of L_n and that $Y_{n+1} = Y$, the density function can be represented by

$$Q_{L_n}(Y) = \frac{1}{L_n} e^{-2|Y|/L_n}. \tag{4.60}$$

Hence

$$\lim_{L_n \to 0} Q_{L_n}(0) = \infty. \tag{4.61}$$

Since $Q_{L_n}(Y)$ satisfies the normalisation condition, it can been seen from Equations 4.60 and 4.61 that $Q_{L_n}(Y)$ has the characteristics of a δ function, that is,

$$\lim_{L_n \to 0} \frac{1}{L_n} e^{-2|y|/L_n} = \delta(Y) = \delta(X - p). \tag{4.62}$$

Using the properties

$$\delta(X - p) = \begin{cases} \infty & X = p \\ 0 & X \neq p \end{cases} \tag{4.63}$$

and

$$\int_{p-\varepsilon}^{p+\varepsilon} \delta(X - p)dX = \int_{-\infty}^{+\infty} \delta(X - p)dX = 1, \tag{4.64}$$

it can be shown that for every $\varepsilon > 0$

$$\lim_{n \to \infty} P\left\{|X - p| < \varepsilon\right\} = \lim_{n \to \infty} \int_{-\varepsilon}^{\varepsilon} \frac{1}{L_n} e^{-2|Y|/L_n} dY = \int_{-\varepsilon}^{\varepsilon} \lim_{n \to \infty} \frac{1}{L_n} e^{-2|Y|/L_n} dY$$

$$= \int_{p-\varepsilon}^{p+\varepsilon} \delta(X - p) dX = \int_{-\infty}^{+\infty} \delta(X - p) dX = 1. \qquad (4.65)$$

This completes the proof of sufficient condition.

(ii) Proof of the necessity

If $X_n \xrightarrow{P} p$, as $n \to +\infty$, for every $\varepsilon > 0$

$$\lim_{n \to \infty} P\left\{|X - p| < \varepsilon\right\} = \lim_{t \to \infty} \int_{-\varepsilon}^{\varepsilon} \frac{1}{L_n} e^{-2|Y|/L_n} dY = \int_{-\varepsilon}^{\varepsilon} \lim_{n \to \infty} \frac{1}{L_n} e^{-2|Y|/L_n} dY = 1. \quad (4.66)$$

Consider the following three possible cases.

(a) $L_n \to \infty$, as $n \to +\infty$.

The limit of the probability density function for this case is

$$\lim_{n \to \infty} Q_{L_n}(Y) = \lim_{L_n \to \infty} \frac{1}{L_n} e^{-2|Y|/L_n} = 0. \qquad (4.67)$$

Thus

$$\lim_{n \to \infty} P\{|X - p| < \varepsilon\} = \lim_{n \to \infty} \int_{-\varepsilon}^{\varepsilon} \frac{1}{L_n} e^{-2|Y|/L_n} dY = 0, \qquad (4.68)$$

which contradicts (4.66).

(b) $L_n \to c$, as $n \to +\infty$, where $0 < c < \infty$.

The limit of the density function for this case is

$$\lim_{n \to \infty} Q_{L_n}(Y) = \lim_{L_n \to c} \frac{1}{L_n} e^{-2|Y|/L_n} = \frac{1}{c} e^{-2|Y|/c}. \qquad (4.69)$$

Therefore

$$\lim_{n \to \infty} P\{|X - p| < \varepsilon\} = \lim_{n \to \infty} \int_{-\varepsilon}^{\varepsilon} \frac{1}{L_n} e^{-2|Y|/L_n} dY = \int_{-\varepsilon}^{\varepsilon} \frac{1}{c} e^{-2|Y|/c} dY = 1 - e^{-2\varepsilon/c} < 1,$$

(4.70)

which also contradicts (4.66).

(c) $L_n \to 0$, as $n \to +\infty$.

The limits of the density function for this case is

$$\lim_{L_n \to 0} \frac{1}{L_n} e^{-2|Y|/L_n} = \delta(Y) = \delta(X - p).$$

(4.71)

One obtains

$$\lim_{n \to \infty} P\{|X - p| < \varepsilon\} = \lim_{n \to \infty} \int_{-\varepsilon}^{\varepsilon} \frac{1}{L_n} e^{-2|Y|/L_n} dY = \int_{-\varepsilon}^{\varepsilon} \lim_{n \to \infty} \frac{1}{L_n} e^{-2|Y|/L_n} dY = 1. \quad (4.72)$$

This completes the proof of Theorem 4.3.　　　　■

4.3.4 Discussion

Apart from the δ potential well model, there are many other quantum potential models that can be established for PSO. Consider a quantum harmonic oscillator (QHO) field for a PSO system with the same methodology as described in previous sections. The normalised wave function of the ground state for QHO model is given by

$$\psi_0(Y) = \frac{\sqrt{\alpha}}{\pi^{1/4}} e^{-1/2\alpha^2 Y^2}.$$

(4.73)

The corresponding probability density function is

$$Q(Y) = |\psi_0(Y)| = \frac{\sigma}{\sqrt{\pi}} e^{-\sigma^2 Y^2},$$

(4.74)

and the probability distribution function is given by

$$F(Y) = \int_{-\infty}^{y} \frac{\sigma}{\sqrt{\pi}} e^{-\sigma^2 Y^2} \, dY. \qquad (4.75)$$

Using Monte Carlo simulation, the position of a particle can be measured by

$$Y = \pm \frac{1}{\sigma} \lambda, \qquad (4.76)$$

or

$$X = p \pm \frac{1}{\sigma} \lambda, \qquad (4.77)$$

where
  $1/\sigma$ in the aforementioned formulae is the characteristic length of the harmonic oscillator field
  λ is a random number with standard normal distribution, that is, $\lambda \sim N(0,1)$

σ is an important control parameter like the characteristic length of the δ potential well in QPSO. Note that the difference between Equations 4.49 and 4.77 is that different probability distributions are used. The corresponding probability density functions of these two equations are shown in Figures 4.17 and 4.18 respectively.

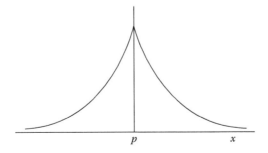

FIGURE 4.17 The probability density function for $Q = (1/L)e^{-2|x-p|/L}$.

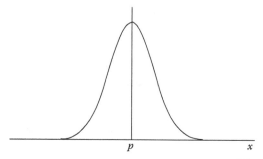

FIGURE 4.18   The probability density function for $Q = (\alpha/\sqrt{\pi})e^{-\alpha^2(x-p)^2}$.

It is observed from Figures 4.17 and 4.18 that the curve of Equation 4.72 is relatively narrower than that of Equation 4.47, showing that the particle converges much faster in the PSO system with the QHO model than in the QPSO system with the same parameter control method. Such convergence speed is attained with the sacrifice of the global search ability of the algorithm. The QHO model of PSO was tested by using some benchmark functions [2]. Results suggest that the PSO algorithm with QHO model is prone to encounter premature convergence.

4.4   QPSO ALGORITHM

4.4.1   Two Search Strategies of QPSO

The evolutionary equation in Equation 4.55 has an open problem of determining the value of $L_{i,n}^j$. A simple method is by fixing the value of $L_{i,n}^j$ during the search process. This approach would most certainly guarantee the boundedness of the particle, but convergence of the algorithm becomes slower at the later stage of search. Another intuitive approach is to decrease the value of $L_{i,n}^j$ according to Theorem 4.3 so that each particle can move towards its local focus. Note that its performance is very sensitive to problems. For example, decreasing $L_{i,n}^j$ linearly from the size of the search domain to zero may generate a good solution to the sphere function but a poor solution to Rosenbrock function.

There are two efficient strategies for determining the value of $L_{i,n}^j$ leading to good performance of the QPSO algorithm. In [2], it was suggested that the value of $L_{i,n}^j$ in Equation 4.55 can be determined by

$$L_{i,n}^j = 2\alpha \, | \, X_{i,n}^j - p_{i,n}^j \, |. \tag{4.78}$$

Upon substituting Equation 4.78 into Equation 4.53 leads to the evolution equation,

$$X_{i,n+1}^{j} = p_{i,n}^{j} \pm \alpha \, | \, X_{i,n}^{j} - p_{i,n}^{j} \, | \ln\left(\frac{1}{u_{i,n+1}^{j}}\right), \qquad (4.79)$$

where α is a positive real number known as the CE coefficient. The PSO defined in Equation 4.79 is known as the quantum delta potential well-based PSO (QDPSO). Although QDPSO shows satisfactory performance on several widely used benchmark functions, its sensitivity to the values of α has been noted by many researchers [3].

In order to overcome the shortcomings of QDPSO, one strategy for controlling the value of $L_{i,n}^{j}$ by scaling the distance from the current position to a global point C_n was proposed in [3]. In this strategy, $L_{i,n}^{j}$ is calculated as follows:

$$L_{i,n}^{j} = 2\alpha \, | \, X_{i,n}^{j} - C_n^{j} \, |, \qquad (4.80)$$

where $C_n = (C_n^1, C_n^2, \ldots, C_n^N)$ is known as the mean best position which can be defined by the average of the personal best positions of all particles, that is,

$$C_n^{j} = \left(\frac{1}{M}\right)\sum_{i=1}^{M} P_{i,n}^{j} \qquad (1 \le j \le N). \qquad (4.81)$$

Accordingly, the position of the particle can be updated by using the equation

$$X_{i,n+1}^{j} = p_{i,n}^{j} \pm \alpha \, | \, X_{i,n}^{j} - C_n^{j} \, | \ln\left(\frac{1}{u_{i,n+1}^{j}}\right). \qquad (4.82)$$

The parameter α in the previous equation is the same as the one defined in Equation 4.79. In this book, the PSO with Equation 4.79 or Equation 4.82 is referred to as the QPSO algorithm. To distinguish them, the QPSO with Equation 4.79 is denoted as QPSO-Type 1 and that with Equation 4.82 as QPSO-Type 2, particularly when the behaviour of the particle with either of the two evolution equations is analysed in Chapter 5. In Equations 4.79 and 4.82, $p_{i,n}^j$ is may be obtained by Equation 4.21 and is rewritten as follows:

$$p_{i,n}^j = \varphi_{i,n}^j P_{i,n}^j + (1 - \varphi_{i,n}^j) G_n^j, \quad \varphi_{i,n}^j \sim U(0,1). \tag{4.83}$$

In QPSO, the CE coefficient α is the most important algorithmic parameter that requires to be adjusted to balance the local and global search of the algorithm during the search process. Chapter 5 provides an analysis of the impact of the CE coefficient on the behaviour of the particle in both types of the QPSO algorithms and discusses how to select its value for practical usage.

4.4.2 Procedure of QPSO

The procedure of QPSO is similar to that of the PSO algorithm, except that they have different evolution equations. In the QPSO algorithm, there is no velocity vector for each particle, and the position of the particle updates directly according to Equation 4.79 or 4.82. For ease of implementation, here is a description on the flow of the procedure. The QPSO algorithm starts with the initialisation of the current positions of the particles and their personal best positions by setting $P_{i,0} = X_{i,0}$, followed with the iterative update of the particle swarm using the update equations provided earlier. In each iteration of the procedure, the mean best position of the particle swarm is computed (for QPSO-Type 2), and the current position of each particle is updated according to Equation 4.79 or 4.82 with the coordinates of its local focus evaluated by Equation 4.83. After each particle updates its current position, its fitness value is evaluated together with an update of the personal best position and the current global best position. In Equation 4.79 or 4.82, the probability of using either '+' operation or '−' operation is equal to 0.5. The iteration continues until the termination condition is met.

The procedure of the QPSO algorithm is outlined in Figure 4.19. Note that rand$i(\cdot)$, $i = 1, 2, 3$, is used to denote random numbers generated uniformly and distributed on $(0,1)$. When the procedure is used for solving an optimisation problem, the value of α must be determined. Methods of selecting the value of α are important in QPSO and are addressed in Chapter 5.

4.4.3 Properties of QPSO

In this section, properties of QPSO are discussed. The authors discuss and examine the significance of these properties in terms of the solution process in three different aspects.

```
Begin
Initialize the current position X_{i,0}^{j} and the personal best position P_{i,0}^{j}
of each particle, evaluate their fitness values and find the global
best position G_0; Set n = 0.
While (termination condition = false)
Do
  Set n = n + 1;
  Compute mean best position C (for QPSO-Type 2);
  Select a suitable value for α;
  for (i = 1 to M)
    for j = 1 to N
      φ_{i,n}^{j} = rand1 (·);
      p_{i,n}^{j} = φ_{i,n}^{j}P_{i,n}^{j} + (1 - φ_{i,n}^{j})G_n^j;
      u_{i,n}^{j} = rand2 (·);
      if (rand3(·) < 0.5)
        X_{i,n+1}^{j} = p_{i,n}^{j} + α|X_{i,n}^{j} - p_{i,n}^{j}|ln(1/u_{i,n+1}^{j}) (for QPSO-Type 1);
        (or X_{i,n+1}^{j} = p_{i,n}^{j} + α|X_{i,n}^{j} - C_n^j|ln(1/u_{i,n+1}^{j}) (for QPSO-Type 2));
      else
        X_{i,n+1}^{j} = p_{i,n}^{j} - α|X_{i,n}^{j} - p_{i,n}^{j}|ln(1/u_{i,n+1}^{j}) (for QPSO-Type 1);
        (or X_{i,n+1}^{j} = p_{i,n}^{j} - α|X_{i,n}^{j} - C_n^j|ln(1/u_{i,n+1}^{j}) (for QPSO-Type 2));
      end if
    end for
    Evaluate the fitness value of X_{i,n+1}, that is, the objective func-
    tion value f(X_{i,n+1});
    Update P_{i,n} and G_n
  end for
end do
end
```

FIGURE 4.19 The procedure of the QPSO algorithm.

4.4.3.1 Advantages of Using the Quantum Model

In the QPSO algorithm, the quantum δ potential well model is used to simulate the thinking model of human beings. Individual particles with quantum behaviour are being assembled to form a particle swarm system which simulates a human society. It is possible to make the analogy of the search process of the particle swarm to the discovery and development of knowledge in the human society. The quantum model for PSO can be used to overcome the shortcomings of the original PSO for the following reasons.

First, the principle of state superposition applies in a quantum system. As a result, a quantum system has far more states than a classical linear dynamical system of PSO. In essence, a quantum system is a system of uncertainty and is very different from classical dynamical systems. Such uncertainty means that a particle in the system can appear in any position, before actual measurement takes place, according to certain probability distribution because there is no determined trajectory. This property resembles many different ideas in the mind of a human before an actual decision is made.

Second, the particles in PSO must be in bound states in order to guarantee collectiveness of the particle swarm. This property allows the PSO algorithm to converge to optima or suboptima. In the canonical PSO, a bound state provides constraints to the particles ensuring that they are in a finite zone. In QPSO, a particle of the swarm can appear at any position in the whole feasible search space with a certain probability, even at a position far from the global best position. Such a position may be superior to the current global best position of the swarm.

4.4.3.2 Knowledge Seeking of Particles in QPSO

In terms of knowledge evolvement of a social organism, there are two kinds of important cognition associated with the knowledge discovering of an individual (a particle) in the PSO algorithm. One is the personal best position which represents the best previous cognition excogitated by the individual. The other is the global best position which represents the best cognition across the whole population. Each particle would search in the direction of its current position to a point $p_{i,n}$ that lies in between the personal best and the global best positions. The point $p_{i,n}$ has its coordinates given by Equation 4.83 resulting from interaction between self-study and teaching when the individual is learning. $p_{i,n}$ is referred to as the 'learning inclination point' (LIP) of the individual [3]. Tendentiousness of learning makes it inevitable that the individual seeks new knowledge around its LIP, which is determined by the existing knowledge in the global best and the personal best positions.

In the canonical PSO, learning tendentiousness of an individual particle is reflected by its oscillation about LIP taken as the equilibrium point (or focus). However, a social organism is far from such a linear system, and the thinking process of each individual is of uncertainty to a great extent. As a result, the trajectory movement is inadequate to describe such tendentiousness. Moreover, the original PSO model cannot explain the reasons for a talented person like Einstein in the human society which is a social organism. In fact, collectiveness of particles in the canonical PSO confines the search scope of the particle. As a consequence, a particle may not appear at a position which is far from the particle swarm and could well be a better solution than the current global best position.

At every iteration of the QPSO algorithm, each particle records its personal best position and compares its personal best position with all of those particles in its neighbourhood or population resulting in the global best position. The LIP, $p_{i,n}$, is given by Equation 4.83. The process is followed by establishing a δ potential well at the point $p_{i,n}$ in order to simulate the tendentiousness of the particle. To begin the next knowledge-seeking step, parameter $L_{i,n}^j$ must be evaluated. In fact, the parameter $L_{i,n}^j$ reflects creativity or imagination and characterises the knowledge-seeking scope of a particle. The larger the value of $L_{i,n}^j$, the more likely the particle would find out new knowledge. The creativity of a particle in QPSO is evaluated by the gap between the current position of the particle and its LIP or the one between the current position and the mean best position. Finally by taking state transformation from the search space to the solution space, the new position can be obtained by Equation 4.79 or 4.82. If the new position is of better knowledge than the personal best position, then it will be replaced by the new knowledge. It is at this point that a particle in QPSO is more creative and more intelligent because the new position of the particle can appear at any point in the search space provided it has better fitness value.

4.4.3.3 Wait Effect amongst Particles

An exponential distribution of positions makes QPSO a global convergent algorithm. Besides, a peculiar waiting ability arises due to the mean best position introduced into the QPSO (i.e. QPSO-Type 2), leading to a better particle distribution.

In the canonical PSO, each particle converges to the global best position independently. This becomes the major shortcoming of the method when there are minor numbers of particles that are still far away from the global best position. Figure 4.20 depicts the concept where the personal best positions of several particles, known as the lagged particles, are located far away from the rest of the particles and the global best position, while the rest of the particles are nearer to the global best position. In the QPSO method, the swarm could not gather around the global best position without waiting for the lagged particles. These lagged particles, not being abandoned by the swarm, affect the distribution of particles in the next iteration, that is, the mean best position would be shifted towards the lagged particles. Such distribution of particles will certainly affect the convergence rate, but the idea leads to stronger global search ability than the canonical PSO.

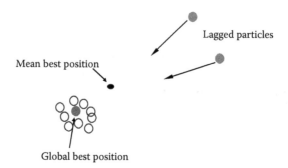

FIGURE 4.20   Lagged particles and wait among particles in QPSO.

4.5 SOME ESSENTIAL APPLICATIONS

This section explores the applicability of QPSO to function optimisation problems, including unconstrained, constrained and multipeak optimisation problems. Since QPSO-Type 2 is the widely used version of QPSO algorithm, this section only investigates the application of QPSO-Type 2 to these problems. For simplicity, QPSO-Type 2 is referred to as QPSO in this section as well as in Chapter 6.

4.5.1 Applications to Unconstrained Function Optimisation

The application of the QPSO algorithm to unconstrained function optimisation problems in this section involves optimising the benchmark functions as described in Chapter 3. Only the sphere function with dimensionality 30 is used in this section as an example problem to illustrate the implementation of the QPSO algorithm for unconstrained optimisation problems. A comprehensive set of performance comparison between QPSO and other PSO variants is given in Chapter 5.

Assuming that the readers have foundation knowledge in MATLAB®, sphere.m implements the sphere function, and QPSO.m contains the source code of QPSO implementation. The content of QPSO.m is shown in Figure 4.21 and is explained as follows.

By and large, the source codes of QPSO.m as listed in Figure 4.21 are similar to those of PSO.m shown in Chapter 2. The two methods differ in the evolution equations used in QPSO as listed in lines from line 27 to line 33. Here the value of the CE coefficient α needs to be fixed. Two ways of fixing the values may be used. First the value of α is a constant throughout the search process. As an example, in the previous code, α is set as 0.75 in line 9. Second, the value of α decreases linearly during the course of the search process. As an example, in the previous code α decreases linearly

```
%%%%%%%QPSO.m (Source Code of QPSO with Sphere Function as a
Benchmark) %%%%%%%%%
1. popsize=20;  % population size
2. MAXITER=3000;    % Maximum number of iterations
3. dimension=30;   % Dimensionality of the problem
4. irange_l=-100;  % Lower bound of initialization scope
5. irange_r=100;   % Upper bound of initialization scope
6. xmax=100;    % Upper bound of the search scope
7. xmin=-100;   % Lower bound of the search scope
8. M=(xmax-xmin)/2;  % The middle point of the search cope on
each dimension
9. alpha=0.75;  % A fixed value is used. It can be changed.
10. runno=50;    % Number of times the algorithm runs
11. data=zeros(runno, MAXITER); % The matrix recording the
fitness value of gbest position at each iteration %%%%%%%%%%%%
Running for runno times%%%%%%%%%%%
12. for run=1:runno
%%%%%%%%%%%%%%%% Initialization of the particle swarm %%%%%%%%%%%%
%%%%%
13.    x=(irange_r-irange_l)*rand(popsize,dimension,1) +
irange_l; % Initialize the particle position
14.    pbest=x; %Set the pbest position as the current position
of the particle
15.    gbest=zeros(1,dimension); % Initialize the gbest poistion
vector

16.    for i=1:popsize
17.      f_x(i)=sphere(x(i,:)); %Calculate the fitness value of
the current position of the particle
18.      f_pbest(i)=f_x(i);% Set the fitness value of the pbest
position to be that of the current position
19.    end

20.    g=min(find(f_pbest==min(f_pbest(1:popsize)))); % Find
index of the particle with gbest position
21.    gbest=pbest(g,:); % Determine the gbest position
22.    f_gbest=f_pbest(g); % Determine the fitness value of the
gbest position

%%%%%%%%%%%%%% The following is the loop of the QPSO's search
process %%%%%%%%%%%
23.    for n=1:MAXITER
24. % alpha=(1.0-0.5)*(MAXITER-n)/MAXITER+0.5; % Determine the
value of alpha
25.      mbest=sum(pbest)/popsize; % Calculate the mbest position

26.      for i=1:popsize %The following is the update of the
particle's position
27.        fi=rand(1,dimension); % Generate a vector of random
numbers with distribution U(0,1)
```

FIGURE 4.21   MATLAB source code for the QPSO algorithm.

(*continued*)

```
28.         p=fi.*pbest(i,:)+(1-fi).*gbest; % Determine the vector
local focus of the particle
29.         u=rand(1,dimension);
30.         x(i,:)=p+((-1).^ceil(0.5+rand(1,dimension))).*(alpha.
*abs(mbest-x(i,:)).*log(1./u));
31.         x(i,:)=x(i,:)-(xmax+xmin)/2;% These tree lines are to
restrict the position in search scopes
32.         x(i,:)=sign(x(i,:)).*min(abs(x(i,:)),M);
33.         x(i,:)=x(i,:)+(xmax+xmin)/2;

34.         f_x(i)=sphere(x(i,:)); % Calculate the fitness value
of the particle's current position
35.         if (f_x(i)<f_pbest(i))
36.            pbest(i,:)=x(i,:);% Update the pbest position of the
particle
37.            f_pbest(i)=f_x(i);% Update the fitness value of the
particle's pbest position
38.         end
39.         if f_pbest(i)<f_gbest
40.            gbest=pbest(i,:); % Update the gbest position
41.            f_gbest=f_pbest(i); % Update the fitness value of
the gbest position
42.         end
43.       end
44.    data(run, n)=f_gbest; % Record the fitness value of the
gbest at each iteration at this run
45.    end
46. end
```

FIGURE 4.21 (continued)

from 1.0 to 0.5 as a function of the iteration number n as shown in line 24. The code QPSO.m implements QPSO-Type 2 which means the mean best position must be determined at each iteration as shown in line 25. Note that the velocity for the particle in QPSO is not required.

From line 27 to line 33, the position of a particle and the restriction of the position within the search scope on each dimension are computed. Noted that line 30 implements the update of the position of a particle in such a way that selection between operations '+' and '−' is executed simultaneously. These lines can be replaced by the codes in Figure 4.22, which update the particle's position in a loop for dimensions instead of in a whole matrix.

In a typical experiment, the source code in Figure 4.21 using either way of defining α with 20 particles was run for 50 times, and each run executed 3000 iterations. The fitness value at each iteration during every run of the algorithm was recorded in the matrix 'data'. These fitness values in 'data' were averaged over the 50 runs resulting in an array of averaged fitness

```
%%%%%The following source codes are used to update the
particle's position in each dimension%%%%%
1.   for d=1: dimension
2.     fi=rand; % Generate the random number □~U(0,1)
3.     p=fi*pbest(i,d)+(1-fi)*gbest(d); % Determine the component
of the local focus on each dimension
4.     u=rand; % Generate the random number u.~U(0,1)
5.     if rand>0.5
6.        x(i,d)=p+alpha*abs(mbest(d)-x(i,d))*log(1/u); % Use "+"
operation in the evolution equation
7.     else
8.        x(i,d)=p-alpha*abs(mbest(d)-x(i,d))*log(1/u); % Use "-"
operation in the evolution equation
9.     end
10.    if x(i,d)>xmax;
11.       x(i,d)=xmax; % If the component of the position larger
than xmax, set it to be xmax
12.    end
13.    if x(i,d)<-xmin;
14.       x(i,d)=-xmin; % If the component of the position
smaller than xmin, set it to be xmin
15.    end
16. end
```

FIGURE 4.22   MATLAB source code to update the position of a particle in each dimension.

TABLE 4.1   Mean and Standard Deviation of the Best Fitness Values Obtained with 50 Runs of Each Algorithm

Algorithms	Mean Best Fitness Value	Standard Deviation of the Best Fitness Values
PSO	3.289×10^4	3.7464×10^3
PSO-In	3.5019×10^{-14}	4.8113×10^{-14}
PSO-Co	8.6167×10^{-38}	3.1475×10^{-37}
QPSO ($\alpha = 0.75$)	1.6029×10^{-71}	7.6101×10^{-71}
QPSO ($\alpha = 1.0 \rightarrow 0.5$)	2.9682×10^{-22}	1.6269×10^{-21}

values at each iteration. The length of the array in this study was 3000. The last value in the array is the mean best fitness value over 50 runs and is given in Table 4.1. The last column of the matrix 'data' records the best fitness values obtained after 50 runs of QPSO. The standard deviation of the 50 best fitness values is also listed in Table 4.1 with the corresponding data for PSO, PSO-In and PSO-Co as a comparison. The array of averaged fitness values at each iteration can be used to plot the convergence history. Figure 4.23 presents the convergence histories of QPSO using both choices of α, PSO, PSO-In and PSO-Co.

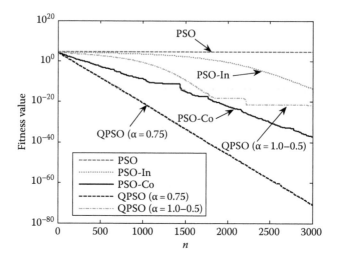

FIGURE 4.23   Convergence histories for the sphere function.

4.5.2 Solutions to Constrained NPL

4.5.2.1 Problem Description

This section is devoted to numerical experiments of several constrained non-linear programming (NLP) problems by using the QPSO method. In particular, the QPSO-Type 2 method was tested. Most of the results in this section are extracted from the publication in [43] with some modifications. For simplicity, only the principles or the procedures of algorithms are presented.

The mathematical description of a constrained NLP problem is given by

$$\min_{x} f(x), \quad x \in S \subset R^{N}. \tag{4.84}$$

$$
\begin{aligned}
\text{subject to} \quad & g_i(x) \leq 0, & i = 1, 2, \ldots, m, \\
& h_j(x) = 0, & j = 1, 2, \ldots, k, \\
& a_i \leq x_i \leq b_i, & 1 \leq i \leq N, \\
& x = (x_1, x_2, \ldots, x_N),
\end{aligned}
\tag{4.85}
$$

where
    $f(x)$ is the objective function
    $g_i(x)$ and $h_j(x)$ represent equality and inequality constraints respectively
    a_i and b_i are the upper and lower bounds of the search domain for x_i

The formulation of the constraints in (4.85) is not restrictive, since an inequality constraint of the form $g_i(x) \geq 0$ can also be represented as $-g_i(x) \leq 0$, and the equality constraint $h_i(x) = 0$ is equivalent to two inequality constraints $h_i(x) \geq 0$ and $h_i(x) \leq 0$. Thus, the restrictive form of (4.85) without equality constraints can be written as

$$g_i(x) \leq 0, \qquad i = 1,2,\ldots,m,$$

$$a_i \leq x_i \leq b_i, \qquad 1 \leq i \leq N, \qquad (4.86)$$

$$x = (x_1, x_2, \ldots, x_N).$$

The most common approach of solving the constrained optimisation problems is to use a penalty function. The purpose of a penalty function is to transform the constrained NLP problem to an unconstrained optimisation problem by building a single objective function and penalising the constraints. The new single objective function can then be minimised by using an unconstrained optimisation algorithm. This is most probably the reason behind the popularity of the penalty function approach when EAs are used to address constrained NLP problems only.

However, it is very difficult to select a suitable penalty. If the penalty values are high, the minimisation algorithms usually would be trapped in local minima. On the other hand, if penalty values are low, there are hardly any feasible optimal solutions that can be detected. In this section, a nonstationary multistage assignment penalty function is adopted as in [108]. The penalty values are dynamically modified according to the inequality constraints $g_i(x)$ and the equality constraints $h_j(x)$. The penalty function is defined by

$$F(x) = f(x) + \lambda(n)H(x), \quad x \in S \subset R^N, \qquad (4.87)$$

where
$f(x)$ is the original objective function of the constrained NLP problem in Equation 4.84
$\lambda(n)$ is a dynamically modified penalty value and n is the number of iterations

$H(x)$ is known as a penalty factor and is defined as

$$H(x) = \sum_{i=1}^{m} \theta(q_i(x)) q_i(x)^{\gamma(q_i(x))} \qquad (4.88)$$

where $q_i(x) = \max\{0, g_i(x)\}$, $i = 1,...,m$. The function $q_i(x)$ is a relative violated function of the constraints, $\theta(q_i(x))$ is a multistage assignment function, $\gamma(q_i(x))$ is the power of the penalty function and $g_i(x)$ is the constraints described in Equation 4.86. The functions $\lambda(\cdot)$, $\theta(\cdot)$ and $\gamma(\cdot)$ are problem dependent. Details of the penalty function used in this study are described in the following section.

4.5.2.2 Experiments of Benchmark Problems for Constrained Optimisation

4.5.2.2.1 Constrained Benchmark Functions   QPSO, PSO-In and PSO-Co were tested on six benchmark problems; all concerning constrained optimisation is described as follows.

Problem 1

$$f_1(x) = (x_1 - 2)^2 + (x_2 - 1)^2$$

$$\text{s.t.} \quad x_1 = 2x_2 - 1, \quad \frac{x_1^2}{4} + x_2^2 - 1 \le 0,$$

The best known solution of the problem is $f^* = 1.3934651$.

Problem 2

$$f_2(x) = (x_1 - 10)^3 + (x_2 - 20)^3$$

$$\text{s.t.} \quad \begin{cases} 100 - (x_1 - 5)^2 - (x_2 - 5)^2 \le 0, (x_1 - 6)^2 + (x_2 - 5)^2 - 82.81 \le 0, \\ 13 \le x_1 \le 100, 0 \le x_2 \le 100. \end{cases}$$

The best known solution of the problem is $f^* = -6961.81381$.

Problem 3

$$f_3(x) = (x_1 - 10)^2 + 5(x_2 - 12)^2 + x_3^4 + 3(x_4 - 11)^2 + 10x_5^6$$

$$+ 7x_6^2 + x_7^4 - 4x_6 x_7 - 10x_6 - 8x_7$$

$$\text{s.t.} \quad \begin{cases} -127 + 2x_1^2 + 3x_2^4 + x_3 + 4x_4^2 + 5x_5 \le 0 \\ -282 + 7x_1 + 3x_2 + 10x_3^2 + x_4 - x_5 \le 0 \\ -196 + 23x_1 + x_2^2 + 6x_6^2 \\ -8x_7 \le 0, \ 4x_1^2 + x_2^2 - 3x_1 x_2 + 2x_3^2 + 5x_6 - 11x_7 \le 0 \\ -10 \le x_i \le 10, \quad i = 1,\ldots,7 \end{cases} \quad ,$$

The best known solution of the problem is $f^* = 680.630057$.

Problem 4

$$f_4(x) = 5.3578547x_3^2 + 0.83568911x_1 x_5 + 37.293239x_1 - 40,792.141$$

$$\text{s.t.} \quad \begin{cases} 0 \le 85.334407 + 0.0056858T_1 + T_2 x_1 x_4 - 0.0022053x_3 x_5 \le 92 \\ 90 \le 80.51249 + 0.0071317x_2 x_5 + 0.0029955x_1 x_2 + 0.0021813x_3^2 \le 110 \\ 20 \le 9.300961 + 0.0047026x_2 x_5 + 0.0012547x_1 x_3 + 0.0019085x_3 x_4 \le 25 \\ 78 \le x_1 \le 102, \ 33 \le x_2 \le 45, \ 27 \le x_i \le 45, \quad i = 3,4,5, \end{cases}$$

where
$T_1 = x_2 x_5$
$T_2 = 0.0006262$

The best known solution of Problem 4 is $f^* = -30,665.538$.

Problem 5

$$f_5(x) = 5.3578547x_3^2 + 0.83568911x_1 x_5 + 37.293239x_1 - 40,792.141$$

$$
\text{s.t.}
\begin{cases}
0 \le 85.334407 + 0.0056858 T_1 + T_2 x_1 x_4 - 0.0022053 x_3 x_5 \le 92 \\[2mm]
90 \le 80.51249 + 0.0071317 x_2 x_5 + 0.0029955 x_1 x_2 + 0.0021813 x_3^2 \le 110 \\[2mm]
20 \le 9.300961 + 0.0047026 x_2 x_5 + 0.0012547 x_1 x_3 + 0.0019085 x_3 x_4 \le 25 \\[2mm]
78 \le x_1 \le 102,\ 33 \le x_2 \le 45,\ 27 \le x_i \le 45, \quad i = 3,4,5,
\end{cases}
$$

where
$$T_1 = x_2 x_3$$
$$T_2 = 0.00026$$

The best solution of Problem 5 is unknown.

Problem 6

$$
f_6(x,y) = -10.5 x_1 - 7.5 x_2 - 3.5 x_3 - 2.5 x_4 - 1.5 x_5 - 10 y - 0.5 \sum_{i=1}^{5} x_i^2
$$

$$
\text{s.t.}
\begin{cases}
6 x_1 + 3 x_2 + 3 x_3 + 2 x_4 + x_5 - 6.5 \le 0, \\[2mm]
10 x_1 + 10 x_3 + y \le 20,\ 0 \le x_i \le 1, \quad i = 1,\ldots,5,\ 0 \le y.
\end{cases}
$$

The best known solution of the problem is $f^* = -213.0$.

4.5.2.2.2 Experimental Configurations   The parameter α for QPSO was made to vary linearly from 1.0 to 0.5 during a single run. The population size for any of variants of PSO was set to 100. For each problem, each algorithm ran 10 times, with each run executing 1000 iterations. A violation tolerance, $g_i(x) > 10^{-5}$, was used for the constraints. The same values for the penalty parameters reported in [108] were used in the experiments here to obtain results which could then be compared to those obtained using a different algorithm. For the purpose of clarity, these penalty parameters are listed here. If $q_i(x) < 1$, then $\gamma(q_i(x)) = 1$, otherwise $\gamma(q_i(x)) = 2$. Moreover, if $q_i(x) < 0.001$, then $\theta(q_i(x)) = 10$, else if $q_i(x) < 0.1$ then $\theta(q_i(x)) = 20$, else if $q_i(x) \le 1$ then $\theta(q_i(x)) = 100$, otherwise $\theta(q_i(x)) = 300$. Finally $\lambda(n) = \sqrt{n}$ for Problem 1 and $\lambda(n) = n\sqrt{n}$ for all other problems.

TABLE 4.2   Mean Objective Function Values and the Objective Function Value
of the Best Solution Obtained over 10 Runs of Each Algorithm on Each Problem

Benchmark Problem	Algorithms	The Mean Objective Function Value	The Best Objective Function Value	The Best Known Objective Function Value
1	PSO-In	1.394006	1.393431	1.3934651
	PSO-Co	1.393431	1.393431	
	QPSO	1.39346498	1.39346498	
2	PSO-In	−6,960.866	−6,961.798	−6,961.81381
	PSO-Co	−6,961.836	−6,961.837	
	QPSO	−6,961.7274	−6,961.80434	
3	PSO-In	680.671	680.639	680.630057
	PSO-Co	680.663	680.635	
	QPSO	680.646034	680.635235	
4	PSO-In	−31,493.190	−31,543.484	−30,665.538
	PSO-Co	−31,528.289	−31,542.578	
	QPSO	−30,665.535	−30,665.5382	
5	PSO-In	−31,523.859	−31,544.036	No known solution
	PSO-Co	−31,526.308	−31,543.312	
	QPSO	−31,026.428	−31,026.4277	
6	PSO-In	−213.0	−213.0	−213.0
	PSO-Co	−213.0	−213.0	
	QPSO	−213.0	−213.0	

The mean objective function value and the best objective function value obtained after 10 runs of each algorithm for each problem are listed in Table 4.2. In most cases, QPSO outperformed the PSO-In and PSO-Co algorithms. In particular, the result reaches its theoretical value for Problem 4. The solution of Problem 5 obtained by QPSO algorithm is closer to its unknown theoretical optima. It was also observed during the experiment that proper fine-tuning parameters of QPSO may result in better solutions.

4.5.3 Solving Multipeak Optimisation Problems by QPSO
4.5.3.1 Introduction
There are many optimisation problems in practical applications which involve functions with multipeaks. However, PSO and QPSO are meant to guide the swarm (i.e. the population) to a single optimum in the search space. In such situation, the entire swarm can be potentially misled to one of the local or global optima. This section presents a recent method

[109] which is based on QPSO using a form of speciation allowing the development of parallel subpopulations. This algorithm is also known as the species-based algorithm and is similar to that developed in work by Li et al. on a GA for multimodal optimisation [110]. In the species-based QPSO (SQPSO), the swarm is divided into species subpopulations based on their similarity properties. Each species is grouped around a dominating particle called the species seed. At each iteration step, species seeds are identified from the entire swarm population and then adopted as neighbourhood best for these individual species groups separately. Species are formed adaptively at each step based on the feedback obtained from the multimodal fitness landscape. Over successive iterations, species are able to simultaneously optimise towards multiple optima. SQPSO encourages particles to converge upon local optima rather than all converging to a single global optimum. This idea when applied to the QPSO algorithm may be used to develop multiple subpopulations in parallel so that it can efficiently search the peaks of multipeak functions.

4.5.3.2 SQPSO Algorithm

4.5.3.2.1 Speciation Algorithm   A species is a class of individuals with common characteristics. These species are centred on the best-known position of the fittest particle known as the 'species seed'. All particles within the species adopt the personal best position of the 'species seed' as their global best position. A swarm may be classified into groups according to their similarities between the particles measured by Euclidean distance. The smaller the Euclidean distance between two individuals, the more similar they are. The distance between two individual particles, $X_{i,n} = (X_{i,n}^1, X_{i,n}^2, \ldots, X_{i,n}^N)$ and $X_{j,n} = (X_{j,n}^1, X_{j,n}^2, \ldots, X_{j,n}^N)$, at the nth iteration of the algorithm is defined by

$$d(X_{i,n}, X_{j,n}) = \sqrt{\sum_{d=1}^{N} (X_{i,n}^d - X_{j,n}^d)}. \tag{4.89}$$

The properties of a species also depend on another parameter, r_s, which is known as the speciation radius. All particles that fall within a distance r_s from the species seed are classified as the same species. If a particle is a candidate member of two species, it will be allocated to the species with the fitter species seed.

```
Begin
  S = Φ
  while not reaching the end of F_sorted do
    found ← FALSE;
    for all p ∈ S do
      if d(s,p)≤r_s then
        found ← TRUE;
        break;
      end if
    end for
    if (not found) then
      let S ← S ∪ {s};
    end if
  end while
end
F_sorted: all particles sorted in decreasing order fitness
p: existing species seed
s: unprocessed particles in population
```

FIGURE 4.24 The procedure of speciation algorithm.

The algorithm for determining species seeds introduced by Li et al. [110] is adopted here. At each iterative step, different species seeds are identified for multiple species, and the personal best position of the species seeds is used as the global best position for the particles in different species accordingly.

The method for determining species seeds is summarised in Figure 4.24. The set of species seeds, S, is set as the empty set initially. All particles are checked successively, in decreasing order of fitness, against the latest species seeds found. If the distance between a particle and all the seeds of S does not fall within the radius r_s, the particle will be added to S as a new seed. Figure 4.25 illustrates the procedure of this algorithm. In this case, S_1, S_2 and S_3 are identified as the species seeds by applying the algorithm.

4.5.3.2.2 SQPSO Algorithm   Once the species seeds have been chosen for each species from the population, the personal best position of a seed is set as the global best position for all the other particles belong to the same species at each iteration step. As a result, an identified species seed is always the best-fit individual in that species. The set of species seeds acquired over a run of the algorithm is the solutions to multipeak functions. If the function has only single solution, then the first seed is the global optimum. The SQPSO can be summarised as in Figure 4.26.

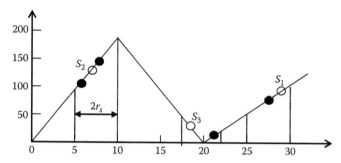

FIGURE 4.25 Determining the species seeds from the population at each iteration step.

Step 1: Generate an initial population with randomly generated particles;

Step 2: Calculate fitness value of all particle individuals in the population;

Step 3: Sort all particles in descending order of their fitness values;

Step 4: Identify the species seeds for the current population;

Step 5: Assign each species seed as the global best position to all individuals identified in the same species;

Step 6: Updating particle positions according to Equations 4.81 and 4.82;

Step 7: Go back to Step 2, unless maximum number of iterations or minimum error criteria is attained.

FIGURE 4.26 Procedure of the SQPSO algorithm.

4.5.3.3 Numerical Results

4.5.3.3.1 Test Functions   In this part, six problems suggested by Beasley et al. [111] are used to test the performance of SQPSO in order to locate a single or multiple maxima and to compare with the species-based PSO (SPSO) [112]. Problem 6 is the sum of square of the first four test functions and has one global optimal solution and six local ones.

Problem 1

$$f_1(x) = \sin^6(5\pi x) \quad (0 \le x \le 1)$$

Problem 2

$$f_2(x) = \exp\left(-2\log(2) \cdot \left(\frac{x - 0.1}{0.8}\right)^2\right) \cdot \sin^6(5\pi x) \quad (0 \le x \le 1)$$

Problem 3

$$f_3(x) = \sin^6(5\pi(x^{3/4} - 0.05)) \quad (0 \le x \le 1)$$

Problem 4

$$f_4(x) = \exp\left[-2\log(2)\cdot\left(\frac{x-0.08}{0.854}\right)^2\right]\cdot\sin^6(5\pi(x^{3/4} - 0.05)) \quad (0 \le x \le 1)$$

Problem 5

$$f_5(x, y) = 200 - (x^2 + y - 11) - (x + y^2 - 7) \quad (-6 \le x, y \le +6)$$

Problem 6

$$f_6(x, y) = \sum_{i=1}^{4} f_i(x)^2 \quad (0 \le x \le 1)$$

Figure 4.27 shows the graphs of these six functions. It can easily be seen that f_1 has five evenly spaced maxima with a function value of 1.0; f_2 has five peaks decreasing exponentially in height, with only one peak as the global maximum; f_3 and f_4 are similar to f_1 and f_2 but the peaks are unevenly spaced; f_5 is known as the Himmelblau function and has four global optima; f_6 has one global maximum, but has six local optima.

4.5.3.3.2 Experimental Settings   SQPSO and SPSO were tested against each of the previous problems. The population size used in each test was set to 50. Each algorithm was run 30 times for every test function, and each run executed a maximum of 100 iterations for f_1 to f_4 and f_6 while 200 iterations for f_5. The default speciation radius r_s was set to a value between 1/20 and 1/10 except for f_5 as shown in Table 4.3.

For SPSO, the evolution equation of PSO-Co was used, and the parameters c_1 and c_2 were both set as 2.05. The constriction factor χ was set as 0.729844. V_{max} was set to be 1.1 times the difference between the upper bounds and the lower and upper bounds of the search domain.

For SQPSO, α was linearly scaled from 1.0 to 0.5 over a run.

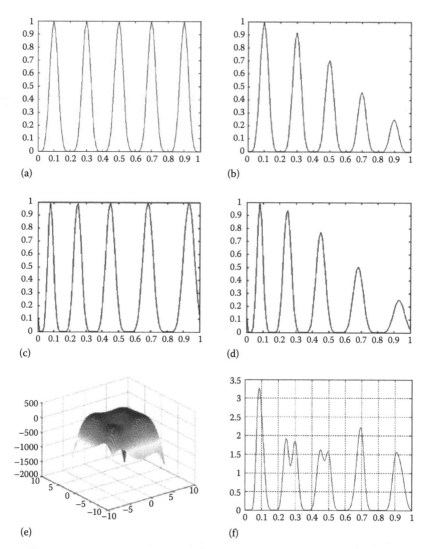

FIGURE 4.27 Graphs of the six test functions. (a) f_1. (b) f_2. (c) f_3. (d) f_4. (e) f_5 the Himmelblau function. (f) f_6.

4.5.3.3.3 Experimental Results The performance of SQPSO and SPSO was evaluated according to four criteria: number of iterations required to locate the optima, accuracy obtained by measuring the closeness to the optima, convergence speed and success rate. The success rate is the percentage of runs in which all global optima are successfully located. A global optimum is found by checking whether the seed of each species in S is close enough to the known global optima. A solution acceptance

TABLE 4.3   Performance of SQPSO and SPSO for Different Problems

Test Function	Number of Global Optimum	Speciation Radius r_s	Number of Iteration Steps to Achieve Convergence		Success Rate (%)	
			SQPSO	SPSO	SQPSO	SPSO
f_1	5	0.05	30	95	100	100
f_2	1	0.05	20	70	100	100
f_3	2	0.05	15	55	100	100
f_4	1	0.05	15	70	100	100
f_5	4	2	40	115	100	100
f_6	1	0.04	20	110	100	100

tolerance $0 < \varepsilon \leq 1$ is defined to check whether the solution is close enough to a global optimum. The acceptance tolerance is defined as

$$E_p = \left| f_{max} - f_p \right| \leq \varepsilon, \tag{4.90}$$

where

E_p is the error

f_{max} is the known maximal (highest) fitness value for a test function

f_p is the fitness value of seed p in S

If the number of global optima is larger than one, then all global optima are checked for the required accuracy using Equation 4.90 before a run is terminated. In the experiment, ε was set to 0.0001 for the acceptance of a solution as a global optimum.

In order to check the accuracy of the algorithm acquired, the mean error, $\bar{E}$, can be computed as

$$\bar{E} = \frac{\sum_{p=1}^{|S|} E_p}{|S|}, \tag{4.91}$$

where $|S|$ is the number of seeds in S. The smaller the mean error after the computation completes for a maximum number of iterations, the better the accuracy it achieves.

Table 4.3 also lists the average number of iterations required and the success rate for finding all global optima over 30 runs. Both SQPSO and SPSO converged to the required accuracy of 0.0001, with 100% success rate for functions f_1 to f_6. In particular, SQPSO found all the global optima in all runs with less number of iterations compared with SPSO. This means that SQPSO achieved the solutions with less computational time. The convergence histories of SQPSO and SPSO for functions f_1 to f_6 are presented in Figure 4.28. It can be seen that SQPSO converged faster than SPSO for all test functions. Finally, Table 4.4 lists the average of the mean errors

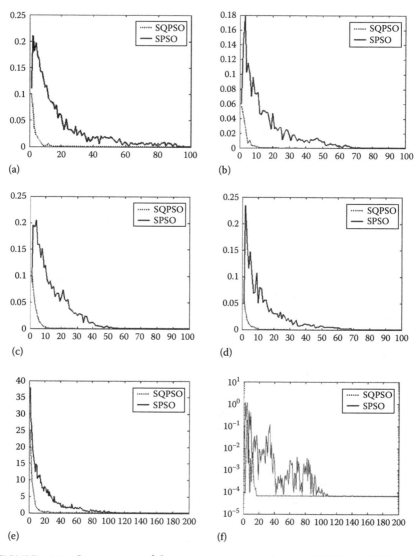

FIGURE 4.28 Comparison of the convergence speed using SQPSO and SPSO for different problems. (a) f_1. (b) f_2. (c) f_3. (d) f_4. (e) f_5. (f) f_6.

TABLE 4.4   The Average of the Mean Errors for
Each Problem Using SQPSO and SPSO

Test Function	AME	
	SQPSO	SPSO
f_1	0.0	4.1×10^{-5}
f_2	0.0	4.0×10^{-6}
f_3	0.0	1.1×10^{-5}
f_4	0.0	3.0×10^{-6}
f_5	4.0×10^{-6}	8.4×10^{-4}
f_6	2.08×10^{-3}	2.56×10^{-2}

AME, Average of the mean error for each problem.

obtained over 30 runs. It can be seen that SQPSO shows better accuracy
compared with SPSO for all test functions in the experiments.

Figure 4.29 shows one typical simulation result for f_1 using SQPSO,
which took only 24 iterations to search for all global optima. All particles of the swarm converged to the five peak values as observed from

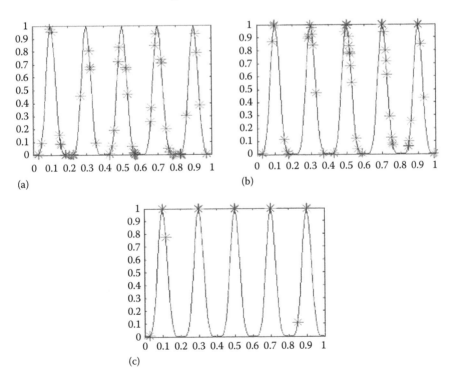

FIGURE 4.29   The history of finding the optima of f_1 using SQPSO (with five
same optimum). (a) The 1st iteration. (b) The 5th iteration. (c) The 24th iteration.

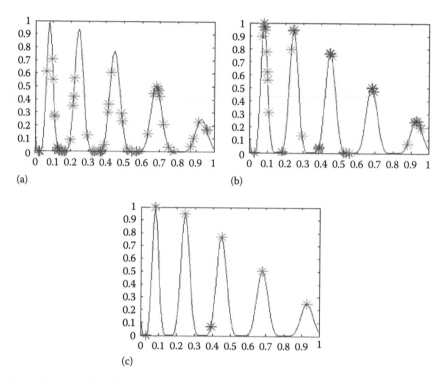

FIGURE 4.30 The history of finding the optima of f_4 using SQPSO. (a) The 1st iteration. (b) The 6th iteration. (c) The 70th iteration.

the graph depicted. Figure 4.30 shows another typical simulation results for f_4. Figures 4.31 and 4.32 are simulation results for f_6 using SQPSO and SPSO respectively. It can be observed from Figures 4.30 and 4.31 that SQPSO has achieved the global optimum and successfully found other local optima in later iterations. From Figure 4.32, one can observe that SPSO is able to find the global optimum but fails to find any of the other local optima. In general, one might suggest that SQPSO may be used to find all peak values of the function irrespective of global or local optima. The simulation results for f_2 and f_3 are similar to f_1 and f_4. For simplicity, these results are not presented here. Figure 4.33 displays a brief convergence history of the solution for f_5 by SQPSO. Initially, many seeds (based on speciation radius) are produced by SQPSO. After several iterations, particles begin to move towards four subspecies where maximum exists. At the 70th iteration, almost all particles have converged to the four global optima.

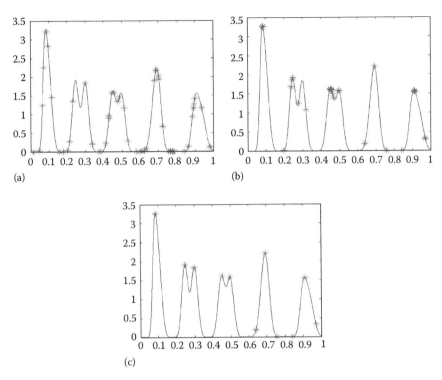

FIGURE 4.31 The history of finding the optima of f_6 using SQPSO. (a) The 1st iteration. (b) The 15th iteration. (c) The 30th iteration.

4.5.4 Solving Systems of Non-Linear Equations

This section provides some numerical experiments of applying the SQPSO algorithm to systems of non-linear equations which could represent problems in science and engineering. The traditional numerical methods for solving a system of linear or non-linear equations can be direct methods or iterative methods. If a problem needs to be solved in real time under time constraints, existing numerical solutions may not be able to deliver solutions in real time. On the other hand, EAs have been studied as an alternative class of methods for difficult and complex problems to which traditional numerical methods have difficulties in obtaining solutions. If there is more than one solution in the system of equations, the objective function is a multipeak function with multiple optima. Such problems would be best transformed into the optimisation of multipeak functions.

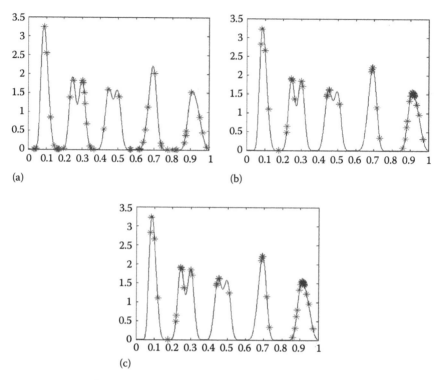

FIGURE 4.32 The history of finding the optima of f_6 using SPSO. (a) The 1st iteration. (b) The 30th iteration. (c) The 110th iteration.

4.5.4.1 Objective Function of Systems of Equations
Given the system of non-linear equations

$$F(X) = [f_1(X), f_2(X), \ldots, f_K(X)]^T = 0, \tag{4.92}$$

where $X = (x_1, x_2, \ldots, x_N)^T$. In general, the fitness function for solving systems of equations can be expressed as follows:

$$f(X) = \sum_{k=1}^{K} (f_k(X))^2, \tag{4.93}$$

where $f_k(X)$ represents the kth equation of the system. The objective of solving a system of equations becomes the minimisation of the error function defined in Equation 4.93.

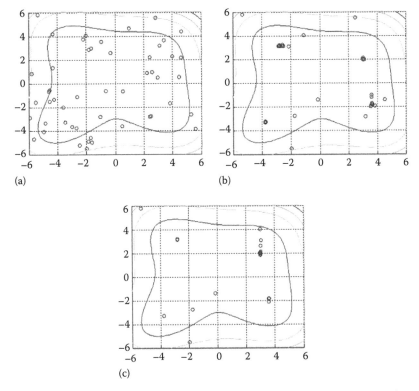

FIGURE 4.33 The history of finding the optima of f_5 using SQPSO. (a) The 1st iteration. (b) The 27th iteration. (c) The 70th iteration.

4.5.4.2 Experimental Results

This section provides a study of five non-linear systems as follows:

F1:

$$
\begin{cases}
f_1(x, y) = x + 2y - 3 \\
f_2(x, y) = 3x - 4y + 1
\end{cases}
$$

F2:

$$
\begin{cases}
f_1(x, y) = e^x - y \\
f_2(x, y) = 2x - y + 2
\end{cases}
$$

F3:

$$\begin{cases} f_1(x,y) = x^2 - y \\ f_2(x,y) = 2x - y + 2 \end{cases}$$

F4:

$$\begin{cases} f_1(x,y) = \cos x \ln x - y \\ f_2(x,y) = \tan x - y \end{cases}$$

F5:

$$\begin{cases} f_1(x_1,x_2) = 0.05x_1^2 - 0.05x_2^2 - 0.125x_1 + 0.1x_2 - 3.8 \\ f_2(x_1,x_2) = 0.1x_1x_2 - 0.1x_1 - 0.125x_2 + 0.125 \end{cases}$$

SQPSO, SPSO, PSO with global best model (PSO-gbest) and PSO with local best model (PSO-lbest) were applied to solve each of the previous systems of non-linear equations. For each problem, the experiments were run for 2000 iterations. Thirty runs were performed for each algorithm on each problem, and the results averaged over the runs. The number of particles was set to 30.

Parameter settings for PSOs (including SPSO, PSO-gbest and PSO-lbest): Parameters c_1 and c_2 were both set to 2. The inertia weight w was linearly scaled from 0.5 to 0.1 over a run. Other data employed in the experiments are shown in Table 4.5.

Parameter settings of SQPSO: Parameter α was linearly scaled from 1.2 to 0.5 over one single run. The default settings of speciation radius r_s are

TABLE 4.5 Parameters Settings for Each Problem

Systems of Equations	X_{min}	$V_{max} = X_{max}$	r_s
F1	−10	10	5
F2	−10	10	5
F3	−10	10	5
F4	0.1	10	2
F5	−15	15	2.5

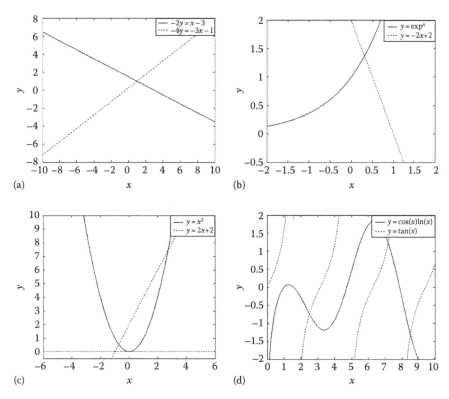

FIGURE 4.34   Systems of equations used in the numerical tests. (a) F1. (b) F2. (c) F3. (d) F4.

shown in Table 4.5. Here $[X_{min}, X_{max}]$ denotes the domain of the problem, while V_{max} is the largest velocity update that is allowed. In PSOs, these limits were chosen since all solutions are within the defined search domain in the solution space (Figure 4.34).

Table 4.6 summarises the number of solutions found by using SQPSO, SPSO, PSO-gbest and PSO-lbest for each of the systems of equations.

TABLE 4.6   Number of Solutions Found by Each Optimisation Algorithm

Systems of Equations	PSO-gbest	PSO-lbest	SPSO	SQPSO	Actual Solutions
F1	1	1	1	1	1
F2	1	1	1	1	1
F3	1	1	2	2	2
F4	1	1	3	3	3
F5	1	1	2	2	2

TABLE 4.7   Comparisons of Mean
Errors (Mean Best Fitness Values) over
All the Problems Obtained by the
Algorithms

Algorithms	Mean Error
PSO-gbest	2.15×10^{-5}
PSO-lbest	2.86×10^{-2}
SPSO	5.78×10^{-8}
SQPSO	8.43×10^{-16}

TABLE 4.8   Comparisons of Mean Errors
(Mean Best Fitness Value) on Each Problem
Obtained by SQPSO and SPSO

Systems of Equations	Mean Error	
	SPSO	SQPSO
F1	0.00	0.00
F2	0.00	0.00
F3	2.53×10^{-7}	9.33×10^{-16}
F4	9.34×10^{-16}	9.34×10^{-16}
F5	9.44×10^{-15}	2.22×10^{-16}

As shown in Table 4.6, both the PSO-gbest and PSO-lbest algorithms always converged to a single solution and cannot find all solutions even when multiple solutions exist. The SQPSO and SPSO both succeeded in finding all solutions for all problems. In these cases, all experiments led to good average of the mean errors (i.e. mean best fitness values) over all problems close to zero, as shown in Table 4.7. Table 4.8 shows that, for each problem, the mean error due to SQPSO is smaller than that due to SPSO.

4.6 SOME VARIANTS OF QPSO

Several important variants of QPSO are introduced in this section. For simplicity, only motivations, principles and procedures of these variants are presented. Readers who are interested in the detail of results and discussions should refer to the corresponding references.

4.6.1 QPSO with Diversity Control Strategy

Recall that the major issue with PSO and other EAs in multimodal optimisation is their premature convergence resulting in significant performance loss and suboptimal solutions. The effect would make particles in the PSO algorithm difficult to escape from local optima. The main reason for this is that the collectiveness of particles leads to fast information flow between particles and in turn rapid declination of the diversity of the particle swarm. During the later stage of the search process, the low diversity may result in a stagnation of the search as an overall outcome. Although the search space, in each iterative step, of an individual particle of the QPSO algorithm is the whole feasible solution space of the problem, diversity loss of the whole population becomes inevitable due to the collectiveness.

One way to circumvent the aforementioned shortcoming of EAs is to apply diversity-decreasing operators (selection, recombination) and diversity-increasing operators (mutation) to alternate between two modes based on a distance-to-average-point measure [113]. The technique was proposed by Ursem in 2002 and is known as the diversity-guided evolutionary algorithm (DGEA). Performance presented by Ursem clearly shows the potential of DGEA in multimodal optimisation.

Riget et al. [13] used the diversity-guided concept described earlier in the canonical PSO. The decreasing and increasing diversity operators are then used to control the population. In the actual implementation, a control strategy for the diversity measure, using the attraction and repulsion mode, is included into the PSO so that the swarm alternates between exploring and exploiting behaviour. The improved PSO algorithm is now known as the attraction and repulsion PSO (ARPSO) algorithm.

A similar diversity control strategy was also incorporated into the QPSO algorithm, and the new algorithm is known as diversity-controlled QPSO (DCQPSO) [14,16]. The diversity measure at the nth iteration, $D_n(S)$, can be calculated by using the formula

$$D_n(S) = \frac{1}{|S| \times |A|} \cdot \sum_{i=1}^{|S|} \sqrt{\sum_{j=1}^{N} (X_{i,n}^j - \bar{X}_n^j)^2} , \qquad (4.94)$$

where

S is the swarm

$|S|$ is the swarm size

$|A|$ is the length of the longest diagonal in the search space

N is the dimensionality of the problem

$X_{i,n}^j$ is the jth component of the position of the ith particle at the nth iteration

$\bar{X}_n^j$ is the jth component of the mean of the all current positions of the particles

The difference between the model used by Sun et al. and the other two models is that only a low bound, denoted as d_{low}, to the diversity of the population is preassigned. As reported in [5] through stochastic simulation, the particles converge when $\alpha < 1.781$ but otherwise diverge. This is also proved mathematically in Chapter 5. The attraction mode is obtained by varying α from 1.0 to 0.5. If the diversity measure of the swarm declines to below the threshold value d_{low}, the particles are set to explosion mode in order to increase the diversity until it is larger than d_{low}. Two methods of making the particle to explode were proposed in [14] when the diversity measure is smaller than the threshold value d_{low}. One method is to control the CE coefficient α by taking the value α_0, where $\alpha_0 > 1.781$. In the procedure presented in Figure 4.35, this is denoted as DCQPSO1. Another method of increasing the diversity is to reinitialise the mean

```
Begin
Initialize the current position Xⱼᵢ,₀ and the personal best position
Pⱼᵢ,₀ of each particle, evaluate their fitness values and find the
global best position G₀ = P_g,₀;
Pre-assign α₀ > 1.781; Pre-set d_low;
for n = 1 to n_max
    Compute the mean best position and calculate Dₙ(S) by using (4.94);
    if (Dₙ(S) < d_low)
      Explosion mode:
    For DCQPSO1: Set α = α₀;
      For DCQPSO2: Reinitialize the mean best position;
      else
      Attraction mode: Compute α = (1.0 - 0.5)*(n - n_max)/n_max + 0.5;
    endif
    for (i = 1 to M)
      for j = 1 to N
        φⱼᵢ,ₙ = rand1(·);
        pⱼᵢ,ₙ = φⱼᵢ,ₙPⱼᵢ,ₙ + (1 - φⱼᵢ,ₙ)Gⱼₙ;
        uⱼᵢ,ₙ = rand2(·);
        if (rand3(·) < 0.5)
            Xⱼᵢ,ₙ₊₁ = pⱼᵢ,ₙ + α|Xⱼᵢ,ₙ - Cⱼₙ|ln(1/uⱼᵢ,ₙ₊₁);
        else
            Xⱼᵢ,ₙ₊₁ = pⱼᵢ,ₙ - α|Xⱼᵢ,ₙ - Cⱼₙ|ln(1/uⱼᵢ,ₙ₊₁);
        end if
      end for
      Evaluate the objective function value f(Xᵢ,ₙ);
      Update Pᵢ,ₙ and Gₙ
    end for
  endfor
end
```

FIGURE 4.35 The procedure of the DCQPSO algorithm.

is that the collectiveness of particles leads to fast information flow between particles and in turn rapid declination of the diversity of the particle swarm. During the later stage of the search process, the low diversity may result in a stagnation of the search as an overall outcome. Although the search space, in each iterative step, of an individual particle of the QPSO algorithm is the whole feasible solution space of the problem, diversity loss of the whole population becomes inevitable due to the collectiveness.

One way to circumvent the aforementioned shortcoming of EAs is to apply diversity-decreasing operators (selection, recombination) and diversity-increasing operators (mutation) to alternate between two modes based on a distance-to-average-point measure [113]. The technique was proposed by Ursem in 2002 and is known as the diversity-guided evolutionary algorithm (DGEA). Performance presented by Ursem clearly shows the potential of DGEA in multimodal optimisation.

Riget et al. [13] used the diversity-guided concept described earlier in the canonical PSO. The decreasing and increasing diversity operators are then used to control the population. In the actual implementation, a control strategy for the diversity measure, using the attraction and repulsion mode, is included into the PSO so that the swarm alternates between exploring and exploiting behaviour. The improved PSO algorithm is now known as the attraction and repulsion PSO (ARPSO) algorithm.

A similar diversity control strategy was also incorporated into the QPSO algorithm, and the new algorithm is known as diversity-controlled QPSO (DCQPSO) [14,16]. The diversity measure at the nth iteration, $D_n(S)$, can be calculated by using the formula

$$D_n(S) = \frac{1}{|S| \times |A|} \cdot \sum_{i=1}^{|S|} \sqrt{\sum_{j=1}^{N} (X_{i,n}^j - \bar{X}_n^j)^2}, \qquad (4.94)$$

where
    S is the swarm
    $|S|$ is the swarm size
    $|A|$ is the length of the longest diagonal in the search space
    N is the dimensionality of the problem
    $X_{i,n}^j$ is the jth component of the position of the ith particle at the nth iteration
    $\bar{X}_n^j$ is the jth component of the mean of the all current positions of the particles

The difference between the model used by Sun et al. and the other two models is that only a low bound, denoted as d_{low}, to the diversity of the population is preassigned. As reported in [5] through stochastic simulation, the particles converge when $\alpha < 1.781$ but otherwise diverge. This is also proved mathematically in Chapter 5. The attraction mode is obtained by varying α from 1.0 to 0.5. If the diversity measure of the swarm declines to below the threshold value d_{low}, the particles are set to explosion mode in order to increase the diversity until it is larger than d_{low}. Two methods of making the particle to explode were proposed in [14] when the diversity measure is smaller than the threshold value d_{low}. One method is to control the CE coefficient α by taking the value α_0, where $\alpha_0 > 1.781$. In the procedure presented in Figure 4.35, this is denoted as DCQPSO1. Another method of increasing the diversity is to reinitialise the mean

```
Begin
Initialize the current position X^j_{i,0} and the personal best position
P^j_{i,0} of each particle, evaluate their fitness values and find the
global best position G_0 = P_{g,0};
Pre-assign α_0 > 1.781; Pre-set d_low;
for n = 1 to n_max
   Compute the mean best position and calculate D_n(S) by using (4.94);
   if (D_n(S) < d_low)
     Explosion mode:
   For DCQPSO1: Set α = α_0;
     For DCQPSO2: Reinitialize the mean best position;
   else
     Attraction mode: Compute α = (1.0 - 0.5)*(n - n_max)/n_max + 0.5;
   endif
   for (i = 1 to M)
     for j = 1 to N
        φ^j_{i,n} = rand1(·);
        p^j_{i,n} = φ^j_{i,n}P^j_{i,n} + (1 - φ^j_{i,n})G^j_n;
        u^j_{i,n} = rand2(·);
        if (rand3(·) < 0.5)
            X^j_{i,n+1} = p^j_{i,n} + α|X^j_{i,n} - C^j_n|ln(1/u^j_{i,n+1});
        else
            X^j_{i,n+1} = p^j_{i,n} - α|X^j_{i,n} - C^j_n|ln(1/u^j_{i,n+1});
        end if
     end for
     Evaluate the objective function value f(X_{i,n});
     Update P_{i,n} and G_n
   end for
 endfor
end
```

FIGURE 4.35 The procedure of the DCQPSO algorithm.

best position of the swarm across the search space. The reason for the initialisation of the mean best position is that when the diversity is low, the distance between the particle and the mean best position is too small for the particle to escape the local optima. Therefore re-initialisation of the mean best could enlarge the gaps between particles and the mean best position, consequently making the particles explode temporarily. In the procedure in Figure 4.35, this operation is denoted as DCQPSO2.

Note that previous experiences show that DCQPSO runs the attraction mode in most of the iterations. Only when the diversity falls below the d_{low} do the particles perform the explosion mode which is transitory. Once diversity measure returns above the threshold, the population will return to the attraction mode again.

There is an alternative method, in addition to the two mentioned earlier, of diversifying the swarm which is to exert the following mutation operation on the global best position G_n and the global best particle $P_{g,n}$ as soon as the diversity measure is smaller than d_{low} [15]. The mutation operation can be defined as

$$G_n^j = G_n^j + \gamma |A| \varepsilon, \quad j = 1, 2, \ldots, N, \tag{4.95a}$$

$$P_{g,n}^j = G_n^j, \quad j = 1, 2, \ldots, N \tag{4.95b}$$

where
$\varepsilon \sim N(0,1)$
γ is a user-specified parameter which is generally set to lie between 0.001 and 0.1

When the mutation operation is exerted, the displacement of the global best position $P_{g,n}$ increases the value of $|P_{g,n}^j - P_{i,n}^j|$ for each j generally and pulls the mean best position C_n away from its original position, consequently extending the search domain of the particles and gaining greater diversity. The procedure of this version of DCQPSO can be obtained by replacing DCQPSO1 and DCQPSO2 with mutation in (4.95a) and (4.95b).

4.6.2 QPSO with Selection Operator

Particles in QPSO ideally would converge to the global best position G_n. The convergence rate of QPSO algorithms is in general faster than GAs from experience. Consider a general situation during the iterative process in Figure 4.20. Most of the particles cluster around the global best position and the lagged particles are pulled towards it. Suppose the global optima lies in the area where the lagged particles are located, the optimal solutions could well be missed by the particles since G_n is relatively far from the global optima. Under such situation, the particles around G_n have little opportunity to hit the global optima. The lagged particles could also miss the global optima because they are rushing to G_n so that they appear in the promising area with little probability. Hence the probability with which the global optima may be found is decreasing, causing premature convergence of the algorithm as an overall result. The premature convergence is one shortcoming of this algorithm, and such situation is due to the social cognition of particles.

To overcome this shortcoming, two selection operations on global best position were introduced into QPSO to enhance the global search ability of the algorithm [27,114].

4.6.2.1 QPSO with Tournament Selection

TS is one selection method commonly used in GAs. The idea is to run a 'tournament' among a few individuals chosen at random from the population and select the one with the best fitness for crossover. The tournament size may be adjusted to suit various situations. If the tournament size is larger, weaker individuals have a smaller chance to be selected.

The way of introducing a tournament into QPSO described in Figure 4.36 is very different from that in [115]. The QPSO with Tournament Selection (QPSO-TS) here does not use the global best solution to determine the local focus $p_{i,n}$ of a particle. It is determined by selecting G_n randomly among the personal best positions which have better fitness value than that of the particle. However, the personal best position of the global best particle, $P_{g,n}$, has more chance to be selected as G_n.

4.6.2.2 QPSO with Roulette-Wheel Selection

An alternative selection method is the RS [114], which is a genetic operation used in many GAs for selecting potentially useful solutions for recombination. A fitness value, which is used to associate a probability of selection with each individual chromosome, to a possible solution is assigned by using the fitness function. If f_i is the fitness value of an individual i in the population, its probability of being selected is

```
Begin
Initialize the current position X_{i,0}^{j} and the personal best position
P_{i,0}^{j} of each particle; evaluate their fitness values and find the
global best position G_0 = P_{g,0};
for n = 1 to n_max
   Compute the mean best position and determine the value of α;
   for (i = 1 to M)
      Randomly select a particle k
      if f(P_{k,n}) < f(P_{i,n})
         G_n = P_{k,n};
      else
         G_n = P_{g,n} (g is the index of the global best particle);
      end if
      for j = 1 to N
         φ_{i,n}^{j} = rand1(·);
         p_{i,n}^{j} = φ_{i,n}^{j}P_{i,n}^{j} + (1 - φ_{i,n}^{j})G_n^{j};
         u_{i,n}^{j} = rand2(·);
         if (rand3(·) < 0.5)
            X_{i,n+1}^{j} = p_{i,n}^{j} + α | X_{i,n}^{j} - C_n^{j} | ln(1/u_{i,n+1}^{j});
         else
            X_{i,n+1}^{j} = p_{i,n}^{j} - α | X_{i,n}^{j} - C_n^{j} | ln(1/u_{i,n+1}^{j});
         end if
      end for
      Evaluate the fitness value of X_{i,n+1}, that is, the objective
      function value f(X_{i,n+1});
      Update P_{i,n} and P_{g,n};
   end for
end for
end
```

FIGURE 4.36 The procedure of QPSO-TS.

$f_i \Big/ \sum_{i=1}^{M} f_i$, where M is the number of individuals in the population. Candidate solutions with a higher fitness will be less likely to be eliminated. A less sophisticated selection algorithm, such as the truncation selection, eliminates a fixed percentage of the weakest candidates, while the fitness proportionate selection might allow some weaker solutions to survive the selection process. This could be viewed as an advantage when such weaker solutions appear in the recombination process.

In QPSO with Roulette-Wheel Selection (QPSO-RS), the selection is also exerted on the global best position. The procedure of QPSO-RS looks similar to that in Figure 4.36 if the pseudo codes for selecting the global best position with TS are replaced by those of RS for the global best position.

4.6.3 QPSO with Mutation Operators

It is interesting to note that the loss of diversity in the population in QPSO is inevitable due to the collective behaviour of the swarm. It is particularly true at a later stage of the search as happened in PSO and other population-based algorithm. This is the major reason for the particles to be trapped in local minima. Mutation could be one way out. In [24], Liu introduced a mutation mechanism into QPSO for the global best and the mean best positions.

In Table 4.9, mutation operations from A to H mean that the mutated position of a particle is generated from the different position of the particles randomly selected from the swarm. Here, Xr_1, Xr_2 and Xr_3 are the positions of different selected particles and N is the dimensionality of the given optimisation problem. Mutation operations I and J are Gaussian mutation and Cauchy mutation respectively, where $G(0, \delta)$ is the Gaussian distribution with mean 0 and standard deviation δ. Both mutation operations are represented as a function of X_i, that is, $m(X_i)$, which means that the resulting position of X_i is obtained by exerting the mutation on X_i.

The function of the local focus is to attract the members of the swarm. It is possible to lead the swarm away from a current location by mutating

TABLE 4.9   Mutation Operators

Mutation Operations

A	Single-dimension mutation	$X_i = Xr_1 + 0.01Xr_2$
B	Different mutation	$X_i = Xr_3 + 0.01(Xr_1 - Xr_2)$
C	Log mutation	$X_i = Xr_3 + \log_{10}|(Xr_1 - Xr_2) + Xr_3|$
D	Exponential mutation	$X_i = Xr_3 + e^{-(|Xr_1 - Xr_2|)}$
E	Algebraic mutation	$X_i = Xr_3 + \dfrac{Xr_1 - Xr_2}{N}$
F	Average mean mutation	$X_i = \dfrac{Xr_1 + Xr_2}{2}$
G	Geometric mean mutation	$X_i = \pm\sqrt{|Xr_1 + Xr_2|}$
H	Harmonic mean mutation	$X_i = \dfrac{2}{1/Xr_1 + 1/Xr_2}$
I	Gaussian mutation	$m(X_i) = X_i + G(0, \delta)$
J	Cauchy mutation	$m(X_i) = \dfrac{a}{\pi(X_i^2 + a^2)}$

```
Begin
Initialize the current position X^j_{i,0} and the personal best position
P^j_{i,0} of each particle;
Evaluate their fitness values and find the global best position
G_0 = P_{g,0};
for n = 1 to n_{max}
   Compute the man best position;
   Determine the value of α;
   for (i = 1 to M)
      for j = 1 to N
         φ^j_{i,n} = rand1 (·);
         p^j_{i,n} = φ^j_{i,n}P^j_{i,n} + (1 - φ^j_{i,n})G^j_n;
         u^j_{i,n} = rand2 (·);
         if (rand3(·) < 0.5)
            X^j_{i,n+1} = p^j_{i,n} + α | X^j_{i,n} - C^j_n | ln (1/u^j_{i,n+1});
         else
            X^j_{i,n+1} = p^j_{i,n} - α | X^j_{i,n} - C^j_n | ln (1/u^j_{i,n+1});
         end if
      end for
      Evaluate the fitness value of X^j_{i,n+1}, that is, the objective
      function value f(X^j_{i,n+1});
      Update P_{i,n};
         Use a mutation operator in Table 4.9 to perform mutation on
         P_{i,n}, and revaluate P_{i,n};
         If the new P_{i,n} has better fitness value then update P_{i,n} and G_n;
   end for
end for
end
```

FIGURE 4.37 The procedure of the QPSO with mutation.

a single individual if the mutated individual takes a better position. This mechanism can potentially provide a means of escaping local optima which speeds up the search. It is possible to exert one of the mutation operators in Table 4.8 on the personal best position for all particles in the swarm [26]. The result would influence the local focus. If a new particle is better than the original one after mutation, it replaces its original personal best position. This method not only diversifies the swarm but also finds better solutions. The procedure of QPSO with mutation operators is presented in Figure 4.37.

4.6.4 QPSO with Hybrid Distribution

The particle in QPSO flies in the δ potential well and its position is subject to double exponential distribution. It has been shown by many researchers through numerical experiments that the exponential distribution may be the best unitary distribution for QPSO. It is therefore thought that a hybrid distribution may improve QPSO [8].

Consider the evolution equation of QPSO (Equation 4.53) which is rewritten as

$$X_{i,n+1}^{j} = p_{i,n}^{j} + A_{i,n+1}^{j}, \tag{4.96}$$

where $A_{i,n}^{j}$ is a random sequence which converges to zero to ensure that $X_{i,n}$ reaches to the local focus $p_{i,n}$. The random sequence $A_{i,n}^{j}$ takes exponential distribution in QPSO algorithm. Assuming $A_{i,n}^{j}$ is the sum of two random sequences, that is,

$$A_{i,n+1}^{j} = a_{i,n+1}^{j} + b_{i,n+1}^{j}, \tag{4.97}$$

where $a_{i,n}^{j}$ and $b_{i,n}^{j}$ are two random sequences generated by two different probability distributions. One example choice for $a_{i,n}^{j}$ and $b_{i,n}^{j}$ as used in [8] is the normal distribution and exponential distribution respectively, that is,

$$a_{i,n+1}^{j} = \pm \alpha \left| X_{i,n}^{j} - p_{i,n}^{j} \right| \ln\left(\frac{1}{u_{i,n+1}^{j}}\right), \tag{4.98}$$

$$b_{i,n+1}^{j} = \beta \left| X_{i,n}^{j} - C_{n}^{j} \right| \lambda_{i,n+1}^{j}, \tag{4.99}$$

where
$u_{i,n}^{j}$ is a sequence of random numbers uniformly distributed random number on (0,1)

$\lambda_{i,n}^{j}$ is a sequence of random numbers with standard normal distribution

C_{n}^{j} is the jth component of the mean best position at the nth iteration

α and β are two parameters known as the CE coefficients and are used to balance the local and global search of the particle

Therefore the iterative Equation 4.97 can be written as

$$X_{i,n+1}^{j} = p_{i,n}^{j} \pm \alpha \left| X_{i,n}^{j} - p_{i,n}^{j} \right| \ln\left(\frac{1}{u_{i,n+1}^{j}}\right) + \beta \left| X_{i,n}^{j} - C_{n}^{j} \right| \lambda_{i,n+1}^{j}. \tag{4.100}$$

The QPSO algorithm with the previous evolution equation is called QPSO-HPD which is outlined as follows in Figure 4.38.

```
Begin
Initialize the current position X_{i,0}^j and the personal best position
P_{i,0}^j of each particle, evaluate their fitness values and find the
global best position G_0 = P_{g,0};
for n = 1 to n_{max}
  Compute the mbest position;
  Determine the value of α;
  for (i = 1 to M)
    for j = 1 to N
      φ_{i,n}^j = rand1 (·) ;
      p_{i,n}^j = φ_{i,n}^j P_{i,n}^j + (1 − φ_{i,n}^j)G_n^j;
      u_{i,n}^j = rand2 (·);
      if (rand3(·) < 0.5)
        X_{i,n+1}^j = p_{i,n}^j + α | X_{i,n}^j − p_{i,n}^j | ln (1/u_{i,n+1}^j) + β | X_{i,n}^j − C_n^j | λ_{i,n+1}^j;
      else
        X_{i,n+1}^j = p_{i,n}^j − α* | X_{i,n}^j − p_{i,n}^j | * ln (1/u_{i,n+1}^j) + β | X_{i,n}^j − C_n^j | λ_{i,n+1}^j;
      end if
    end for
    Evaluate the fitness value of X_{i,n+1}^j that is, the objective func-
    tion value f(X_{i,n+1}^j)
    Update P_{i,n} and G_n;
  end for
end for
end
```

FIGURE 4.38   The procedure of the QPSO with hybrid distribution.

4.6.5 Cooperative QPSO

Before looking at cooperative swarms in depth, let us first consider some issues encountered by PSO and QPSO. The general principle [30,116] involved in PSO or QPSO is to update each particle which is represented by a vector containing N components, where N is the dimensionality of the objective function, as a potential solution. Each iterative update step is performed in a component-wise fashion. It happens that some components in the vector may have moved closer to the solution, while others actually moved away from the solution. The accuracy is judged through some metrics measuring the overall performance of the vector. Therefore, it is significantly harder to find the global optimum of a high-dimensional problem. One way to reduce the computational effort is to employ the concept of a cooperative method [30] into the QPSO algorithm. This algorithm is known as the CQPSO, in which all the particles are partitioned into N swarms of 1D vectors. Each swarm attempts to optimise a single component of solution vector, essentially a 1D problem.

Since each particle in CQPSO intends to optimise each dimension of the solution, it is clearly not possible to directly compute the fitness of a particle because its contribution in each dimension is not implicit described. A *context vector* is required to provide a suitable context so that the particle in each dimension can be evaluated in a fair environment. The simplest scheme for constructing such a context vector is to take the global best particle from each of the N swarms and concatenate them to form such an N-dimensional vector. To calculate the fitness for all particles in the jth swarm, the other $N - 1$ components in the context vector are kept constant (with their values set to the global best particles from other $N - 1$ swarms), while the jth component of the context vector is replaced in turn by each particle from the jth swarm represented by q_j.

The CQPSO algorithm is presented in Figure 4.39. A swarm containing M particles has its search space being split into exactly N subspaces, each of which represents one dimension of the solution. $q_j X_{i,n}$ now becomes $q_j P_{i,n}$, which refers to the position and personal best position of particle i in the jth swarm at iteration n, which can therefore be substituted into the jth component of the *context vector* if necessary. Each of the N swarms now has a global best position denoted as $q_j G_n$. The function $b(j, z)$ returns an

```
Define  b(j, z) ≡ (q₁Gₙ, q₂Gₙ, ..., qⱼ₋₁Gₙ, qⱼ₊₁Gₙ, ..., q_NGₙ)
Begin
  Create and initialize N one-dimensional QPSOs: qⱼ, j ∈ {1,2,..., N};
  Set n = 0;
  While (termination condition = false)
Do
  Set n = n + 1;
    for (i = 1 to M)
      for j = 1 to N
        if f(b(j, qⱼ Xᵢ,ₙ)) < f(b(j, qⱼPᵢ,ₙ))
            qⱼPᵢ,ₙ = qⱼXᵢ,ₙ
        end if
        if f (b(j, qⱼPᵢ,ₙ)) < f (b(j, qⱼGₙ))
            qⱼGₙ = qⱼPᵢ,ₙ
        end if
      end for
      Perform particle updates according to the evolution equation of
      QPSO as in Equation 4.82
    end for
  end do
end
```

FIGURE 4.39 The procedure of CQPSO.

N-dimensional vector which is formed by concatenating all the global best vectors across all swarms, except for the jth component which is replaced by the position z of a particle from the jth swarm. Each swarm in the group only contains information regarding a specific component of the solution vector; the rest of the vector is provided by the other $N - 1$ swarms. This promotes cooperation between different swarms, as they all contribute to b. This is the reason why the algorithm is called the CQPSO.

By using the cooperative method among the particles in QPSO, each particle can potentially make contribution to the searching of the solution. The diversity of solutions generated by CQPSO certainly exceeds that of the QPSO. This ensures that the search space is sampled more thoroughly.

4.7 SUMMARY

A brief review is given of the improvements of QPSO algorithm and its applications. While QPSO is effective and adaptable to various applications, the improved QPSO algorithms aim to enhance the performance. Methods of improvements range from parameter selection to control swarm diversity, from cooperative methods to using different probability distribution functions, from novel search methods to hybrid methods. Application areas of QPSO include clustering and classification, control, electronics and electromagnetics, biomedical, graphics and image processing, signal processing, power systems, neural network, fuzzy, modelling, antennas, combinatorial optimisation, etc.

The development of QPSO is motivated from two aspects, one is the dynamical analysis of the particle in PSO and the other is quantum mechanics. The fundamental model for and pseudo-codes procedure of QPSO are included. The basic evolution equation of QPSO for 1D case is derived by solving the stationary Schrödinger equation and by performing Monte Carlo inverse transformation on the obtained distribution function.

Two search strategies are presented for QPSO resulting in two versions of QPSO algorithms. The procedure of the algorithm is provided, and some interesting properties of QPSO are exposed. In particular, the QPSO with the mean best position is emphasised due to its good performance reported in the literature.

Some detailed experimental results using the QPSO algorithm are described. These problems include unconstrained function optimisation problem, constrained optimisation problem, multipeak function optimisation and systems of non-linear equations. The source codes in MATLAB for QPSO are included for practical usage.

Finally, several variants of QPSO, proposed by the authors and their collaborators, are introduced. It is anticipated that the review and the exposition in this chapter provides the readers a clearer view on the QPSO and its variants. The work here is by no means complete, but it is a first attempt to gather these materials in a short space.

REFERENCES

1. M. Clerc, J. Kennedy. The particle swarm—Explosion, stability, and convergence in a multidimensional complex space. *IEEE Transactions on Evolutionary Computation*, 2002, 6(1): 58–73.
2. J. Sun, B. Feng, W. Xu. Particle swarm optimization with particles having quantum behavior. In *Proceedings of the 2004 IEEE Congress on Evolutionary Computation*, Piscataway, NJ, 2004, pp. 325–331.
3. J. Sun, B. Feng, W. Xu. A global search strategy of quantum-behaved particle swarm optimization. In *Proceedings of the 2004 IEEE Conference on Cybernetics and Intelligent Systems*, Singapore, 2004, pp. 111–116.
4. J. Kennedy. Some issues and practices for particle swarms. In *Proceedings of the 2007 IEEE Swarm Intelligence Symposium*, Honolulu, HI, 2007, pp. 162–169.
5. J. Sun, W. Xu, J. Liu. Parameter selection of quantum-behaved particle swarm optimization. In *Proceedings of the 2005 International Conference on Natural Computation*, Changsha, China, 2005, pp. 543–552.
6. J. Sun, W. Fang, X. Wu, V. Palade, W. Xu. Quantum-behaved particle swarm optimization: Analysis of the individual particle and parameter selection. *Evolutionary Computation,* in press.
7. J. Sun, W. Xu, B. Feng. Adaptive parameter control for quantum-behaved particle swarm optimization on individual level. In *Proceedings of the 2005 IEEE International Conference on Systems, Man and Cybernetics*, Waikoloa, HI, 2005, pp. 3049–3054.
8. J. Sun, W. Xu, W. Fang. Quantum-behaved particle swarm optimization with a hybrid probability distribution. In *Proceedings of the Ninth Pacific Rim International Conference on Artificial Intelligence*, Guilin, China, 2006, pp. 737–746.
9. L.S. Coelho. Novel Gaussian quantum-behaved particle swarm optimiser applied to electromagnetic design. *IET Science, Measurement and Technology*, 2007, 1(5): 290–294.
10. L.S. Coelho, N. Nedjah, L.D.M. Mourelle. Gaussian quantum-behaved particle swarm optimization applied to fuzzy pid controller design. *Studies in Computational Intelligence*, 2008, 121: 1–15.
11. L.S. Coelho. Gaussian quantum-behaved particle swarm optimization approaches for constrained engineering design problems. *Expert Systems with Applications*, 2010, 37(2): 1676–1683.

12. J. Sun, C. Lai, W. Xu, Y. Ding, Z. Chai. A Modified quantum-behaved particle swarm optimization. In *Proceedings of the 2007 International Conference on Computational Science*, Kuala Lumpur, Malaysia, 2007, pp. 294–301.

13. J. Riget, J. Vesterstroem. A diversity-guided particle swarm optimizer—The ARPSO: Department of Computer Science, University of Aarhus, Aarhus, Denment, 2002.

14. J. Sun, W. Xu, W. Fang. Quantum-behaved particle swarm optimization algorithm with controlled diversity. In *Proceedings of the 2006 International Conference on Computational Science*, Reading, MA, 2006, pp. 847–854.

15. J. Sun, W. Xu, W. Fang. A diversity-guided quantum-behaved particle swarm optimization algorithm. In *Proceedings of the 2006 International Conference on Simulated Evolution and Learning*, Hefei, China, 2006, pp. 497–504.

16. J. Sun, W. Xu, W. Fang. Enhancing global search ability of quantum-behaved particle swarm optimization by maintaining diversity of the swarm. In *Proceedings of the 2006 International Conference on Rough Sets and Current Trends in Computing*, Kobe, Japan, 2006, pp. 736–745.

17. J. Wang, Y. Zhou. Quantum-behaved particle swarm optimization with generalized local search operator for global optimization. In *Proceedings of the 2007 International Conference on Intelligent Computing*, Lecture Notes in Computer Science, Qingdao, China, Vol. 4682, 2007, pp. 851–860.

18. Z. Huang, Y. Wang, C. Yang, C. Wu. A new improved quantum-behaved particle swarm optimization model. In *Proceedings of the 2009 IEEE Conference on Industrial Electronics and Applications*, Taipei, Taiwan, 2009, pp. 1560–1564.

19. Y. Kaiqiao, N. Hirosato. Quantum-behaved particle swarm optimization with chaotic search. *IEICE: Trans Information Systems*, 2008, E91-D(7): 1963–1970.

20. J. Liu, J. Sun, W. Xu. Improving quantum-behaved particle swarm optimization by simulated annealing. In *Proceedings of the 2006 International Conference on Intelligent Computing*, Kunming, China, 2006, pp. 130–136.

21. J. Liu, J. Sun, W. Xu, X. Kong. Quantum-behaved particle swarm optimization based on immune memory and vaccination. In *Proceedings of the 2006 IEEE International Conference on Granular Computing*, San Jose, CA, 2006, pp. 453–456.

22. J. Liu, J. Sun, W. Xu. Quantum-behaved particle swarm optimization with immune operator. In *Proceedings of the 16th International Symposium on Foundations of Intelligent Systems*, Bari, Italy, 2006, pp. 77–83.

23. W. Fang, J. Sun, W. Xu. Improved quantum-behaved particle swarm optimization algorithm based on differential evolution operator and its application. *Journal of System Simulation*, 2008, 20(24): 6740–6744.

24. J. Liu, J. Sun, W. Xu. Quantum-behaved particle swarm optimization with adaptive mutation operator. In *Proceedings of the 2006 International Conference on Natural Computing*, Hainan, China, 2006, pp. 959–967.

25. L.S. Coelho. A quantum particle swarm optimizer with chaotic mutation operator. *Chaos, Solitons and Fractals*, 2008, 37(5): 1409–1418.

26. W. Fang, J. Sun, W. Xu. Analysis of mutation operators on quantum-behaved particle swarm optimization algorithm. *New Mathematics and Natural Computation*, 2009, 5(2): 487–496.

27. J. Sun, C. Lai, W. Xu, Z. Chai. A novel and more efficient search strategy of quantum-behaved particle swarm optimization. In *Proceedings of the 2007 International Conference on Adaptive and Natural Computing Algorithms*, Warsaw, Poland, 2007, pp. 394–403.

28. M. Xi, J. Sun, W. Xu. Quantum-behaved particle swarm optimization with elitist mean best position. In *Proceedings of the 2007 International Conference on Complex Systems and Applications–Modeling, Control and Simulations*, Jinan, China, 2007, pp. 1643–1647.

29. M. Xi, J. Sun, W. Xu. An improved quantum-behaved particle swarm optimization algorithm with weighted mean best position. *Applied Mathematics and Computation*, 2008, 205(2): 751–759.

30. H. Gao, W. Xu, T. Gao. A cooperative approach to quantum-behaved particle swarm optimization. In *Proceedings of the 2007 IEEE International Symposium on Intelligent Signal Processing*, Madrid, Spain, 2007, pp. 1–6.

31. S. Lu, C. Sun. Quantum-behaved particle swarm optimization with cooperative-competitive coevolutionary. In *Proceedings of the 2008 International Symposium on Knowledge Acquisition and Modeling*, Wuhan, China, 2008, pp. 593–597.

32. S. Lu, C. Sun. Coevolutionary quantum-behaved particle swarm optimization with hybrid cooperative search. In *Proceedings of the 2008 Pacific-Asia Workshop on Computational Intelligence and Industrial Application*, Washington, DC, 2008, pp. 109–113.

33. Q. Baida, J. Zhuqing, X. Baoguo. Research on quantum-behaved particle swarms cooperative optimization. *Computer Engineering and Applications*, 2008, 44(7): 72–74.

34. R. Poli. An analysis of publications on particle swarm optimization applications. Department of Computer Science, University of Essex, Essex, U.K., 2007.

35. S. Mikki, A.A. Kishk. Infinitesimal dipole model for dielectric resonator antennas using the QPSO algorithm. In *Proceedings of the 2006 IEEE Antennas and Propagation Society International Symposium*, Albuquerque, NM, 2006, pp. 3285–3288.

36. Y. Cai, J. Sun, J. Wang, Y. Ding, N. Tian, X. Liao, W. Xu. Optimizing the codon usage of synthetic gene with QPSO algorithm. *Journal of Theoretical Biology*, 2008, 254(1): 123–127.

37. L. Liu, J. Sun, D. Zhang, G. Du, J. Chen, W. Xu. Culture conditions optimization of hyaluronic acid production by *Streptococcus zooepidemicus* based on radial basis function neural network and quantum-behaved particle swarm optimization algorithm. *Enzyme and Microbial Technology*, 2009, 44(1): 24–32.

38. K. Lu, R. Wang. Application of PSO and QPSO algorithm to estimate parameters from kinetic model of glutamic acid batch fermentation. In *Proceedings of the Seventh World Congress on Intelligent Control and Automation*, Chongqing, China, 2008, pp. 8968–8971.

39. Y. Chi, X. Liu, K. Xia, C. Su. An intelligent diagnosis to type 2 diabetes based on QPSO algorithm and WLS-SVM. In *Proceedings of the 2008 Intelligent Information Technology Application Workshops*, Washington, DC, 2008, pp. 117–121.

40. H. Liu, S. Xu, X. Liang. A modified quantum-behaved particle swarm optimization for constrained optimization. In *Proceedings of the 2008 International Symposium on Intelligent Information Technology Application Workshops*, Shanghai, China, 2008, pp. 531–534.

41. J. Wang, Y. Zhang, Y. Zhou, J. Yin. Discrete quantum-behaved particle swarm optimization based on estimation of distribution for combinatorial optimization. In *Proceedings of the 2008 IEEE World Congress on Computational Intelligence*, Hong Kong, China, 2008, pp. 897–904.

42. J. Liu, J. Sun, W. Xu. Quantum-behaved particle swarm optimization for integer programming. In *Proceedings of the 2006 International Conference on Neural Information Processing*, Hong Kong, China, 2006, pp. 1042–1050.

43. J. Sun, J. Liu, W. Xu. Using quantum-behaved particle swarm optimization algorithm to solve non-linear programming problems. *International Journal of Computer Mathematics*, 2007, 84(2): 261–272.

44. B. Xiao, T. Qin, D. Feng, G. Mu, P. Li, G.M. Xiao. Optimal planning of substation locating and sizing based on improved QPSO algorithm. In *Proceedings of the Asia-Pacific Power and Energy Engineering Conference*, Shanghai, China, 2009, pp. 1–5.

45. S.N. Omkar, R. Khandelwal, T.V.S. Ananth, G.N. Naik, S. Gopalakrishnan. Quantum behaved particle swarm optimization (QPSO) for multi-objective design optimization of composite structures. *Expert Systems with Applications*, 2009, 36(8): 11312–11322.

46. J. Sun, J. Liu, W. Xu. QPSO-based QoS multicast routing algorithm. In *Proceedings of the 2006 International Conference on Simulated Evolution and Learning*, Hefei, China, 2006, pp. 261–268.

47. D. Zhao, K. Xia, B. Wang, J. Gao. An approach to mobile IP routing based on QPSO algorithm. In *Proceedings of the Pacific-Asia Workshop on Computational Intelligence and Industrial Application*, Wuhan, China, 2008, pp. 667–671.

48. R. Ma, Y. Liu, X. Lin, Z. Wang. Network anomaly detection using RBF neural network with hybrid QPSO. In *Proceedings of the IEEE International Conference on Networking, Sensing and Control*, Chicago, IL, 2008, pp. 1284–1287.

49. R. Ma, Y. Liu, X. Lin. Hybrid QPSO based wavelet neural networks for network anomaly detection. In *Proceedings of the Second Workshop on Digital Media and its Application in Museum and Heritages*, Qingdao, China, 2007, pp. 442–447.

50. R. Wu, C. Su, K. Xia, Y. Wu. An approach to WLS-SVM based on QPSO algorithm in anomaly detection. In *Proceedings of the 2008 World Congress on Intelligent Control and Automation*, Chongqing, China, 2008, pp. 4468–4472.

51. C. Yue, Z. Dongming, X. Kewen, W. Rui. Channel assignment based on QPSO algorithm. *Communications Technology*, 2009, 42(2): 204–206.

52. S. Jalilzadeh, H. Shayeghi, A. Safari, D. Masoomi. Output feedback UPFC controller design by using Quantum Particle Swarm Optimization. In *Proceedings of the Sixth International Conference on Electrical Engineering/ Electronics, Computer, Telecommunications and Information Technology*, Pataya, Thailand, 2009, pp. 28–31.

53. M. Xi, J. Sun, W. Xu. Quantum-behaved particle swarm optimization for design H infinite structure specified controllers. In *Proceedings of the 2006 International Symposium on Distributed Computing and Its Applications in Business, Engineering and Science*, Wuxi, China, 2006, pp. 1016–1019.

54. M. Xi, J. Sun, W. Xu. Parameter optimization of PID controller based on quantum-behaved particle swarm optimization. In *Proceedings of the 2007 International Conference on Computational Science and Applications*, Kuala Lumpur, Malaysia, 2007, pp. 603–607.

55. J. Liu, Q. Wu, D. Zhu. Thruster fault-tolerant for UUVs based on quantum-behaved particle swarm optimization. Opportunities and challenges for next-generation applied intelligence. *Studies in Computational Intelligence*, 2009, 214: 159–165.

56. F Gao, H. Tong. Parameter estimation for chaotic system based on particle swarm optimization. *Acta Physica Sinica*, 2006, 2: 577–582

57. J. Sun, W. Xu, B. Ye. Quantum-behaved particle swarm optimization clustering algorithm. In *Proceedings of the 2006 International Conference on Advanced Data Mining and Applications*, Xi'an, China, 2006, pp. 340–347.

58. W. Chen, J. Sun, Y. Ding, W. Fang, W. Xu. Clustering of gene expression data with quantum-behaved particle swarm optimization. In *Proceedings of the 21st International Conference on Industrial, Engineering and Other Applications of Applied Intelligent Systems*, Wroclaw, Poland, 2008, pp. 388–396.

59. K. Lu, K. Fang, G. Xie. A hybrid quantum-behaved particle swarm optimization algorithm for clustering analysis. In *Proceedings of the Fifth International Conference on Fuzzy Systems and Knowledge Discovery*, Shandong, China, 2008, pp. 21–25.

60. X. Peng, Y. Zhang, S. Xiao, Z. Wu, J. Cui, L. Chen, D. Xiao. An alert correlation method based on improved cluster algorithm. In *Proceedings of the 2008 Pacific-Asia Workshop on Computational Intelligence and Industrial Application*, Wuhan, China, 2008, pp. 342–247.

61. X. Zhang, H. Zhang, Y. Zhu, Y. Liu, T. Yang, T. Zhang. Using IACO and QPSO to solve spatial clustering with obstacles constraints. In *Proceedings of the 2009 IEEE International Conference on Automation and Logistics*, Shenyang, China, 2009, pp. 1699–1704.

62. H. Wang, S. Yang, W. Xu, J. Sun. Scalability of hybrid fuzzy c-means algorithm based on quantum-behaved PSO. In *Proceedings of the Fourth International Conference on Fuzzy Systems and Knowledge Discovery*, Hainan, China, 2007, pp. 261–265.

63. L. Tao. Text topic mining and classification based on quantum-behaved particle swarm optimization. *Journal of Southwest University for Nationalities*, 2009, 35(3): 603–607.

64. L. Tao, Y. Feng, C. Jianying, H. Weilin. Acquisition of classification rule based on quantum-behaved particle swarm optimization. *Application Research of Computers*, 2009, 26(2): 496–499.

65. H. Zhu, X. Zhao, Y. Zhong. Feature selection method combined optimized document frequency with improved RBF network. In *Proceedings of the 2009 International Conference on Advanced Data Mining and Applications*, Beijing, China, 2009, pp. 796–803.

66. S.-Y. Lv, X.-M. Zheng, X.-D. Wang. Attribute reduction based on quantum-behaved particle swarm optimization. *Computer Engineering*, 2008, 34(18): 65–69.
67. J.-Y. Wang, Y. Xie. Minimal attribute reduction algorithm based on quantum particle swarm optimization. *Computer Engineering*, 2009, 35(12): 148–150.
68. S. Mikki, A.A. Kishk. Investigation of the quantum particle swarm optimization technique for electromagnetic applications. In *Proceedings of the 2005 IEEE Antennas and Propagation Society International Symposium*, Honolulu, HI, 2005, 2A: 45–48.
69. S. Mikki, A.A. Kishk. Quantum particle swarm optimization for electromagnetics. *IEEE Transactions on Antennas and Propagation*, 2006, 54(10): 2764–2775.
70. S. Mikki, A.A. Kishk. Theory and applications of infinitesimal dipole models for computational electromagnetics. *IEEE Transactions on Antennas and Propagation*, 2007, 55(5): 1325–1337.
71. L.S. Coelho, P. Alotto. Global optimization of electromagnetic devices using an exponential quantum-behaved particle swarm optimizer. *IEEE Transactions on Magnetics*, 2008, 44(6): 1074–1077.
72. R. Wu, J. Wang, K. Xia, R. Yang. Optimal design on CMOS operational amplifier with QPSO algorithm. In *Proceedings of the 2008 International Conference on Wavelet Analysis and Pattern Recognition*, Hong Kong, China, 2008, pp. 821–825.
73. N. Liu, K. Xia, J. Zhou, C. Ge. Numerical simulation on transistor with CQPSO algorithm. In *Proceedings of the Fourth IEEE Conference on Industrial Electronics and Applications*, Xi'an, China, 2009, pp. 3732–3736.
74. J. Sun, W. Xu, W. Fang. Solving multi-period financial planning problem via quantum-behaved particle swarm algorithm. In *Proceedings of the 2006 International Conference on Intelligent Computing*, Lecture Notes in Computer Science, Kunming, China, 2006, Vol. 4114, pp. 1158–1169.
75. J. Sun, M. Xi, W. Xu. Empirical study on stock through GARCH based on QPSO algorithm. *Journal on Numerical Methods and Computer Applications*, 2007, 28(4): 260–266.
76. L. Tang, F. Xue. Using data to design fuzzy system based on quantum-behaved particle swarm optimization. In *Proceedings of the 2008 International Conference on Machine Learning and Cybernetics*, Kunming, China, 2008, pp. 624–628.
77. Y. Xue, J. Sun, W. Xu. QPSO algorithm for rectangle-packing optimization. *Journal of Computer Applications*, 2006, 9: 2068–2070.
78. Z. Di, J. Sun, W. Xu. Polygonal approximation of curves using binary quantum-behaved particle swarm optimization. *Journal of Computer Applications*, 2007, 27(8): 2030–2032.
79. J. Huang, J. Sun, W. Xu, H. Dong. Study on layout problem using quantum-behaved particle swarm optimization algorithm. *Journal of Computer Applications*, 2006, 12: 3015–3018.

80. J. Sun, W. Xu, W. Fang, Z. Chai. Quantum-behaved particle swarm optimization with binary encoding. In *Proceedings of the Eighth International Conference on Adaptive and Natural Computing Algorithms*, Warsaw, Poland, 2007, pp. 376–385.

81. S. Li, R. Wang, W. Hu, J. Sun. A new QPSO based BP neural network for face detection. In *Proceedings of the Second International Conference on Fuzzy Information and Engineering*, Piscataway, NJ, 2007, pp. 355–363.

82. X. Lei, A. Fu. Two-dimensional maximum entropy image segmentation method based on quantum-behaved particle swarm optimization algorithm. In *Proceedings of the Fourth International Conference on Natural Computation*, Jinan, China, 2008, pp. 692–696.

83. B. Feng, W. Zhang, J. Sun. Image threshold segmentation with Ostu based on quantum-behaved particle swarm algorithm. *Computer Engineering and Design*, 2008, 29(13): 3429–3431.

84. Y. Zhao, Z. Fang, K. Wang, H. Pang. Multilevel minimum cross entropy threshold selection based on quantum particle swarm optimization. In *Proceedings of the Eighth ACIS International Conference on Software Engineering, Artificial Intelligence, Networking, and Parallel/Distributed Computing*, Qingdao, China, 2007, pp. 65–69.

85. L. Yang, Z.W. Liao, W.F. Chen. An automatic registration framework using quantum particle swarm optimization for remote sensing images. In *Proceedings of the 2007 International Conference on Wavelet Analysis and Pattern Recognition*, Beijing, China, 2007, pp. 484–488.

86. W. Xu, W. Xu, J. Sun. Image interpolation algorithm based on quantum-behaved particle swarm optimization. *Journal of Computer Applications*, 2007, 27(9): 2147–2149.

87. L.S. Coelho, V.C. Mariani. Particle swarm approach based on quantum mechanics and harmonic oscillator potential well for economic load dispatch with valve-point effects. *Energy Conversion and Management*, 2008, 49(11): 3080–3085.

88. J. Sun, W. Fang, D. Wang, W. Xu. Solving the economic dispatch problem with a modified quantum-behaved particle swarm optimization method. *Energy Conversion and Management*, 2009, 50(12): 2967–2975.

89. L. Zhou, H. Yang, C. Liu. QPSO-based hyper-parameters selection for LS-SVM regression. In *Proceedings of the Fourth International Conference on Natural Computation*, Jinan, China, 2008, pp. 130–133.

90. X. Li, L. Zhou, C. Liu. Model selection of least squares support vector regression using quantum-behaved particle swarm optimization algorithm. In *Proceedings of 2009 International Workshop on Intelligent Systems and Applications*, Wuhan, China, 2009, pp. 1–5.

91. J. Wang, Z. Liu, P. Lu. Electricity load forecasting based on adaptive quantum-behaved particle swarm optimization and support vector machines on global level. In *Proceedings of the 2008 International Symposium on Computational Intelligence and Design*, Wuhan, China, 2008, pp. 233–236.

92. Q. Zhang, Z. Che. A novel method to train support vector machines for solving quadratic programming task. In *Proceedings of the Seventh World Congress on Intelligent Control and Automation*, Chongqing, China, 2008, pp. 7917–7921.

93. C. Lin, P. Feng. Parameters selection and application of support vector machines based on quantum delta particle swarm optimization algorithm. *Automation and Instrumentation*, 2009, 1: 5–8.

94. S.L. Sabat, L.S. Coelho, A. Abraham. MESFET DC model parameter extraction using quantum particle swarm optimization. *Microelectronics Reliability*, 2009, 49(6): 660–666.

95. J. Liu, W. Xu, J. Sun. Nonlinear system identification of hammerstien and wiener model using swarm intelligence. In *Proceedings of the 2006 IEEE International Conference on Information Acquisition*, Shandong, China, 2006, pp. 1219–1223.

96. F. Gao, H. Gao, Z. Li, H. Tong, J. Lee. Detecting unstable periodic orbits of nonlinear mappings by a novel quantum-behaved particle swarm optimization non-Lyapunov way. *Chaos, Solitons and Fractals*, 2009, 42(4): 2450–2463.

97. J. Sun, W. Xu, J. Liu. Training RBF neural network via quantum-behaved particle swarm optimization. In *Proceedings of the 2006 International Conference on Neural Information Processing*, Hong Kong, China, 2006, pp. 1156–1163.

98. S. Tian, T. Liu. Short-term load forecasting based on RBFNN and QPSO. In *Proceedings of the 2009 Asia-Pacific Power and Energy Engineering Conference*, Wuhan, China, 2009, pp. 1–4.

99. Q. Tan, Y. Song. Sidelobe suppression algorithm for chaotic FM signal based on neural network. In *Proceedings of the Ninth International Conference on Signal Processing*, Beijing, China, 2008, pp. 2429–2433.

100. G.-H. Yang, B.-Y. Wen. Identification of power quality disturbance based on QPSO-ANN. *Proceedings of the CSEE*, 2008, 28(10): 123–129.

101. X. Kong, J. Sun, B. Ye, W. Xu. An efficient quantum-behaved particle swarm optimization for multiprocessor scheduling. In *Proceedings of the 2007 International Conference on Computational Science*, Kuala Lumpur, Malaysia, 2007, pp. 278–285.

102. W. Fang, J. Sun, W. Xu, J. Liu. FIR digital filters design based on quantum-behaved particle swarm optimization. In *Proceedings of the First International Conference on Innovative Computing, Information and Control*, Dalian, China, 2006, pp. 615–619.

103. W. Fang, J. Sun, W. Xu. FIR filter design based on adaptive quantum-behaved particle swarm optimization algorithm. *Systems Engineering and Electronics*, 2008, 30(7): 1378–1381.

104. W. Fang, J. Sun, W. Xu. Design IIR digital filters using quantum-behaved particle swarm optimization. In *Proceedings of the 2006 International Conference on Natural Computation*, Dalian, China, 2006, pp. 637–640.

105. W. Fang, J. Sun, W. Xu. Analysis of adaptive IIR filter design based on quantum-behaved particle swarm optimization. In *Proceedings of the Sixth World Congress on Intelligent Control and Automation*, Dalian, China, 2006, pp. 3396–3400.

106. W. Fang, J. Sun, W. Xu. Design of two-dimensional recursive filters by using quantum-behaved particle swarm optimization. In *Proceedings of the 2006 International Conference on Intelligent Information Hiding and Multimedia Signal Processing*, Pasadena, CA, 2006, pp. 240–243.

107. J. Sun, W. Fang, W. Chen, W. Xu. Design of two-dimensional IIR digital filters using an improved quantum-behaved particle swarm optimization algorithm. In *Proceedings of the 2008 American Control Conference*, Seattle, Washington, USA, 2008, pp. 2603–2608.

108. K.E. Parsopoulos, M.N. Vrahatis. Particle swarm optimization method for constrained optimization problems. In *Proceedings of the 2002 Euro-International Symposium on Computational Intelligence*, Kosice, Slovakia, 2002, pp. 214–220.

109. J. Zhao, J. Sun, C.-H. Lai, W. Xu. An improved quantum-behaved particle swarm optimization for multi-peak optimization problems. *International Journal of Computer Mathematics*, 2011, 88(3): 517–532.

110. J. Li, M.E. Balazs, G.T. Parks. P.J. Clarkson. A species conserving genetic algorithm for multimodal function optimization. *Evolutionary Computation*, 2002, 10(3): 207–234.

111. D. Beasley, D.R. Bull, R.R. Martin. A sequential niche technique for multimodal function optimization. *Evolutionary Computation*, 1993, 1(2): 101–125.

112. X.D. Li. Adaptively choosing neighbourhood bests using species in a particle swarm optimizer for multimodal function optimization. In *Proceedings of the 2004 Genetic and Evolutionary Computation Conference*, Seattle, WA, 2004, pp. 105–116.

113. R.K. Ursem. Diversity-guided evolutionary algorithms. In *Proceedings of the 2011 Parallel Problem Solving from Nature Conference*, Paris, France, 2001, pp. 462–471.

114. H. Long, J. Sun, X. Wang, C. Lai, W. Xu. Using selection to improve quantum-behaved particle swarm optimisation. *International Journal of Innovative Computing and Applications*, 2009, 2(2): 100–114.

115. P.J. Angeline. Using selection to improve particle swarm optimization. In *Proceedings of the 1998 IEEE International Conference on Evolutionary Computation*, Anchorage, AK, 1998, pp. 84–89.

116. F. van den Bergh, A.P. Engelbrech. A cooperative approach to particle swarm optimization. *IEEE Transactions on Evolutionary Computation*, 2004, 8(3): 225–238.

CHAPTER 5

Advanced Topics

THIS CHAPTER INVOLVES SOME advanced topics on the QPSO algorithm. Readers who are not interested in the theory of QPSO should skip this chapter except the last section. First, the behaviour of an individual particle is analysed using probability theory and stochastic simulation to validate the theoretical results. Second, the global convergence of QPSO is proved by using probability measures, Markov processes and fixed-point theorems in probabilistic metric (PM) spaces. Third, the definitions of computational complexity and the rate of convergence of a stochastic search algorithm are introduced and applied to the QPSO algorithm, and tested experimentally against the sphere function. Finally, parameter selection of the QPSO algorithm is discussed, and numerical tests of several benchmark functions are performed and analysed.

5.1 BEHAVIOUR ANALYSIS OF INDIVIDUAL PARTICLES

This section addresses the dynamical behaviour of individual particles in QPSO. The aim is to provide a mathematical framework suitable for analysing QPSO. Most analysis relies on the work in [1] with several improvements and modifications. After providing some mathematical preliminaries and fundamental results, the iterative update equations are discussed. This follows with the convergence results for Type-1 particles and the boundedness for Type-2 particles. Numerical simulation results are also included in this section to demonstrate the main results.

5.1.1 Some Preliminaries

Definition 5.1

The set Ω containing all possible outcomes of a random experiment is called the *space of elementary events* or *the sample space*. Each element, or a point, ω in Ω is an outcome known as an *elementary event* or a *sample point*.

Definition 5.2

The space Ω together with a σ-algebra $\mathcal{F}$ of its subsets is a measurable space which is denoted as $(\Omega, \mathcal{F})$.

Definition 5.3

An ordered triple $(\Omega, \mathcal{F}, P)$ where

  (a) Ω is a set of points ω
  (b) $\mathcal{F}$ is a σ-algebra of subsets of Ω
  (c) P is a probability measure on $\mathcal{F}$

is called a *probabilistic model* or a *probability space*. Here Ω is the sample space or the space of elementary events, the set A in $\mathcal{F}$ is known as an event, and $P(A)$ is the probability of the event A.

Let $R = (-\infty, +\infty)$ be the real line. $(R, \mathcal{B}(R))$ or $(R, \mathcal{B})$ is a measure space. $\mathcal{B}(R)$ or $\mathcal{B}$ is called the Borel algebra of subsets.

Definition 5.4

A real function $\xi = \xi(\omega)$ defined on $(\Omega, \mathcal{F}, P)$ is an $\mathcal{F}$-measurable function or a random variable, if $\{\omega : \xi(\omega) \in B\} \in \mathcal{F}$ for every $B \in \mathcal{B}(R)$; or equivalently, if the inverse image $\xi^{-1}(B) \equiv \{\omega : \xi(\omega) \in B\}$ is a measurable set in Ω.

Definition 5.5

Let $\xi_1, \xi_2, \dots$ be random variables defined on the probability space $(\Omega, \mathcal{F}, P)$. The sequence of random variables $\{\xi_n\}$ defined on the probability space $(\Omega, \mathcal{F}, P)$ is called a sequence of *independent identically distributed (i.i.d.)*

random variables, if each random variable ξ_i has the same probability distribution as the others and all are mutually independent.

There are several different concepts of convergence of random variables in probability theory. Four of these are of particular importance in this work: *with probability one, in probability, in distribution, in mean of order r* [2]. In the subsequent definitions, $\xi_1, \xi_2, \dots$ and $\{\xi_n\}$ always refer to the previous definition.

Definition 5.6

The sequence $\{\xi_n\}$ converges *with probability one (almost surely, almost everywhere)* to the random variable ξ if

$$P\{\lim_{n\to\infty} \xi_n(\omega) = \xi\} = 1. \tag{5.1}$$

On the other hand, the set of sample points ω for which $\xi_n(\omega)$ does not converge to ξ has probability zero. In the case of convergence, it is usually denoted as $\xi_n(\omega) \to \xi$, $\xi_n(\omega) \xrightarrow{a.s.} \xi$ or $\xi_n(\omega) \xrightarrow{a.e.} \xi$.

Definition 5.7

The sequence $\{\xi_n\}$ converges *in probability* to the random variable ξ (usually denoted as $\xi_n \xrightarrow{P} \xi$) if for every $\varepsilon > 0$,

$$\lim_{n\to\infty} P\{|\xi_n(\omega) - \xi| > \varepsilon\} = 0, \tag{5.2}$$

or

$$\lim_{n\to\infty} P\{|\xi_n(\omega) - \xi| < \varepsilon\} = 1. \tag{5.3}$$

Definition 5.8

Let $F_n(x)$ and $F(x)$ be distribution functions for $\{\xi_n\}$, respectively. The sequence $\{\xi_n\}$ converges *in distribution* to the random variable ξ (usually denoted as $\xi_n \xrightarrow{d} \xi$) if for every continuous point $x \in R$,

$$\lim_{n\to\infty} F_n(x) = F(x). \tag{5.4}$$

Definition 5.9

Suppose $E|\xi_n|^r < \infty$, $0 < r < \infty$, holds on $(\Omega, \mathscr{F}, P)$. The sequence $\{\xi_n\}$ converges *in mean of order r* to the random variable ξ (usually denoted as $\xi_n \xrightarrow{r} \xi$) if

$$\lim_{n \to \infty} E\,|\xi_n - \xi|^r = 0. \tag{5.5}$$

The following theorem indicates the relationships between the aforementioned four types of convergence. It should be noted that the converses of (5.6), (5.7) and (5.8) are false in general.

Theorem 5.1

There are three implications of the four types of convergence:

$$\xi_n \xrightarrow{a.s.} \xi \Rightarrow \xi_n \xrightarrow{P} \xi, \tag{5.6}$$

$$\xi_n \xrightarrow{P} \xi \Rightarrow \xi_n \xrightarrow{d} \xi, \tag{5.7}$$

$$\xi_n \xrightarrow{r} \xi \Rightarrow \xi_n \xrightarrow{P} \xi, \quad r > 0. \tag{5.8}$$

The following two theorems present the strong and weak laws of large numbers, respectively.

Theorem 5.2 (Kolmogorov's Strong Law of Large Numbers)

Let $\xi_1, \xi_2, \ldots$ be a sequence of i.i.d. random variables with $E|\xi_1| < \infty$. Then

$$\frac{1}{n} \sum_{i=1}^{n} \xi_i \xrightarrow{a.s.} E\,|\xi_1|. \tag{5.9}$$

Theorem 5.3 (Khintchine's Weak Law of Large Numbers)

Let $\xi_1, \xi_2,\ldots$ be a sequence of i.i.d. random variables with $E\,|\xi_1| < \infty$. Then for every $\varepsilon > 0$

$$\lim_{n\to\infty} P\left\{ \left| \frac{1}{n}\sum_{i=1}^{n}\xi_i - E\,|\xi_1| \right| < \varepsilon \right\} = 1, \qquad (5.10)$$

or

$$\frac{1}{n}\sum_{i=1}^{n}\xi_i \xrightarrow{P} E\,|\xi_1|. \qquad (5.11)$$

5.1.2 Simplification of the Iterative Update Equations

It is obvious that the convergence or boundedness of the position of an individual particle must be consistent with the convergence or boundedness of its component in each dimension. For QPSO-Type-1, the necessary and sufficient condition for $X_{i,n}$ to converge to $p_{i,n}$ in any kind of convergence is that $X_{i,n}^j$ converges to $p_{i,n}^j$ for each $1 \le j \le N$ in relevant kind of convergence. For QPSO-Type-2, the necessary and sufficient condition for $X_{i,n}$ to be probabilistic bounded is that $X_{i,n}^j$ is probabilistic bounded for each $1 \le j \le N$.

It appears from Equation 4.79 or 4.82 that the update in each dimension of the position of a particle is done independently following the same equation. The only link between the dimensions of the problem space relies on the objective function and, in turn, through the locations of the personal and global best positions and the mean best position among particles found so far. As a result, the convergence or boundedness of an individual particle would rely on the component-wise knowledge. Without loss of generality, the study of convergence or boundedness for a particle in the N-dimensional space can be reduced to the one for a single particle in the one-dimensional (1D) space using the iterative update equation given by

$$X_{n+1} = p \pm \alpha\,|X_n - p|\ln\left(\frac{1}{u_{n+1}}\right), \qquad u_{n+1} \sim U(0,1). \qquad (5.12)$$

or

$$X_{n+1} = p \pm \alpha |X_n - C| \ln\left(\frac{1}{u_{n+1}}\right), \quad u_{n+1} \sim U(0,1). \tag{5.13}$$

In the previous equations, the local attractor of the particle and the mean best position are denoted as p and C. These are probabilistic bounded random variables rather than constants as used in [3]. The probabilistic boundedness of p and C means that $P\{\sup|p| < \infty\} = 1$ and $P\{\sup|C| < \infty\} = 1$, respectively. $\{u_n\}$ is a sequence of independent random variables such that $u_n \sim U(0,1)$ for all $n \geq 0$ and thus the position sequence $\{X_n\}$ in either of the previous two equations is a sequence of random variables. For convenience, the particle moving according to Equation 4.79 or 5.12 is denoted as Type-1 particle and the one according to Equation 4.82 or 5.13 as Type-2 particle. The remainder of this section focuses on the behaviour of these two types of particles.

5.1.3 Convergence Behaviour of Type-1 Particle

Rewrite Equation 5.12 as

$$|X_{n+1} - p| = \alpha |X_n - p| \ln\left(\frac{1}{u_{n+1}}\right) = \lambda_{n+1} |X_n - p|, \quad u_{n+1} \sim U(0,1), \tag{5.14}$$

where $\lambda_{n+1} = \alpha \ln(1/u_{n+1})$, one obtains

$$|X_n - p| = |X_0 - p| \prod_{i=1}^{n} \lambda_i, \tag{5.15}$$

where X_0 is the initial position of the particle. As such, the convergence analysis of $\{X_n\}$ can be reduced to that of the infinite product

$$\beta_n = \prod_{i=1}^{n} \lambda_i = \alpha^n \prod_{i=1}^{n} \ln\left(\frac{1}{u_i}\right). \tag{5.16}$$

The following theorem gives an integral which is of significance to the subsequent analysis of the convergence of β_n and X_n.

Lemma 5.1

The improper integral

$$\int_0^1 \ln\left[\ln\left(\frac{1}{x}\right)\right] dx = -\gamma \tag{5.17}$$

holds with the Euler constant $\gamma \approx 0.577215665$.

Proof: Let $s = 1/x$, one has

$$\int_0^1 \ln\left[\ln\left(\frac{1}{x}\right)\right] dx = \int_0^\infty e^{-s} \ln s\, ds.$$

Since $\Gamma(m) = \int_0^1 x^{m-1} e^{-x} dx$, where $\Gamma(\cdot)$ is the gamma function,

$$\Gamma'(m) = \int_0^1 x^{m-1} e^{-x} \ln x\, dx.$$

From [4], one obtains $\Gamma'(1) = \int_0^1 e^{-x} \ln x\, dx = -\gamma$, which implies that

$$\int_0^1 \ln\left[\ln\left(\frac{1}{x}\right)\right] dx = \int_0^\infty e^{-s} \ln s\, ds = \Gamma'(1) = -\gamma.$$

This completes the proof of the theorem. ∎

With the aforementioned preliminaries, the sufficient and necessary condition for $\{X_n\}$ to converge to random variable p almost surely is derived.

Lemma 5.2

If $\{u_n\}$ is a sequence of i.i.d. random variables with $n \sim U(0,1)$ for all n and $\zeta_n = \ln\left[\ln(1/u_n)\right]$, then

$$\frac{1}{n}\sum_{i=1}^n \zeta_i \xrightarrow{a.s.} -\gamma. \tag{5.18}$$

Proof: Since $\{u_n\}$ is a sequence of i.i.d. random variables, $\{\zeta_n\}$ is also a sequence of i.i.d. random variables. Lemma 5.1 implies that

$$E(\zeta_n) = E\left\{\ln\left[\ln\left(\frac{1}{u_n}\right)\right]\right\} = \int_0^1 \ln\left[\ln\left(\frac{1}{x}\right)\right] dx = -\gamma.$$

Thus, by Kolmogorov's strong law of large numbers, one obtains

$$\frac{1}{n}\sum_{i=1}^{n} \zeta_n \xrightarrow{a.s.} E(\zeta_n) = -\gamma. \qquad \blacksquare$$

Theorem 5.4

Let $\lambda_i = \alpha \ln(1/u_i)$ and $\beta_n = \prod_{i=1}^{n} \lambda_i$, where $u_i \sim U(0,1)$. The necessary and sufficient condition for β_n to converge to zero almost surely is $\alpha < e^\gamma \approx 1.781$. The necessary and sufficient condition for β_n to be probabilistic bounded, that is, $P\{\sup \beta_n < \infty\} = 1$, is $\alpha \leq e^\gamma \approx 1.781$.

Proof: Let $\zeta_i = \ln [\ln(1/u_i)]$ and consider the following three possible cases.

(a) *Case 1*: $\alpha < e^\gamma$

    (i) From Lemma 5.2, $\forall m \in Z^+$, $\exists K_1 \in Z^+$ (the set of all positive integers) such that whenever $k \geq K_1$

$$P\left\{\ln\alpha - \gamma - \frac{1}{m} < \ln\alpha + \frac{1}{k}\sum_{i=1}^{k} \zeta_i < \ln\alpha - \gamma + \frac{1}{m}\right\} = 1. \qquad (5.19)$$

Since $\alpha < e^\gamma$, $\ln\alpha < \gamma$, one obtains

$$\ln\alpha - \gamma + \frac{1}{m} < \frac{1}{m},$$

and therefore

$$\left\{\ln\alpha + \frac{1}{k}\sum_{i=1}^{k} \zeta_i < \frac{1}{m}\right\} \supset \left\{\ln\alpha + \frac{1}{k}\sum_{i=1}^{k} \zeta_i < \ln\alpha - \gamma + \frac{1}{m}\right\}$$

$$\supset \left\{\ln\alpha - \gamma - \frac{1}{m} < \ln\alpha + \frac{1}{k}\sum_{i=1}^{k} \zeta_i < \ln\alpha - \gamma + \frac{1}{m}\right\}. \qquad (5.20)$$

From (5.20), one obtains

$$P\left\{\ln\alpha+\frac{1}{k}\sum_{i=1}^{k}\zeta_i<\frac{1}{m}\right\}\geq P\left\{\ln\alpha+\frac{1}{k}\sum_{i=1}^{k}\zeta_i<\ln\alpha-\gamma+\frac{1}{m}\right\}$$

$$\geq P\left\{\ln\alpha-\gamma-\frac{1}{m}<\ln\alpha+\frac{1}{k}\sum_{i=1}^{k}\zeta_i<\ln\alpha-\gamma+\frac{1}{m}\right\}=1,$$

and thus

$$P\left\{\ln\alpha+\frac{1}{k}\sum_{i=1}^{k}\zeta_i<\frac{1}{m}\right\}=1. \tag{5.21}$$

Since $(-m/k)<1/m$, one obtains

$$\left\{\ln\alpha+\frac{1}{k}\sum_{i=1}^{k}\zeta_i<-\frac{m}{k}\right\}=\left\{\ln\alpha+\frac{1}{k}\sum_{i=1}^{k}\zeta_i<\frac{1}{m}\right\}$$

$$-\left\{-\frac{m}{k}\leq\ln\alpha+\frac{1}{k}\sum_{i=1}^{k}\zeta_i<\frac{1}{m}\right\}$$

which leads to the fact

$$P\left\{\ln\alpha+\frac{1}{k}\sum_{i=1}^{k}\zeta_i<-\frac{m}{k}\right\}$$

$$=P\left\{\ln\alpha+\frac{1}{k}\sum_{i=1}^{k}\zeta_i<\frac{1}{m}\right\}-P\left\{-\frac{m}{k}\leq\ln\alpha+\frac{1}{k}\sum_{i=1}^{k}\zeta_i<\frac{1}{m}\right\}$$

$$=1-P\left\{-\frac{m}{k}\leq\ln\alpha+\frac{1}{k}\sum_{i=1}^{k}\zeta_i<\frac{1}{m}\right\} \tag{5.22}$$

(ii) $\forall m \in Z^+$, $\exists K_2 = m^2$ such that whenever $k \geq K_2$ and $(-m/k) > (-1/m)$, the following is true

$$\left\{ -\frac{m}{k} \leq \ln\alpha + \frac{1}{k}\sum_{i=1}^{k}\zeta_i < \frac{1}{m} \right\} \subset \left\{ -\frac{1}{m} < \ln\alpha + \frac{1}{k}\sum_{i=1}^{k}\zeta_i < \frac{1}{m} \right\}.$$

This leads to

$$P\left\{ -\frac{m}{k} \leq \ln\alpha + \frac{1}{k}\sum_{i=1}^{k}\zeta_i < \frac{1}{m} \right\} \leq P\left\{ -\frac{1}{m} < \ln\alpha + \frac{1}{k}\sum_{i=1}^{k}\zeta_i < \frac{1}{m} \right\}.$$

(5.23)

From (5.22) and (5.23), $\forall m \in Z^+$, $\exists K = \max(K_1, K_2)$ such that whenever $k \geq K$, one obtains

$$P\left\{ \ln\alpha + \frac{1}{k}\sum_{i=1}^{k}\zeta_i < -\frac{m}{k} \right\} = 1 - P\left\{ -\frac{m}{k} \leq \ln\alpha + \frac{1}{k}\sum_{i=1}^{k}\zeta_i < \frac{1}{m} \right\}$$

$$\geq 1 - P\left\{ -\frac{1}{m} \leq \ln\alpha + \frac{1}{k}\sum_{i=1}^{k}\zeta_i < \frac{1}{m} \right\},$$

which is equivalent to

$$P\left\{ \bigcap_{m=1}^{\infty}\bigcup_{n=1}^{\infty}\bigcap_{k=n}^{\infty}\left(\ln\alpha + \frac{1}{k}\sum_{i=1}^{k}\zeta_i < -\frac{m}{k} \right) \right\}$$

$$\geq 1 - P\left\{ \bigcap_{m=1}^{\infty}\bigcup_{n=1}^{\infty}\bigcap_{k=n}^{\infty}\left(-\frac{1}{m} < \ln\alpha + \frac{1}{k}\sum_{i=1}^{k}\zeta_i < \frac{1}{m} \right) \right\}$$

$$= 1 - P\left\{ \lim_{n\to\infty}\frac{1}{n}\sum_{i=1}^{n}\zeta_i = -\ln\alpha \right\}.$$

(5.24)

Since $1/n \sum_{i=1}^{n} \zeta_i \xrightarrow{a.s.} -\gamma$, $P\left\{ \lim_{n\to\infty} \frac{1}{n} \sum_{i=1}^{n} \zeta_i = -\ln\alpha \right\} = P\{\ln\alpha = \gamma\}$.

The condition that $\alpha < e^\gamma$ implies that $P\{\ln\alpha = \gamma\} = 0$. Hence,

$$P\left\{ \lim_{n\to\infty} \frac{1}{n} \sum_{i=1}^{n} \zeta_i = -\ln\alpha \right\} = 0. \tag{5.25}$$

From inequality (5.24), one obtains

$$P\left\{ \bigcap_{m=1}^{\infty} \bigcup_{n=1}^{\infty} \bigcap_{k=n}^{\infty} \left(\ln\alpha + \frac{1}{k}\sum_{i=1}^{k} \zeta_i < -\frac{m}{k} \right) \right\}$$

$$\geq 1 - P\left\{ \lim_{n\to\infty} \frac{1}{n}\sum_{i=1}^{n} \zeta_i = -\ln\alpha \right\} = 1 - 0 = 1,$$

and thus

$$P\left\{ \bigcap_{m=1}^{\infty} \bigcup_{n=1}^{\infty} \bigcap_{k=n}^{\infty} \left(\ln\alpha + \frac{1}{k}\sum_{i=1}^{k} \zeta_i < -\frac{m}{k} \right) \right\} = 1,$$

which implies

$$P\left\{ \lim_{n\to\infty} \left(n\ln\alpha + \sum_{i=1}^{n} \zeta_i \right) = -\infty \right\} = 1,$$

and thus

$$P\left\{ \lim_{n\to\infty} \ln\left[\alpha^n \prod_{i=1}^{n} \ln\left(\frac{1}{u_i}\right) \right] = -\infty \right\} = 1.$$

Hence,

$$P\{\lim_{n\to\infty} \beta_n = 0\} = 1 \quad \text{or} \quad \beta_n \xrightarrow{a.s.} 0.$$

Convergence in probability implies convergence in distribution and in probability, and hence, $\beta_n \xrightarrow{d} 0$ and $\beta_n \xrightarrow{P} 0$.

(b) *Case 2*: $\alpha = e^{\gamma}$, $\ln \alpha = \gamma$

Since $1/n \sum_{i=1}^{n} \zeta_i \xrightarrow{a.s.} -\gamma$, one obtains

$$P\left\{\lim_{n \to \infty}\left|\frac{1}{n}\sum_{i=1}^{n}\zeta_i + \gamma\right| = 0\right\} = 1,$$

and

$$P\left\{\lim_{n \to \infty}\left|\frac{1}{n}\sum_{i=1}^{n}\zeta_i + \ln\alpha\right| = 0\right\} = 1.$$

In other words, $\forall m \in Z^+$, $\exists K \in Z^+$, such that whenever $k \geq K$,

$$P\left\{\left|\frac{1}{k}\sum_{i=1}^{k}\zeta_i + \ln\alpha\right| < \frac{1}{m}\right\} = 1.$$

Hence,

$$P\left\{\left|\sum_{i=1}^{k}\zeta_i + k\ln\alpha\right| < \frac{k}{m}\right\} = P\left\{\left|\ln\alpha^k\prod_{i=1}^{k}\ln\left(\frac{1}{u_i}\right)\right| < \frac{k}{m}\right\}$$

$$= P\left\{\left|\ln\beta_k\right| < \frac{k}{m}\right\} = 1. \tag{5.26}$$

Equation 5.26 is equivalent to

$$P\left\{\bigcap_{m=1}^{\infty}\bigcup_{n=1}^{\infty}\bigcap_{k=n}^{\infty}\left(\left|\ln\beta_k\right| < \frac{k}{m}\right)\right\} = P\left\{\bigcap_{m=1}^{\infty}\bigcup_{n=1}^{\infty}\bigcap_{k=n}^{\infty}\left(-\frac{k}{m} < \ln\beta_k < \frac{k}{m}\right)\right\}$$

$$= P\left\{-\infty < \lim_{n \to \infty}\beta_n < \infty\right\} = 1.$$

Note that $P\{-\infty < \lim_{n\to\infty} \beta_n < \infty\} = 1$, that is, when $n \to \infty$, the value of β_n

is bounded. This implies that β_n does not converge to 0 in distribution.

(c) *Case 3*: $\alpha > e^\gamma$, $\ln \alpha > \gamma$

Since $\quad 1/n \sum_{i=1}^{n} \zeta_i \xrightarrow{a.s.} -\gamma, \quad$ it $\quad$ is $\quad$ obvious $\quad$ that

$$P\left\{\lim_{n\to\infty} \left|\frac{1}{n}\sum_{i=1}^{n} \zeta_i + \ln\alpha\right| > 0\right\} = 1. \text{ In other words, } \exists b > 0, \text{ such that}$$

$$P\left\{\lim_{n\to\infty} \left|\frac{1}{n}\sum_{i=1}^{n} \zeta_i + \ln\alpha\right| = b\right\} = 1. \tag{5.27}$$

It is easy to deduce that $\forall m \in Z^+$, $\exists K \in Z^+$, such that whenever $k \geq K$,

$$P\left\{\left|\left|\frac{1}{k}\sum_{i=1}^{k} \zeta_i + \ln\alpha\right| - b\right| < \frac{1}{m}\right\} = 1.$$

Hence,

$$P\left\{b - \frac{1}{m} < \left|\frac{1}{k}\sum_{i=1}^{k} \zeta_i + \ln\alpha\right| < \frac{1}{m} + b\right\} = 1.$$

It is also easy to deduce that

$$P\left\{b - \frac{1}{m} < \frac{1}{k}\sum_{i=1}^{k} \zeta_i + \ln\alpha < b + \frac{1}{m}\right\}$$

$$= P\left\{kb - \frac{k}{m} < \sum_{i=1}^{k} \zeta_i + k\ln\alpha < kb + \frac{k}{m}\right\}$$

$$= P\left\{kb - \frac{k}{m} < \ln\left[\alpha^k \prod_{i=1}^{k} \ln\left(\frac{1}{u_i}\right)\right] < kb + \frac{k}{m}\right\}$$

$$= P\left\{kb - \frac{k}{m} < \ln\beta_k < kb + \frac{k}{m}\right\} = 1$$

Since both $kb - \dfrac{k}{m}$ and $kb + \dfrac{k}{m}$ are arbitrary large real numbers, the above proposition leads to

$$P\left\{\bigcap_{m=1}^{\infty}\bigcup_{n=1}^{\infty}\bigcap_{k=n}^{\infty}\left(kb-\frac{k}{m}<\ln\beta_k<kb+\frac{k}{m}\right)\right\}$$

$$= P\left\{\lim_{n\to\infty}\ln\beta_n = +\infty\right\} = 1 \qquad (5.28)$$

This proves that β_n diverges as $n \to +\infty$. Hence, the theorem follows. This completes the proof of the theorem. ■

Theorem 5.5

The necessary and sufficient condition for the position sequence of a Type-1 particle $\{X_n\}$ to almost surely converge to the random variable p, that is, $X_n \xrightarrow{a.s} p$, is $\alpha < e^{\gamma}$.

Proof: Proof of the theorem is the same as in the proof for part (a) of Theorem 5.1. ■

It is also proved in [1] that the sufficient and necessary condition that the position sequence for Type-1 particles to converge to p in any of the four kinds of convergence is $\alpha < e^{\gamma}$. In other words, the four kinds of convergence are equivalent as far as Type-1 particles are concerned. This conclusion has no conflicts with Theorem 5.1. It is shown by (5.6) that almost sure convergence implies convergence in probability, while its converse is not true in general. However, under certain special circumstances, the converse of (5.6) may hold. Kolmogorov's strong law and Khintchine's weak law of large numbers reveal that there is such a case that convergence in probability may imply almost sure convergence, leading to the equivalence between the two kinds of convergence of Type-1 particles.

Due to boundedness of p and its uniform distribution on (P, G) or (G, P), the converse of (5.7) holds for Type-1 particles, as shown in the proof of Theorem 5.4. Because of the equivalence between convergences in probability and in distribution, point p can be treated as a constant as in the stochastic simulations in the next section. Furthermore, as proved in [1], since $\{X_n\}$ is uniformly integrable, the sufficient and necessary condition of its convergence in the mean of order r is the same as that of its convergence in probability.

5.1.4 Boundedness of Type-2 Particle

This section concerns the behaviour of Type-2 particles. Rewrite Equation 5.13 as

$$X_{n+1} - C = p - C + \alpha |X_n - C| \ln\left(\frac{1}{u_{n+1}}\right) = p - C + \lambda_{n+1} |X_n - C|, \quad u_n \sim U(0,1)$$

(5.29)

where $\lambda_{n+1} = \alpha \ln(1/u_{n+1})$. From the earlier equation, one can obtain these two inequalities:

$$|X_n - C| \le |C - p| + \alpha |X_{n-1} - C| \ln\left(\frac{1}{u_n}\right) = |C - p| + \lambda_n |X_{n-1} - C|, \quad (5.30)$$

and

$$|X_n - C| \ge -|C - p| + \alpha |X_{n-1} - C| \ln\left(\frac{1}{u_n}\right) = -|C - p| + \lambda_n |X_{n-1} - C|.$$

(5.31)

The inequalities (Equations 5.30 and 5.31) are used in the proof of Theorem 5.6 which shows the derivation of the necessary and sufficient condition for the probabilistic boundedness of the position of a Type-2 particle and the probabilistic boundedness of $\beta_n = \prod_{i=1}^{n} \lambda_i$.

Theorem 5.6

The necessary and sufficient condition for the sequence of positions for a Type-2 particle, $\{X_n\}$, to be probabilistic bounded, that is, $P\{\sup X_n < \infty\} = 1$, is $\alpha \le e^\gamma$.

Proof: It is obvious that $P\{\lambda_n = 1\} = 0$ because λ_n is a continuous random variable and that $P\{\sup |C - p| < \infty\} = 1$ implies that $P\{\sup[|C - p|/(1 - \lambda_n)] < \infty\} = 1$.

Let $r = \sup[|C - p|/(1 - \lambda_n)]$, where $0 < r < \infty$. For every $n > 0$, the following inequality is true:

$$|C - p| \le r(1 - \lambda_n) \tag{5.32}$$

Proof of sufficiency: Substituting the inequality (5.32) into Equation 5.30, one obtains the inequality,

$$|X_n - C| - r \le \lambda_n(|X_{n-1} - C| - r) \quad (n > 0),$$

from which one deduces

$$|X_n - C| - r \le \lambda_n(|X_{n-1} - C| - r) \le \lambda_n \lambda_{n-1}(|X_{n-2} - C| - r)$$

$$\le \lambda_n \lambda_{n-1} \lambda_{n-2}(|X_{n-3} - C| - r) \le \cdots \le (|X_0 - C| - r)\prod_{i=1}^{n} \lambda_i.$$

Thus, the following inequality holds:

$$|X_n - C| \le r + (|X_0 - C| - r)\prod_{i=1}^{n} \lambda_i. \tag{5.33}$$

Since $\beta_n = \prod_{i=1}^{n} \lambda_i > 0$, one obtains

$$\sup|X_n - C| \le \sup\left[r + (|X_0 - C| - r)\prod_{i=1}^{n} \lambda_i\right]$$

$$\le r + \sup(|X_0 - C| - r)\sup\left(\prod_{i=1}^{n} \lambda_i\right) \le r + \sup(|X_0 - C| - r)\sup(\beta_n).$$

Using Theorem 5.4 and taking $\alpha \le e^\gamma$, the following expression is deduced:

$$P\{\sup\beta_n < \infty\} = P\left\{\sup\left(\prod_{i=1}^{n} \lambda_i\right) < \infty\right\} = 1.$$

Note that $P\{\sup(|X_0 - C| - r) < \infty\} = 1$ when $r < \infty$. Therefore,

$$P\left\{\sup|X_n - C| < \infty\right\} \geq P\left\{r + \sup(|X_0 - C| - r)\sup\left(\prod_{i=1}^{n}\lambda_i\right) < \infty\right\}$$

$$= P\left\{\sup(|X_0 - C| - r)\sup\left(\prod_{i=1}^{n}\lambda_i\right) < \infty\right\}$$

$$\geq P\{[\sup(|X_0 - C| - r) < \infty] \cap (\sup\beta_n < \infty)\}$$

$$= P\{\sup(|X_0 - C| - r) < \infty\} + P\{\sup\beta_n < \infty\}$$

$$- P\{[\sup(|X_0 - C| - r) < \infty] \cup (\sup\beta_n < \infty)\}$$

$$= 1 + 1 - P\{[\sup(|X_0 - C| - r) < \infty] \cup (\sup\beta_n < \infty)\} \geq 1,$$

from which one can deduce that $P\{\sup|X_n - C| < \infty\} = 1$, that is, $|X_n - C|$ is probabilistic bounded. This implies that X_n is also probabilistic bounded, that is, $P\{\sup X_n < \infty\} = 1$.

Proof of necessity: Substituting the inequality in (5.32) into Equation 5.31 leads to

$$|X_n - C| + r \geq \lambda_n(|X_{n-1} - C| + r) \quad (n > 0), \tag{5.34}$$

from which the following inequality is deduced.

$$|X_n - C| + r \geq \lambda_n(|X_{n-1} - C| + r) \geq \lambda_n\lambda_{n-1}(|X_{n-2} - C| + r)$$

$$\geq \lambda_n\lambda_{n-1}\lambda_{n-2}(|X_{n-3} - C| + r) \geq \cdots \geq (|X_0 - C| + r)\prod_{i=1}^{n}\lambda_i \tag{5.35}$$

Hence,

$$\sup(|X_n - C| + r) \geq \sup\left((|X_0 - C| + r)\prod_{i=1}^{n}\lambda_i\right)$$

$$= \sup[(|X_0 - C| + r)\beta_n] = \sup(|X_0 - C| + r)\sup\beta_n,$$

and it follows that

$$P\{\sup\left(\mid X_n - C\mid + r\right) < \infty\} \leq P\{\sup(\mid X_0 - C\mid + r)\sup\beta_n < \infty\}$$

$$= P\{[\sup(\mid X_0 - C\mid + r) < \infty] \cap [\sup\beta_n < \infty]\}. \tag{5.36}$$

If X_n is probabilistic bounded, $\mid X_n - C\mid$ is also probabilistic bounded, that is, $P\{\sup\mid X_n - C\mid < \infty\} = 1$. As a result, $\mid X_n - C\mid + r$ is probabilistic bounded if $r < \infty$, that is,

$$P\left\{\sup(\mid X_n - C\mid + r) < \infty\right\} = 1.$$

From the inequality in (5.36), it can easily be seen that

$$P\left\{[\sup(\mid X_0 - C\mid + r) < \infty] \cap [\sup\beta_n < \infty]\right\} \geq 1,$$

which leads to

$$P\{[\sup(\mid X_0 - C\mid + r) < \infty] \cap [\sup\beta_n < \infty]\} = 1. \tag{5.37}$$

Hence,

$$P\{\sup(\mid X_0 - C\mid + r) < \infty\} = 1, \tag{5.38}$$

because of the probabilistic boundedness of $\mid X_0 - C\mid + r$. From Equations 5.37 and 5.38, it can be deduced that $P\{\sup\beta_n < \infty\} = 1$ which means $\alpha \leq e^\gamma$ as shown by Theorem 5.4. This implies that $\alpha \leq e^\gamma$ is the necessary condition for the probabilistic boundedness of X_n.

This completes the proof of the theorem. ■

Theorem 5.6 simply suggests that the sufficient and necessary conditions for the probabilistic boundedness of Type-1 particles and Type-2 particles are the same. In addition, the behaviours of both types of particles are related to the probabilistic boundedness of β_n. Furthermore, the behaviour of Type-2 particles is also influenced by point C. In practice, when the QPSO algorithm is being executed, the personal best positions of all the particles converge to the same point. This implies that $\mid C - p\mid$ converges almost surely to zero, and thus, $\sup[|C - p|/1 - \lambda_n] = r$ also converges almost surely to zero. As a result $P\{\lim_{n\to\infty} \beta_n = 0\} = 1$ if and only if $\alpha < e^\gamma$. According to the inequality in (5.33), either $P\{\lim_{n\to\infty} |X_n - C| = 0\} = 1$ or $P\{\lim_{n\to\infty} |X_n - p| = 0\} = 1$.

5.1.5 Simulation Results

It has been demonstrated that the behaviour of an individual particle in QPSO is affected by the choice of α. In order to validate the aforementioned main analysis results, two sets of stochastic simulations of the behaviour of the particle are carried out, one for Type-1 particles and the other for Type-2 particles.

Table 5.1 consists of a list of values for α selected from a series of numbers in $[0.5, 2.0]$ and their corresponding maximum number of iterations. These values were chosen in such a way that they are sufficiently large for the purpose of checking convergence, boundedness or divergence. In the stochastic simulations, the logarithmic value of the distance between the current position X_n and the point p was recorded as ordinate, and the number of iteration was recorded as abscissa. In the numerical simulations, p was fixed at the origin and the initial position of the particle was set as $X_0 = 1000$ for tests involving Type-1 particles. On the other hand, C was fixed at $X = 0.001$, p was fixed at the origin and the initial position of the particle was set also as $X_0 = 1000$ for tests involving Type-2 particles. Simulation results obtained by using MATLAB® 7.0 are depicted in Figures 5.1 and 5.2.

Figure 5.1 shows the simulation results for the position of Type-1 particles. It is observed that $|X_n - p|$ reaches the minimum or maximum positive value when $\ln|X_n - p| < -700$ or $\ln|X_n - p| > 700$, respectively. Thus, $|X_n - p|$ converges to zero (X_n converges to p) or $|X_n - p|$ diverges to infinity. It can be seen from Figure 5.1 that the position of the particle

TABLE 5.1 The Values of α and the Corresponding Maximum Number of Iterations Chosen for the Simulations

α	0.5	0.7	1.0	1.2	1.5
n_{max}	1,000	1,000	1,500	3,000	5,000
α	1.7	1.76	1.77	1.775	1.785
n_{max}	15,000	80,000	100,000	200,000	200,000
α	1.787	1.79	1.8	1.9	2.0
n_{max}	200,000	100,000	50,000	15,000	7,000

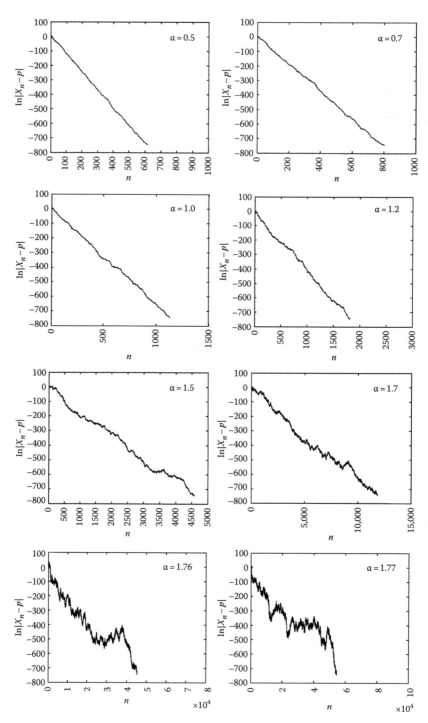

FIGURE 5.1   Simulation results for Type-1 particles at different values of α.

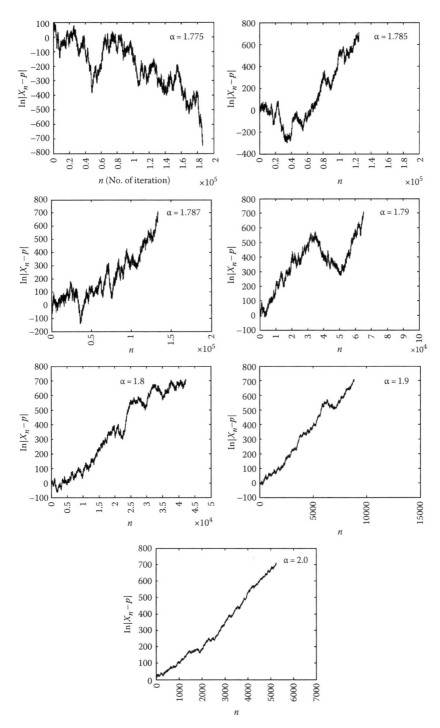

FIGURE 5.1 (continued)

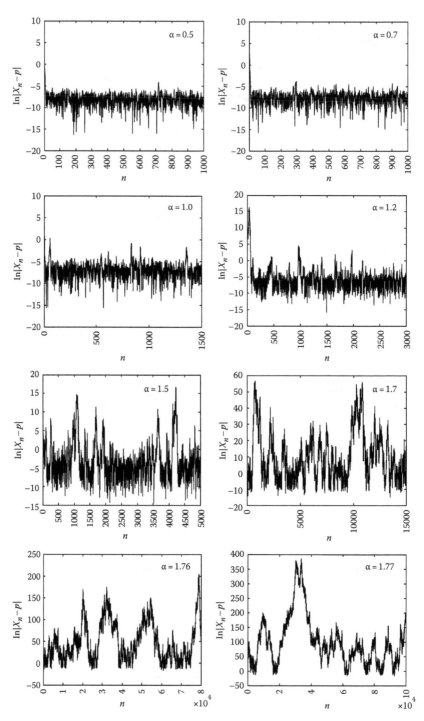

FIGURE 5.2   Simulation results for Type-2 particles at different values of α.

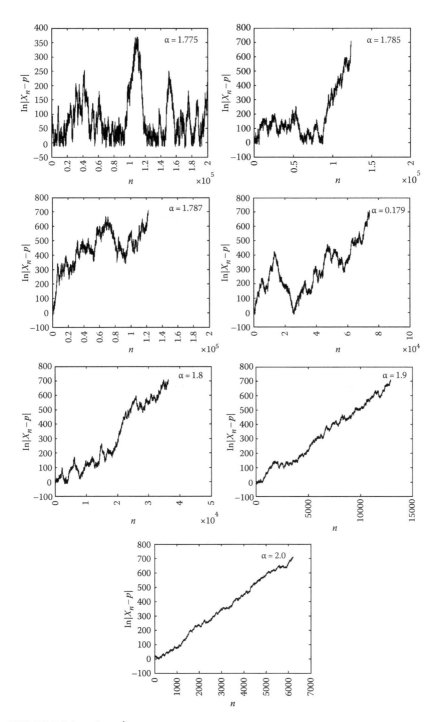

FIGURE 5.2 (continued)

X_n converges to p when $\alpha \leq 1.775$ and diverges when $\alpha \geq 1.785$. When X_n converges to p, the smaller the value of α, the faster X_n converges. On the other hand, when X_n diverges, the larger value of α results in a faster speed at which the position of the particle careens to infinity.

Figure 5.2 shows the simulation results for Type-2 particles. It can be seen that the position of the particle oscillates around p and C when $\alpha \leq 1.775$ and explodes when $\alpha \geq 1.785$. On the other hand, smaller values of α lead to smaller amplitude in the oscillation when X_n is probabilistic bounded. Finally, the larger the value of α, the faster the divergence speed.

From the previous simulation results, it can be concluded that there exists an α_0 within (1.775, 1.785) such that if $\alpha \leq \alpha_0$, the position of a Type-1 particle converges (when $\alpha < \alpha_0$) or is probabilistic bounded (when $\alpha = \alpha_0$), and that of a Type-2 particle is probabilistic bounded (when $\alpha \leq \alpha_0$). If $\alpha > \alpha_0$, the positions of both types of particles would diverge. The earlier numerical results are consistent with the theoretical results in the previous sections, that is, setting $\alpha \leq e^{\gamma} \approx 1.781$ prevents the explosion of the particle.

The fact that smaller values of α result in faster convergence for Type-1 particles and smaller amplitudes of oscillation for Type-2 particles is studied in a different way. Using Equation 5.12, the convergence rate for Type-1 particles at the nth iteration is given by

$$c_n = E\left[\frac{|X_n - p|}{|X_{n-1} - p|}\middle| X_{n-1}\right] = E[-\alpha \ln(u_n)] = \alpha E[-\ln(u_n)] = -\alpha \int_0^1 \ln(u)du = \alpha,$$

$$(5.39)$$

which leads to the result

$$E[|X_n - p| \,\|\, X_{n-1}] = c_n \,|X_{n-1} - p| = \alpha \,|X_{n-1} - p|. \qquad (5.40)$$

Note that using smaller values of c_n or α increases the speed at which $|X_n - p|$ decreases. Similarly using Equation 5.13 leads to the following result for Type-2 particles:

$$E[|X_n - p| \,\|\, X_{n-1}] = \alpha \,|X_{n-1} - C|. \qquad (5.41)$$

This result implies that the smaller the value of α, the smaller the conditional expected value of $|X_n - p|$ and thus the smaller the amplitude of oscillation of X_n. It is obvious that the slower convergence of Type-1 particles or large amplitude of oscillation of Type-2 particles means stronger global search ability of the particle. On the contrary, faster convergence or smaller amplitude of oscillation implies stronger local search ability. Excessive global search may result in slow convergence of the QPSO algorithm. On the other hand, excessive local search may cause premature convergence of the algorithm. It is important to balance the global search (exploration) and the local search (exploitation) to ensure good performance of the algorithm.

5.2 CONVERGENCE ANALYSIS OF THE ALGORITHM

5.2.1 Analysis Using Probability Measures

Solis and Wets have studied the convergence of stochastic search algorithms, in particular pure random search algorithms, and provided criteria and conditions governing whether it is a global search algorithm or a local search algorithm [5]. These definitions are borrowed and used in the following study of the convergence characteristics of QPSO [6]. For convenience, the relevant definitions from [5] are reproduced as follows.

5.2.1.1 Preliminaries

Proposition 5.1

Given a function f from S, a subset of R^N, to R. A point z in S which minimises f on S or at least yields an acceptable approximation of the infimum of f on S, as described by the problem in Equation 1.1, is required.

This proposition provides the definition of the output of a global optimiser with a given function and a search space. The simplest stochastic algorithm designed to perform this task is the basic random search (BRS) algorithm. During the nth iteration of this algorithm, a probability space $(R^N, \mathcal{B}, \mu_n)$, where μ_n is a probability measure (corresponding to a distribution function on R^N) on $\mathcal{B}$ and $\mathcal{B}$ is the σ-algebra of subsets of R^N, is required. The support of the probability measure μ_n is denoted as M_n which is the smallest closed subset of R^N with measure 1 under μ_n. To start the algorithm, it also needs a random initial starting point in S known as z_0. The BRS algorithm is outlined in Figure 5.3.

Step 0: Find z_0 in S and set $n = 0$.

Step 1: Generate a vector ξ_n from the sample space $(R^N,\ \mathcal{B},\ \mu_n)$.

Step 2: Set $z_{n+1} = D(z_n,\ \xi_n)$, choose μ_{n+1}, set $n: = n + 1$ and return to Step 1.

FIGURE 5.3 The procedure of BRS algorithm.

Note that the BRS algorithm is a localised random search as described in Chapter 1. Here $D(.,.)$ is a function taking in two parameters and is used to construct a solution to the problem. The solution obtained by using D guarantees that the newly constructed solution will be no worse than the current solution. Therefore, it needs to satisfy the following condition.

Condition 5.1 (C5.1): $f(D(z, \xi)) \leq f(z)$ and if $\xi \in S$, then $f(D(z, \xi)) \leq f(\xi)$.

Different functions of D lead to different algorithms on the assumption that C5.1 must be satisfied for an optimisation algorithm to work correctly.

Global convergence of any algorithm means that the sequence $\{f(z_n)\}_{n=1}^{\infty}$ converges to the infimum of f on S. A pathological case would be that the function f has a minimum consisting of a single discontinuous point on an otherwise smooth function. To compensate such cases, instead of searching for the infimum, one would search for the *essential infimum* ϕ such that $\phi = \inf(t : v[z \in S|f(z) < t] > 0)$, where $v[A]$ is the Lebesgue measure on the set A. The definition of ϕ means that there must be more than one point in a subset of the search space which yields function values arbitrarily close to ϕ such that ϕ is the infimum of the function values from this nonzero v-measurable set. Typically, $v[A]$ is the N-dimensional volume of the set A. This definition of ϕ avoids the problem of the pathological case mentioned earlier by defining a new infimum so that there is always a non-empty volume surrounding it containing points of the search space. This way it is possible to approach the infimum without having to sample every point in S.

With the earlier discussion in mind an optimal region can be defined as

$$R_\varepsilon = \{z \in S \mid f(z) < \varphi + \varepsilon\}, \tag{5.42}$$

where $\varepsilon > 0$. If the algorithm finds a point in the optimal region, then it has found an acceptable approximation to the global minimum of the function.

It is now possible to consider whether an algorithm can in fact reach the optimal region of the search space. A true global search algorithm satisfies the following assumption:

Condition 5.2 (C5.2): For any (Borel) subset A of S with $v[A] > 0$, the following product holds:

$$\prod_{n=0}^{\infty}(1-\mu_n[A]) = 0, \qquad (5.43)$$

where $\mu_n[A]$ is the probability of A generated by μ_n.

The earlier condition simply says that the probability of repeatedly missing the set A using random samples (e.g. ζ_n in the earlier BRS algorithm) must be zero for any subset A of S with positive measure v. Since $R_\varepsilon \subset S$, this implies that the probability of sampling a point in the optimal region must be nonzero. The following theorem is due to Solis and Wets [5].

Theorem 5.7

Suppose f is a measurable function, S is a measurable subset of R^N and C5.1 and C5.2 are satisfied. Let $\{z_n\}_{n=0}^{\infty}$ be a sequence generated by the BRS algorithm. Then

$$\lim_{n \to +\infty} P[z_n \in R_\varepsilon] = 1, \qquad (5.44)$$

where $P[z_n \in R_\varepsilon]$ is the probability at step n such that the point z_n generated by the algorithm is in R_ε.

By the above global search theorem, one can conclude that an algorithm that satisfies C5.1 and C5.2 is a global optimisation algorithm.

5.2.1.2 Global Convergence of QPSO

The proof presented here suggests that the QPSO algorithm can be treated under the framework of a global stochastic search algorithm allowing the use of Theorem 5.7 to prove convergence. Therefore, it remains to show that QPSO satisfies both C5.1 and C5.2.

Lemma 5.3

The QPSO algorithm satisfies the condition C5.1.

Proof: According to the procedure of QPSO, function D (as introduced in C5.1) is defined for QPSO as

$$D(G_n, X_{i,n}) = \begin{cases} G_n & \text{if } f(g(X_{i,n})) \geq f(G_n), \\ g(X_{i,n}) & \text{if } f(g(X_{i,n})) < f(G_n), \end{cases} \tag{5.45}$$

where f is the objective function of the minimisation problem as stated in Equation 1.1. The notation $g(X_{i,n})$ denotes the application of g, a vector function performing QPSO updates, to $X_{i,n}$, where the subscript n denotes the time step. The sequence $\{G_l\}_{l=0}^{n}$ is a sequence of the global best positions up to and including step n. It is possible to say that

$$X_{i,n+1} = g(X_{i,n}). \tag{5.46}$$

Let $g(X_{i,n})_j$ denote the jth dimension of the function g. According to Equation 4.79 or Equation 4.82, the jth dimension of function g can be represented as

$$g(X_{i,n})_j = p_{i,n}^j \pm \alpha \, | X_{i,n}^j - p_{i,n}^j | \ln\left(\frac{1}{u_{i,n+1}^j}\right), \tag{5.47}$$

or

$$g(X_{i,n})_j = p_{i,n}^j \pm \alpha \, | X_{i,n}^j - C_n^j | \ln\left(\frac{1}{u_{i,n+1}^j}\right). \tag{5.48}$$

The definition of D for QPSO given earlier clearly complies with C5.1, since the sequence $\{f_n = f(G_n), n \geq 0\}$ is monotonic by definition.
   This completes the proof of the lemma. ■

Lemma 5.4

QPSO satisfies C5.2.

Proof: As formulated earlier, at any iterative step n, with $p_{i,n}^j$ being considered as a random variable, the conditional probability density function of the jth dimension of the current position of particle i at step $n + 1$ is given by

$$Q(X_{i,n+1}^j \mid p_{i,n}^j) = \left(\frac{1}{L_{i,n}^j}\right) \exp\left(\frac{-2\mid X_{i,n+1}^j - p_{i,n}^j \mid}{L_{i,n}^j}\right), \qquad (5.49)$$

where $L_{i,n}^j = 2\alpha \mid X_{i,n}^j - p_{i,n}^j \mid$ or $L_{i,n}^j = 2\alpha \mid X_{i,n}^j - C_n^j \mid$. Thus, the probability density function is given by

$$Q(X_{i,n+1}^j) = \int_{-\infty}^{+\infty} Q(X_{i,n+1}^j \mid p_{i,n}^j) Q(p_{i,n}^j) dp_{i,n}^j, \qquad (5.50)$$

where $Q(p_{i,n}^j)$ is the marginal probability density function of $p_{i,n}^j$ and satisfies

$$Q(p_{i,n}^j) \sim U(a_{i,n}^j, b_{i,n}^j), \qquad (5.51)$$

with $a_{i,n}^j = \min(P_{i,n}^j, G_n^j)$ and $b_{i,n}^j = \max(P_{i,n}^j, G_n^j)$. Here, $U(a_{i,n}^j, b_{i,n}^j)$ means the probability uniform distribution on the interval $(a_{i,n}^j, b_{i,n}^j)$. Therefore, the support of $Q(X_{i,n+1}^j \mid p_{i,n}^j)$, $M(X_{i,n+1}^j \mid p_{i,n}^j) = R$ and the support of $Q(X_{i,n+1}^j)$ can be obtained as

$$M(X_{i,n+1}^j) = M(X_{i,n+1}^j \mid p_{i,n}^j) \cup M(p_{i,n}^j) = R \cup [a_{i,n}^j, b_{i,n}^j] = R, \qquad (5.52)$$

where $M(p_{i,n}^j) = [a_{i,n}^j, b_{i,n}^j]$ according to (5.51).

Since the probability density function of particle i can be obtained as

$$Q(X_{i,n+1}) = \prod_{j=1}^{N} Q(X_{i,n+1}^j), \qquad (5.53)$$

the support of $Q(X_{i,n+1})$ is given by

$$M(X_{i,n+1}) = \prod_{j=1}^{N} M(X_{i,n+1}^j) = \underbrace{R \times R \cdots R}_{N} = R^N \supset S. \tag{5.54}$$

Denoting $\mu_{i,n+1}$, the probability measure corresponds to the previous distribution $Q(X_{i,n+1})$. For any Borel subset A of S with $v[A] > 0$, one obtains the expression

$$\mu_{i,n+1}[A] = \int_A Q(X_{i,n+1}) dX_{i,n+1} = \int_A \prod_{j=1}^{N} Q(X_{i,n+1}^j) dX_{i,n+1}, \tag{5.55}$$

because $A \subset M(X_{i,n+1}^j)$ and $Q(X_{i,n+1})$ is Lebesgue integrable for all $n \geq 0$. Therefore, from Equation 5.55 one can conclude that

$$0 < \mu_{i,n+1}[A] < 1, \tag{5.56}$$

for all $n \geq 0$. The union of the supports of all the particles is given by

$$M_{n+1} = \bigcup_{i=1}^{m} M(X_{i,n+1}) = R^N \supset S, \tag{5.57}$$

where M_{n+1} is the support of distribution μ_{n+1}. The probability of A generated by μ_{n+1} can be calculated as

$$\mu_{n+1}[A] = 1 - \prod_{i=1}^{m} (1 - \mu_{i,n+1}[A]). \tag{5.58}$$

From inequality (5.56) and Equation 5.58 it can be seen $0 < \mu_{n+1}[A] < 1$, for $n \geq 0$. At step 0, the particles are initialised within a bounded area M_0, where $M_0 \subset R^N$. If $A \cap M_0 \neq \Phi$ then $0 < \mu_0[A] < 1$, otherwise $\mu_0[A] = 0$. Finally one obtains the following result,

$$\prod_{n=0}^{\infty} (1 - \mu_n[A]) = (1 - \mu_0[A]) \prod_{n=0}^{\infty} (1 - \mu_{n+1}[A]) = (1 - \mu_0[A]) \times 0 = 0, \tag{5.59}$$

which implies that QPSO satisfies C5.2. ■

Theorem 5.8

The QPSO algorithm is a global search algorithm.

Proof: Since QPSO satisfies C5.1 and C5.2, by Theorem 5.7 one can easily see that QPSO is a global search algorithm. ▪

5.2.2 Analysis Using Markov Processes

5.2.2.1 Preliminaries

Definition 5.10

The nonnegative function $p(\cdot,\cdot)$ on $\Omega \times \mathcal{F}$ is called a Markov transfer kernel function in probability measure space $(\Omega, \mathcal{F}, \pi)$, if it satisfies

(1) For all $x \in \Omega$, $p(x, \cdot)$ is a finite σ-measure on $\mathcal{F}$
(2) For all $A \in \mathcal{F}$, $p(\cdot, A)$ is a measurable function about π
(3) For all $x \in \Omega$, $p(x, \Omega) = 1$

Definition 5.11

The Markov transfer kernel function $p(\cdot,\cdot)$ is called a stationary distribution, if there exists a probability measure π such that for all $A \in \mathcal{F}$,

$$\pi(A) = \int_{\Omega} p(x, A)\pi(dx). \tag{5.60}$$

Definition 5.12

Let $p^{(1)}(x, A) = p(x, A)$ and recursively define the Markov transition kernel, that is, transition probability at the nth iteration as

$$p^{(n)}(x, A) = \int_{\Omega} p^{(n-1)}(y, A)p(x, dy). \tag{5.61}$$

The Markov transition kernel $p(\cdot,\cdot)$ is said to be irreducible, if, for all $A \in \Omega$,

$$\sum_{n=1}^{\infty} p^{(n)}(x, A) > 0, \tag{5.62}$$

whenever $\pi(A) > 0$.

Definition 5.13

The Markov transition kernel $p(\cdot,\cdot)$ has a cycle with length k, if there exists a family of disjoint sets $(A_1,\ldots, A_k)$ $(A_i \in \Omega, i = 1, 2,\ldots, k)$ such that

$$\begin{cases} p(x, A_{i+1}) = 1, & x \in A_i, \quad i = 1,\ldots,k-1, \\ p(x, A_1) = 1, & x \in A_k. \end{cases} \tag{5.63}$$

If there is no cycle with length $k > 1$, $p(\cdot,\cdot)$ is said to be non-cyclic.

Definition 5.14

If $\{X_n\}$ is a sequence of random variable with a Markov transition kernel function $p(\cdot,\cdot)$, it is called a discrete Markov process. The discrete Markov process is called a Harris recurrent Markov process whenever $\pi(A) > 0$, if the conditional probability

$$P\left\{\sum_{n=1}^{\infty} 1_A(X_n) \middle| X_0 = x\right\} = 1, \quad \forall x \in \Omega, \tag{5.64}$$

where $1_A(X_n)$ is the indicator function.

5.2.2.2 Markov Process Model for a Stochastic Optimisation Algorithm
Consider the following minimisation problem as stated in Equation 1.1

$$\min_{x \in S} f(x)$$

where f is a real-valued function defined on the region S which is a subset of the N-dimensional Euclidean space R^N. Let V represent the range of f on S, and thus $V \subset R$. Denote $f_* = \min_{x \in S} f(x)$ and $f^* = \max_{x \in S} f(x)$.

During the process of solving the problem as stated in (1.1), a random global search algorithm produces a sequence of random points on S, $\{X_n, n = 1, 2,\ldots\}$, according to a certain distribution. This in turn generates the corresponding sequence of function values $\{Y_n = f(X_n), n = 1, 2, \ldots\}$.

The following mathematical preliminaries [7] are required.

1. Define two probabilistic spaces, $\{S, \mathcal{F}, \mu\}$ for $\{X_n, n = 1, 2,...\}$ and $\{V, \mathcal{B}, P\}$ for $\{Y_n, n = 1, 2, ...\}$.

2. Assuming the probabilistic space of $\{Y_n\}$, $\{V, \mathcal{B}, P\}$, forms an irreducible, non-cyclic discrete Markov process.

3. On $\{V, \mathcal{B}, P\}$, calculate the initial distribution $P\{Y_0 \in B\}$ and the transition probability:

$$p_n(y, B) = P\{Y_{n+1} \in B \mid Y_n = y\}, \tag{5.65}$$

where $y \in V$ and $B \in \mathcal{B}$. If the density function exists, it is called a transition kernel function. $\{Y_n\}$ is called a homogeneous discrete Markov process if $p_n(y, B)$ is independent of n, otherwise $\{Y_n\}$ is inhomogeneous. In the case of homogeneous $p_n(y, B)$ is denoted as $p(y, B)$.

4. Determine the stopping criterion of the random global search algorithm mentioned earlier. Let R_ε ($R_\varepsilon \in \mathcal{B}$) be the stopping domain (i.e. the optimal region) of the algorithm and assume $P(R_\varepsilon) > 0$. If, at a certain step n, $Y_n = f(X_n) \in R_\varepsilon$, the algorithm terminates. Denote $R_\varepsilon^c = V - R_\varepsilon$.

5. Let R_ε be an absorbing set of the Markov process. From the process $\{Y_n\}$ defined in the space $\{V, B, P\}$, a new discrete Markov process $\{\tilde{Y}_n\}$ called the absorbing Markov process is constructed. The one-step transition probability is given by

$$\tilde{p}(y, B) = \begin{cases} 0, & y \in R_\varepsilon, \\ p(y, B), & y \in R_\varepsilon^c, \end{cases} \tag{5.66}$$

and the n-step transition probability can be defined as

$$\tilde{p}^{(0)}(y, B) = \begin{cases} 1, & y \in B, \\ 0, & y \notin B, \end{cases} \tag{5.67}$$

$$\tilde{p}^{(1)}(y, B) = \tilde{p}(y, B), \tag{5.68}$$

$$\tilde{p}^{(n)}(y, B) = \int_S \tilde{p}^{(n-1)}(y, B)\tilde{p}(y, dz), \quad n = 1, 2,... \tag{5.69}$$

In terms of the theoretical framework established earlier, a random global optimisation algorithm can be considered as a discrete Markov process $\{Y_n, n = 1, 2,...\}$, or a discrete absorbing Markov process $\{\tilde{Y}_n, n = 1, 2,...\}$. Hence, the algorithm can be evaluated using the framework.

5.2.2.3 Construction of a Markov Process Model for QPSO

In this discussion, the subscript n is used to denote the iteration number of the QPSO algorithm. The current position and the personal best position of a particle are updated after initialisation of the global best position G_0 and its objective function value $f_0 = f(G_0)$. The QPSO algorithm produces a sequence of global best positions $\{G_n\}$ after n iterations. By computing $f_n = f(G_n)$ a corresponding sequence of random variables $\{f_n\}$ are generated. The discrete Markov process is given by the following procedure.

1. Construct the probability measure space $\{V, \mathcal{B}, P\}$ for the sequence of random variables $\{f_n\}$, where $V = f(S)$, and B is Broel σ-field generated by the interval $[f_*, f]$. According to the operations of the QPSO algorithm, the probability measure for every random variable f_n is given by

$$g_n(f) = P\{f_n \leq f\} = 1 - \prod_{i=1}^{M}\left|1 - \int_{S_f} \theta_{i,n}(G_n, P_{i,n}, x)dx\right|, \qquad (5.70)$$

where $S_f = \{x|f(x) < f\}$. Using the position updating equations of the algorithm, one obtains

$$\theta_{i,n}(G_{n-1}, P_{i,n-1}, X_{i,n}) = \prod_{j=1}^{N}\left|\int_{P_{i,n-1}^{j}}^{G_{n-1}^{j}} \frac{1}{L_{i,n-1}^{j}}\exp\left(\frac{-2|X_{i,n}^{j} - p|}{L_{i,n-1}^{j}}\right)dp\right|. \qquad (5.71)$$

2. Consider the sequence of random variables $\{f_n\}$ that form a homogeneous discrete Markov process. For $z \in V$, $B = [f_*, f]$, the transfer probability is given by

$$p_n(z, B) = 1, \qquad (5.72)$$

whenever $z < f$, and is given by

$$p_n(z, B) = P\{f_{n+1} \in B \mid f_n = z\} = P\{f_{n+1} \leq f \mid f_n = z\} = g_n(f), \quad (5.73)$$

wherever $z \geq f$.

3. Define the stopping domain of the QPSO algorithm as $R_\varepsilon = \{f \mid f \leq f_* + \varepsilon\} = [f_*, f_* + \varepsilon]$, where $\varepsilon > 0$ is sufficiently small such that $f_* + \varepsilon < f^*$. Taking R_ε as the absorbing region, one obtains the discrete absorbing Markov process $\{\tilde{f}_n\}$.

Therefore, the QPSO algorithm can be studied by using the aforementioned Markov process model.

5.2.2.4 Global Convergence of a Stochastic Optimisation Algorithm

For a random optimisation algorithm, the convergence is one of the most important issues in theoretical analysis. By ensuring that the algorithm can be treated as a discrete absorbed Markov process $\{\tilde{Y}_n\}$, the convergence is relevant either to its expectation at any rate or to the distribution of time steps which the algorithm spends on reaching the absorbing region R_ε. A couple of definitions are required before leading to the main result.

Definition 5.15

The discrete random variable

$$K = \inf\left\{n \mid \tilde{Y}_n \in R_\varepsilon\right\}, \quad (5.74)$$

in relation to the stochastic algorithm $\{\tilde{Y}_n, n = 1, 2, \ldots\}$, is called the time of $\{Y_n, n = 1, 2, \ldots\}$ about ε.

In terms of absorbing Markov process, the discrete random variable K is essentially the time when the discrete Markov process $\{\tilde{Y}_n\}$ has first come across R_ε, or the time when $\{\tilde{Y}_n\}$ is absorbed into R_ε.

Definition 5.16

If the series

$$\tilde{Q}(y, B) = \sum_{n=0}^{\infty} \tilde{p}^{(n)}(y, B), \quad (5.75)$$

in relation to the stochastic algorithm $\{\tilde{Y}_n, n = 1, 2,...\}$, is uniformly convergent with regard to $y \in V$, for every $B \in \mathcal{B}$, then the function $\tilde{Q}(y, B)$ forms a potential function from y to B.

Theorem 5.9

For a stochastic algorithm $\{\tilde{Y}_n, n = 0, 1, 2,...\}$, the following conclusions are satisfied. Let $a_n = P\{K = n\}$ represent the probability when the algorithm stops at the nth step, one obtains

(i)

$$a_n = \int_V \tilde{p}^{(n)}(y, R_\varepsilon)P(dy), \quad n = 0, 1, 2,... \quad (5.76)$$

(ii) If the potential function $\tilde{Q}(y, R_\varepsilon) = 1$ holds uniformly in $y \in R_\varepsilon^c$, then the algorithm converges to the global optimum in probability.

Proof:

(i) According to Equations 5.66 through 5.69, one obtains

$$a_0 = P\{K = 0\} = P\{\tilde{Y}_0 \in R_\varepsilon\}$$

$$= \int_V \tilde{p}^{(0)}(y, R_\varepsilon)P(dy),$$

$$a_1 = P\{K = 1\} = P\{\tilde{Y}_0 \in R_\varepsilon^c, \tilde{Y}_1 \in R_\varepsilon\} = P\{\tilde{Y}_1 \in R_\varepsilon \mid \tilde{Y}_0 \in R_\varepsilon^c\}P\{\tilde{Y}_0 \in R_\varepsilon^c\}$$

$$= \int_{R_\varepsilon^c} \tilde{p}(y_0, R_\varepsilon)P(dy_0) = \int_V \tilde{p}^{(1)}(y, R_\varepsilon)P(dy),$$

and

$$a_2 = P\{K = 2\} = P\{\tilde{Y}_0 \in R_\varepsilon^c, \tilde{Y}_1 \in R_\varepsilon^c, \tilde{Y}_2 \in R_\varepsilon\}$$

$$= P\{\tilde{Y}_2 \in R_\varepsilon \mid \tilde{Y}_0 \in R_\varepsilon^c, \tilde{Y}_1 \in R_\varepsilon^c\} \cdot P\{\tilde{Y}_1 \in R_\varepsilon^c \mid \tilde{Y}_0 \in R_\varepsilon^c\}P\{\tilde{Y}_0 \in R_\varepsilon^c\}$$

$$= \int_{R_\varepsilon^c} \int_{R_\varepsilon^c} \tilde{p}(y_1, R_\varepsilon)\tilde{p}(y_0, dy_1)P(dy_0) = \int_V \tilde{p}^{(2)}(y, R_\varepsilon)P(dy).$$

Similarly,

$$a_n = \int_V \tilde{p}^{(n)}(y, R_\varepsilon) P(dy), \quad n = 3, 4, \dots \tag{5.77}$$

(ii) From the earlier results for $a_n (n \geq 0)$, one can obtain

$$\lim_{n \to \infty} P\{\tilde{Y}_n \in R_\varepsilon\} = P\left\{ \bigcup_n^\infty (\tilde{Y}_n \in R_\varepsilon) \right\}$$

$$= \sum_{n=0}^\infty P\{\tilde{Y}_0 \in R_\varepsilon^c, \tilde{Y}_1 \in R_\varepsilon^c, \dots, \tilde{Y}_{n-1}, \tilde{Y}_n \in R_\varepsilon\} = \sum_{n=0}^\infty a_n$$

$$= \sum_{n=1}^\infty \int_V \tilde{p}^{(n)}(y, R_\varepsilon) P(dy) = \int_V \tilde{Q}(y, R_\varepsilon) P(dy) = 1,$$

$$\tag{5.78}$$

which implies that the algorithm converges to the global optima in probability. ∎

5.2.2.5 Global Convergence of QPSO

The following equality is needed in order to show the global convergence of the QPSO algorithm.

Theorem 5.10

Let $\{b_n\}$ be a sequence of real numbers such that

$$g_n = \begin{cases} b_1, & n = 1, \\ b_n(1 - b_{n-1})(1 - b_{n-2}) \cdots (1 - b_1) = b_n \prod_{i=1}^{n-1}(1 - b_i), & n > 1. \end{cases} \tag{5.79}$$

The equality

$$G_n = \sum_{i=1}^n g_i = 1 - \prod_{i=1}^n (1 - b_i), \tag{5.80}$$

holds for all $n \geq 1$.

Proof: Mathematical induction is used to prove the theorem.

If $n = 1$, $G_1 = b_1 = 1 - (1 - b_1)$. It is clear that the equality holds.

If $n = 2$,

$$G_2 = g_1 + g_2 = b_1 + b_2(1-b_1) = b_1 + b_2 - b_1 b_2$$
$$= 1 - 1 + b_1 + b_2 - b_1 b_2 = 1 - (1-b_1)(1-b_2),$$

which implies the equality holds.

Suppose the equality

$$G_k = 1 - \prod_{i=1}^{k}(1-b_i),$$

holds when $n = k$.

When $n = k + 1$, one obtains

$$G_{k+1} = G_k + g_{k+1} = 1 - \prod_{i=1}^{k}(1-b_i) + b_{k+1}\prod_{i=1}^{k}(1-b_k)$$

$$= 1 - (1-b_{k+1})\prod_{i=1}^{k}(1-b_k) = 1 - \prod_{i=1}^{k+1}(1-b_k).$$

Consequently the equality holds for all $n \geq 1$. ■

Theorem 5.11

QPSO converges to the global optima in probability.

Proof: It is only required to prove that $\tilde{Q}(f, R_\varepsilon) = 1$, where $\tilde{Q}(f, R_\varepsilon)$ is the potential function of the discrete absorbing Markov process $\{\tilde{f}_n\}$.

For the case $f \in R_\varepsilon$:

Note that

$$\tilde{p}^{(0)}(f, R_\varepsilon) = \begin{cases} 1 & f \in R_\varepsilon, \\ 0 & f \in R_\varepsilon^c, \end{cases}$$

and $\tilde{p}^{(1)}(f, R_\varepsilon) = 0$. The two expressions imply $\tilde{Q}(f, R_\varepsilon) = 1$.

For the case $f \in R_\varepsilon^c$:

Note that $\tilde{p}^{(0)}(f, R_\varepsilon) = 0$ and $\tilde{p}^{(1)}(f, R_\varepsilon) = p(f, R_\varepsilon) = g_1(f_* + \varepsilon)$. The transition probability of the nth step can be calculated iteratively by

$$\tilde{p}^{(n)}(f, R_\varepsilon) = g_n(f_* + \varepsilon)[1 - g_{n-1}(f_* + \varepsilon)] \cdots [1 - g_1(f_* + \varepsilon)]$$

$$= g_n(f_* + \varepsilon) \prod_{i=1}^{n-1} [1 - g_i(f_* + \varepsilon)].$$

Using Equation 5.80, one obtains

$$\tilde{Q}(f, R_\varepsilon) = \sum_{n=1}^{\infty} \tilde{p}^{(n)}(f, R_\varepsilon) = \lim_{n \to \infty} g_n(f_* + \varepsilon) \prod_{i=1}^{n-1} [1 - g_i(f_* + \varepsilon)]$$

$$= \lim_{n \to \infty} \left\{ 1 - \prod_{i=1}^{n} [1 - g_i(f_* + \varepsilon)] \right\}.$$

Note also that for every $n < \infty$, the characteristic length of δ potential is well used for QPSO

$$0 < L_{i,n}^j < \infty.$$

Consequently $\theta_{i,n}(G_n, P_{i,n}, X_{i,n})$ is Lebesgue integrable. For R_ε, the inequalities $v(R_\varepsilon) > 0$ and $0 < \int_{R_\varepsilon} \theta_{i,n}(G_n, P_{i,n}, x) dx < 1$ hold, where $v(R_\varepsilon)$ is the Lebesgue measure of R_ε, and hence,

$$0 < g_i(f_* + \varepsilon) < 1,$$

from which the following equality can be shown.

$$\tilde{Q}(f, R_\varepsilon) = \lim_{n \to \infty} \left\{ 1 - \prod_{i=1}^{n} [1 - g_i(f_* + \varepsilon)] \right\} = 1. \tag{5.81}$$

The earlier equation essentially says that QPSO is a global convergent algorithm. This completes the proof of the theorem. ∎

5.2.3 Analysis in Probabilistic Metric Spaces

In this section, global convergence of QPSO is analysed by establishing a PM space for the algorithm, in which the algorithm is treated as a self-mapping. By proving that the algorithm is a contraction mapping and its orbit is probabilistic bounded, one can conclude that QPSO converges to the unique fixed point, that is, the global optimum.

5.2.3.1 Preliminaries

Definition 5.17

Denote the set of all real numbers as R, and the set of all non-negative real numbers as R^+. The mapping $f: R \to R^+$ is called a distribution function if it is non-decreasing, left continuous, and $\inf_{t \in R} f(t) = 0$ and $\sup_{t \in R} f(t) = 1$.

Denote the set of all distribution functions by $\mathcal{D}$. Let $H(t)$ be a specific distribution function defined by

$$H(t) = \begin{cases} 1 & t > 0, \\ 0 & t \le 0. \end{cases} \tag{5.82}$$

This distribution function plays an important role in the following study.

Definition 5.18

A PM space is an ordered pair (E, F), where E is a non-empty set and F is a mapping of $E \times E$ into $\mathcal{D}$. The value of the distribution function F at every $x, y \in E$ is denoted by $F_{x,y}$, and $F_{x,y}(t)$ represents the value of $F_{x,y}$ at $t \in R$. The functions $F_{x,y}(t)$, $x, y \in E$, are assumed to satisfy the following conditions:

(PM-1) $F_{x,y}(0) = 0$.
(PM-2) $F_{x,y}(t) = H(t)$ for all $t > 0$ if and only if $x = y$.
(PM-3) $F_{x,y} = F_{y,x}$.
(PM-4) $F_{x,y}(t_1) = 1$ and $F_{y,z}(t_2) = 1$ imply $F_{x,z}(t_1 + t_2) = 1$, $\forall x, y, z \in E$.

The value $F_{x,y}(t)$ of $F_{x,y}$ at $t \in R$ can be interpreted as the probability that the distance between x and y is less than t.

Definition 5.19

A mapping $\Delta: [0,1] \times [0,1] \rightarrow [0,1]$ a triangle norm (briefly t-norm) if it satisfies for every $a, b, c, d \in [0,1]$,

$$(\Delta\text{-}1)\ \Delta(a,1) = a, \Delta(0,0) = 0$$

$$(\Delta\text{-}2)\ \Delta(a,b) = \Delta(b,a)$$

$$(\Delta\text{-}3)\ \Delta(c,d) \geq \Delta(a,b) \quad \text{for } c \geq a, d \geq b$$

$$(\Delta\text{-}4)\ \Delta(\Delta(a,b),c) = \Delta(a,\Delta(b,c))$$

It can be easily verified that $\Delta_1(a, b) = \max\{a + b - 1, 0\}$ is a t-norm.

Definition 5.20

A Menger PM space (briefly, a Menger space) is a triplet (E, F, Δ), where (E, F) is a PM space with t-norm Δ, such that Menger's triangle inequality

$$(\text{PM} - 4)'\ F_{x,z}(t_1 + t_2) \geq \Delta\left(F_{x,y}(t_1), F_{y,z}(t_2)\right)$$

is satisfied for all $x, y, z \in E$ and for all $t_1 \geq 0, t_2 \geq 0$.

Note that (PM-4)' is a generalised version of (PM-4) and may replace it in certain situations.

If (E, F, Δ) is a Menger space with continuous t-norm, then it is a Hausdoff space with the topology $\mathcal{T}$ introduced by the family $\{U_y(\varepsilon, \lambda): y \in E, \lambda > 0\}$, where

$$U_y(\varepsilon,\lambda) = \{x \in E, F_{x,y}(\varepsilon) > 1 - \lambda, \varepsilon > 0, \lambda > 0\}, \tag{5.83}$$

is called an (ε, λ) neighbourhood of $y \in E$. Hence, the following concepts are introduced into (E, F, Δ).

Definition 5.21

Let $\{x_n\}$ be a sequence in a Menger space (E, F, Δ), where Δ is continuous. The sequence $\{x_n\}$ converges to $x_*\in E$ in $\mathcal{T}\left(x_n \xrightarrow{\ \mathcal{T}\ } x_*\right)$, if for every $\varepsilon > 0$ and $\lambda > 0$, there exists a positive integer $K = K(\varepsilon, \lambda)$ such that $F_{x_n,x_*}(\varepsilon) > 1-\lambda$, whenever $n \geq K$.

The sequence $\{x_n\}$ is called a $\mathcal{T}$-Cauchy sequence in E, if for every $\varepsilon > 0$ and $\lambda > 0$, there exists a positive integer $K = K(\varepsilon, \lambda)$ such that $F_{x_n,x_m}(\varepsilon) > 1-\lambda$, whenever $m, n \geq K$. A Menger space (E, F, Δ) is said to be $\mathcal{T}$-Complete if every $\mathcal{T}$-Cauchy sequence in E converges in $\mathcal{T}$ to a point in E. It can be proved that every Menger space with continuous t-norm is $\mathcal{T}$-Complete [8].

Definition 5.22

Let (E, F, Δ) be a Menger space where Δ is continuous. The self-mapping $\mathcal{T}$ that maps E into itself is $\mathcal{T}$-continuous on E, if for every sequence $\{x_n\}$ in E, $Tx_n \xrightarrow{\ \mathcal{T}\ } Tx_*$ whenever $x_n \xrightarrow{\ \mathcal{T}\ } x_* \in E$.

5.2.3.2 Fixed-Point Theorem in PM Space

Definition 5.23

Let T be a self-mapping of a Menger space (E, F, Δ). T is a contraction mapping, if there exists a constant $k \in (0,1)$ and for every $x \in E$ there exists a positive integer $n(x)$ such that for every $y \in E$,

$$F_{T^{n(x)}x,T^{n(x)}y}(t) \geq F_{x,y}\left(\frac{t}{k}\right), \qquad \forall t \geq 0. \tag{5.84}$$

A set $A \subset (E, F, \Delta)$ is called probabilistic bounded if $\sup_{t>0} \inf_{x,y\in A} F_{x,y}(t) = 1$. Denote the orbit generated by T at $x \in E$ by $O_T(x; 0, \infty)$, namely,

$$O_T(x; 0, \infty) = \{x_n = T^n x\}_{n=0}^{\infty}. \tag{5.85}$$

Using the contraction mapping as defined in Definition 5.23, Chang has provided a proof of the following fixed-point theorem (Theorem 5.12) [9].

Theorem 5.12

Let a self-mapping $T: (E, F, \Delta) \rightarrow (E, F, \Delta)$ be the contraction mapping as defined in Definition 5.23. If for every $x \in E$, $O_T(x; 0, \infty)$ is probabilistic bounded, then there exists a unique common fixed point x_* in E for T, and for every $x_0 \in E$, the iterative sequence $\{T^n x_0\}$ converges to x_* in $\mathcal{T}$.

5.2.3.3 Global Convergence of the QPSO Algorithm

5.2.3.3.1 Construction of the PM Space for QPSO   Consider the minimisation problem in Equation 1.1, which is rewritten as follows:

$$\text{Minimize } f(x), \quad \text{s.t. } x \in S \subseteq R^N,$$

where f is a real-valued function defined over region S, a compact subset of the N-dimensional Euclidean space R^N, and continuous almost everywhere. Let V be the range of f over S, and thus, $V \subset R$. Denote $f_* = \min_{x \in S}\{f(x)\}$ and $f^* = \max_{x \in S}\{f(x)\}$.

Theorem 5.13

Consider the ordered pair (V, F), where F is a mapping of $V \times V$ into $\mathfrak{D}$. For every $x, y \in V$, if the distribution function $F_{x,y}$ is defined by $F_{x,y}(t) = P\{|x - y| < t\}, \forall t \in R$, then (V, F) is a PM space.

Proof: It is sufficient just to show the mapping F satisfies conditions (PM-1) to (PM-4).

(1) For every $x, y \in V$, since $|x - y| \geq 0$, $F_{x,y}(0) = P\{|x - y| < 0\} = 0$ implies that F satisfies condition (PM-1).

(2) For every $x, y \in V$, if $F_{x,y}(t) = H(t)$ for every $t > 0$, then $F_{x,y}(t) = P\{|x - y| < t\} = 1$, which implies that for every positive integer m, $P\{|x - y| < 1/m\} = 1$. Therefore,

$$P\left\{\bigcap_{m=1}^{\infty}\left(|x - y| < \frac{1}{m}\right)\right\} = P\{|x - y| = 0\} = 1,$$

namely, $x = y$.

On the contrary, if $x = y$, namely, $|x - y| = 0$, then for every $t > 0$ $F_{x,y}(t) = P\{|x - y| < t\} = 1$; or for every $t < 0$, $F_{x,y}(t) = P\{|x - y| < t\} = 0$. Thus, $F_{x,y}(t) = H(t)$, that is, F satisfies (PM-2).

(3) By using the definition of F, one can easily see that F satisfies (PM-3).

(4) For every $x, y, z \in V$, if $F_{x,y}(t_1) = 1$ and $F_{y,z}(t_2) = 1$, then

$$F_{x,y}(t_1) = P\{|x-y| < t_1\} = 1,$$

and

$$F_{y,z}(t_2) = P\{|y-z| < t_2\} = 1.$$

Using the inequality $|x - z| \leq |x - y| + |y - z|$ leads to

$$F_{x,z}(t_1+t_2) = P\{|x-z| \leq t_1+t_2\} \geq P\{|x-y|+|y-z| < t_1+t_2\}$$
$$\geq P\{(|x-y| < t_1) \cap (|y-z| < t_2)\} = P\{|x-y| < t_1\} + P\{|y-z| < t_2\}$$
$$- P\{(|x-y| < t_1) \cup (|y-z| < t_2)\} = 2 - P\{(|x-y| < t_1) \cup (|y-z| < t_2)\} \geq 1.$$

Hence,

$$F_{x,z}(t_1+t_2) = P\{|x-z| < t_1+t_2\} = 1,$$

which implies that F satisfies (PM-4).

This completes the proof of the theorem. ■

Theorem 5.14

The triplet (V, F, Δ) is a Menger space, where $\Delta = \Delta_1$.

Proof: It is sufficient to prove that (V, F, Δ) satisfies Menger's triangle inequality. For every $x, y, z \in V$ and every $t_1 \geq 0, t_2 \geq 0$, since

$$|x-z| \leq |x-y| + |y-z|,$$

one obtains

$$\{|x-z| < t_1+t_2\} \supset \{|x-y|+|y-z| < t_1+t_2\} \supset \{|x-y| < t_1\} \cap \{|y-z| < t_2\}.$$

Hence,

$$F_{x,z}(t_1+t_2) = P\{|x-z| < t_1 + t_2\} \geq P\{|x-y| + |y-z| < t_1 + t_2\}$$

$$\geq P\{(|x-y| < t_1) \cap (|y-z| < t_2)\}$$

$$= P\{|x-y| < t_1\} + P\{y-z| < t_2\} - P\{(|x-y| < t_1) \cup (|y-z| < t_2)\}$$

$$\geq P\{|x-y| < t_1\} + P\{y-z| < t_2\} - 1 = F_{x,y}(t_1) + F_{y,z}(t_2) - 1$$

$$= \max\{F_{x,y}(t_1) + F_{y,z}(t_2) - 1, 0\} = \Delta_1(F_{x,y}(t_1), F_{y,z}(t_2)).$$

This implies that (V, F, Δ), where $\Delta = \Delta_1$, satisfies Menger's triangle inequality. Therefore, (V, F, Δ) is a Menger space.

This completes the proof of the theorem.   ■

(V, F, Δ) is a Menger space with continuous t-norm Δ_1 and also is a Hausdoff space of the topology $\mathcal{T}$ introduced by the family

$$\{U_y(\varepsilon, \lambda) : y \in V, \lambda > 0\},$$

where $U_y(\varepsilon, \lambda) = \{x \in V, F_{x,y}(\varepsilon) > 1 - \lambda, \varepsilon, \lambda > 0\}$, and consequently is $\mathcal{T}$-Complete.

5.2.3.3.2 Fixed-Point Theorem for the QPSO Algorithm   QPSO may be treated as a mapping, say, denoted by T which is a self-mapping of Menger space (V, F, Δ). When $n = 0$, the global best position G_0 is generated by initialising QPSO, that is, $f_0 = f(G_0)$ and that $f_0 \in V$. A series of iterations using the algorithm generates a sequence of global best positions $\{G_n, n \geq 1\}$ and a sequence of the corresponding non-increasing function values $\{f_n = f(G_n), n \geq 1\}$. Treating $\{f_n, n \geq 1\}$ as a sequence of points generated by T in (V, F, Δ), that is, $f_n = T^n f_0$ and $f_n \in V$, and denoting the orbit generated by T at $f_0 \in V$ as $O_T(f_0; 0, \infty)$ leads to $O_T(f_0; 0, \infty) = \{f_n = T^n f_0\}_{n=0}^{\infty}$. The following theorem proves that T is a contraction mapping on (V, F, Δ).

Theorem 5.15

The mapping T is a contraction mapping on the Menger space (V, F, Δ).

Proof: When $t = 0$, T satisfies the contractive condition in Definition 5.23. Given $f_a \in V$, $\forall f_b \in V$ and $\forall t > 0$, and assume that

$$F_{f_a,f_b}(t) = P\{|\, f_a - f_b\,| < t\} = 1 - \delta, \tag{5.86}$$

where $0 < \delta < 1$. Let $V(t) = \{f : f \in V, f - f_* < t\}$, where f_* is the global minimum of $f(X)$. $V(t)$ is measurable and its Lebesgue measure $v[V(t)] > 0$. Assuming $S(t) = \{x : f(x) \in V(t)\}$, then $v[S(t)] > 0$ due to the continuity of $f(x)$ almost everywhere in S.

Let $a_0(t) = P\{f_0 \in V(t)\} = g_0(t)$ at the starting point. At each subsequent iteration, $f_n = T^n f_0$, and using the update equation of QPSO, one obtains

$$g_n(t) = 1 - \prod_{i=1}^{M}\left[1 - \int_{S(t)} \theta_{X_{i,n}}(G_{n-1}, P_{i,n-1}, x)dx\right],$$

where

$$\theta_{X_{i,n}}(G_{n-1}, P_{i,n}, x) = \prod_{j=1}^{N}\left| \int_{P_{i,n}^j}^{G_n^j} \frac{1}{L_{i,n}^j}\exp\left(\frac{-2\,|\,x - p\,|}{L_{i,n}^j}\right)dp\right|.$$

Thus, one obtains

$$a_n(t) = P\{f_n \in V(t), f_k \notin V(t), k = 1,2,\ldots,n-1\} = g_n(t)\prod_{i=0}^{n-1}[1 - g_i(t)]. \tag{5.87}$$

Hence,

$$F_{f_n,f_*}(t) = P\{|\, f_n - f_*\,| < t\} = P\{f_n \in V(t)\} = \sum_{i=0}^{n} a_n(t) = 1 - \prod_{i=1}^{n}[1 - g_i(t)]. \tag{5.88}$$

For every $0 \le n < \infty$, one has $|X_{i,n}^j| < \infty$, $|C - X_{i,n}^j| < \infty$ or $|p - X_{i,n}^j| < \infty$. Hence, $0 < L_{i,j,n} < \infty$, which implies that $\theta_{X_{i,n}}(G_{n-1}, P_{i,n}, x)$ is Lebesgue integrable and $0 < g_i(t) < 1$. Therefore, $\sup_{n>0} F_{f_n,f_*}(t) = 1$ following Equation

5.88. Hence, for the given δ and f_a, there exists a positive integer $n_1(f_a)$ such that whenever $n \geq n_1(f_a)$,

$$F_{T^n f_a, f_*}(t) = P\left\{|T^n f_a - f_*| < t\right\} = P\left\{T^n f_a \in V(t)\right\} > 1 - \frac{\delta}{2}.$$

There also exists a positive integer $n_2(f_a)$ such that whenever $n \geq n_2(f_a)$,

$$F_{T^n f_b, f_*}(t) = P\left\{|T^n f_b - f_*| < t\right\} = P\left\{T^n f_b \in V(t)\right\} > 1 - \frac{\delta}{2}.$$

Let $n(f_a) \geq \max\{n_1(f_a), n_2(f_a)\}$. Both of the following two inequalities are true:

$$F_{T^{n(f_a)} f_a, f_*}(t) = P\left\{|T^{n(f_a)} f_a - f_*| < t\right\} = P\left\{T^{n(f_a)} f_a \in V(t)\right\}$$

$$= P\left\{T^{n(f_a)} f_a - f_* < t\right\} > 1 - \frac{\delta}{2}. \tag{5.89}$$

$$F_{T^{n(f_a)} f'', f_*}(t) = P\left\{|T^{n(f_a)} f_b - f_*| < t\right\} = P\left\{T^{n(f_a)} f_b \in V(t)\right\}$$

$$= P\left\{T^{n(f_a)} f_a - f_* < t\right\} > 1 - \frac{\delta}{2}. \tag{5.90}$$

Since the diameter of $V(t)$ is t, that is, $\sup_{x, y \in V(t)} |x - y| = t$. If $T^{n(f_a)} f_a \in V(t)$ and $T^{n(f_a)} f_a \in V(t)$, then

$$|T^{n(f_a)} f_a - T^{n(f_a)} f_b| < t.$$

As a result the following equality holds:

$$\left\{|T^{n(f_a)} f_a - T^{n(f_a)} f_b| < t\right\} \supset \left\{(T^{n(f_a)} f_a \in V(t)) \cap T^{n(f_a)} f_b \in V(t)\right\}$$

$$= \left\{(T^{n(f_a)} f_a - f_* < t) \cap (T^{n(f_a)} f_b - f_* < t)\right\}.$$

Hence,

$$F_{T^{n(f_a)}f_a, T^{n(f_a)}f_b}(t)$$

$$= P\{|T^{n(f_a)}f_a - T^{n(f_a)}f_b| < t\} \geq P\{(T^{n(f_a)}f_a - f_* < t) \cap (T^{n(f_a)}f_b - f_* < t)\}$$

$$= P\{T^{n(f_a)}f_a - f_* < t\} + P\{T^{n(f_a)}f_b - f_* < t\}$$

$$- P\{(T^{n(f_a)}f_a - f_* < t) \cup (T^{n(f_a)}f_b - f_* < t)\} > 1 - \frac{\delta}{2} + 1 - \frac{\delta}{2}$$

$$- P\{(T^{n(f_a)}f_a - f_* < t) \cup (T^{n(f_a)}f_b - f_* < t)\} > 2 - \delta - 1 = 1 - \delta,$$

and

$$F_{T^{n(f_a)}f_a, T^{n(f_a)}f_b}(t) > F_{f_a, f_b}(t). \tag{5.91}$$

Since F is monotonically increasing with t, there must exist $k \in (0,1)$ such that

$$F_{T^{n(f_a)}f_a, T^{n(f_a)}f_b}(t) \geq F_{f_a, f_b}\left(\frac{t}{k}\right). \tag{5.92}$$

It implies that T satisfies the contractive condition in Definition 5.23. This completes the proof of the theorem. ∎

Theorem 5.16

f_* is the unique fixed point in V such that for every $f_0 \in V$, the iterative sequence $\{T^n f_0\}$ converges to f_*.

Proof: T is a contraction mapping on Menger space (V, F, Δ) where $\Delta = \Delta_1$ as shown in Theorem 5.15. Given $f_0 \in V$, one can construct $f_n = T^n f_0 \in [f_*, f_0]$ for $n \geq 1$. This implies that $O_T(f_0; 0, \infty) = \{f_n = T^n f_0\}_{n=0}^{\infty} \subset [f_*, f_0]$. Thus, for every $t > f_0 - f_*$, $\inf_{f_a, f_b \in O_T(f_0; 0, \infty)} F_{f_a, f_b}(t) = 1$. As a result, one obtains

$$\sup_{t>0} \inf_{f_a, f_b \in O_T(f_0; 0, \infty)} F_{f_a, f_b}(t) = 1, \tag{5.93}$$

which means that the orbit generated by T at f_0 is probabilistic bounded. By Theorem 5.12, there exists a unique common fixed point in E for T. Since $f_* = Tf_*, f_*$ is the fixed point. Consequently, for every $f_0 \in V$, the iterative sequence $\{T^n f_0\}$ converges to f_* in $\mathcal{T}$.

This completes the proof of the theorem. ■

Denote the optimal region of problem (1.1) by $V_\varepsilon = V(\varepsilon) = \{f: f \in V, f - f_* < \varepsilon\}$. For QPSO in (V, F, Δ), the following theorem shows the equivalence between convergence in $\mathcal{T}$ and convergence in probability.

Theorem 5.17

The sequence of function values $\{f_n, n \geq 0\}$ generated by QPSO converges to f_* in probability.

Proof: Since $\{f_n, n \geq 0\}$ converges to f_* in $\mathcal{T}$, by Definition 5.23 and the definition given in Theorem 5.13, for every $\varepsilon > 0$, $\lambda > 0$, there exists $K = K(\varepsilon, \lambda)$ such that whenever $n \geq K$,

$$F_{f_n, f_*}(\varepsilon) = P\{|f_n - f_*| < \varepsilon\} = P\{f_n \in V_\varepsilon\} > 1 - \lambda. \qquad (5.94)$$

Due to the arbitrariness of λ, (5.94) implies that $f_n \xrightarrow{P} f_*$.

This completes the proof of the theorem. ■

5.3 TIME COMPLEXITY AND RATE OF CONVERGENCE

5.3.1 Measure of Time Complexity

Convergence is an important characteristic of stochastic optimisation algorithms. It is not sufficient to evaluate the efficiency of a stochastic optimisation algorithm by comparing to other optimisation methods. A better approach is to study the distribution of the number of steps required to reach the optimality region $V(\varepsilon)$, more specifically by comparing the expected number of steps and higher moments of this distribution. The number of steps required to reach $V(\varepsilon)$ is defined by $K(\varepsilon) = \inf\{n|\ f_n \in V_\varepsilon\}$. The expected value (time complexity) and the variance of $K(\varepsilon)$, if they exist, can be computed by

$$E[K(\varepsilon)] = \sum_{n=0}^{\infty} na_n, \qquad (5.95)$$

$$Var[K(\varepsilon)] = E[K^2(\varepsilon)] - \{E[K(\varepsilon)]\}^2 = \sum_{n=0}^{\infty} n^2 a_n - \left(\sum_{n=0}^{\infty} n a_n \right)^2, \quad (5.96)$$

where $a_n = a_n(\varepsilon)$. Using (5.87), one obtains

$$a_n = P\{K(\varepsilon) = n\} = P\{f_0 \in V_\varepsilon^c, f_1 \in V_\varepsilon^c, f_2 \in V_\varepsilon^c, \ldots, f_{n-1} \in V_\varepsilon^c, f_n \in V_\varepsilon\} = a_n(\varepsilon).$$
$$(5.97)$$

Using Equation 5.88, one can deduce that

$$F_{f_n, f_*}(\varepsilon) = P\{f_n \in V_\varepsilon\} = \sum_{i=0}^{n} a_n \quad (5.98)$$

The compatibility condition for the algorithm to be global convergent is given by

$$\sum_{i=0}^{\infty} a_n = 1. \quad (5.99)$$

It is obvious that the existence of $E[K(\varepsilon)]$ relies on the convergence of $\sum_{n=0}^{\infty} n a_n$. In general, it is far more difficult to compute a_n analytically than to prove the global convergence of most stochastic optimisation algorithms. This is particularly true for population-based random search techniques. Theoretical determination based on certain particular situations [5] or numerical simulation based on some specific functions [10–15] are usually adapted to evaluate $E[K(\varepsilon)]$ and $Var[K(\varepsilon)]$.

In this chapter, the influences due to the behaviour of each individual particle on the convergence of QPSO, from the perspective of time complexity, are considered. It should be noted that the proof of global convergence of QPSO does not involve the behaviour of the individual particle. It is true that the algorithm may exhibit global convergence even when the particles diverge (i.e. when $\alpha > e^\gamma$). One only needs to ensure the compatibility condition $\sum_{i=0}^{\infty} a_n = 1$ is satisfied or $g_n(\varepsilon) > 0$, for all n, in order to guarantee the global convergence of QPSO.

When the particle diverges, $g_n(\varepsilon)$ decreases continuously, for every $\varepsilon > 0$. The function

$$F_{f_n, f_*}(\varepsilon) = \sum_{i=0}^{n} a_n(\varepsilon) = 1 - \prod_{i=1}^{n} [1 - g_i(\varepsilon)],$$

may still converge to 1 since $0 < g_n(\varepsilon) < 1$ for all $n < \infty$. However, the series $\sum_{n=0}^{\infty} na_n$ may diverge in such cases. The earlier discussion suggests that the divergence of the particles does not affect the global convergence of the algorithm, but results in infinite complexity in general. On the other hand, when the particles converge or are bounded, $g_n(\varepsilon)$ does not decrease continuously but may increase during the search. As a result, for certain $\varepsilon > 0$, the series $\sum_{n=0}^{\infty} na_n$ can converge, which implies that the algorithm has finite time complexity. Therefore, in order to make QPSO converge globally with finite time complexity, the condition $\alpha < e^\gamma$ is required to ensure the convergence of the particle. Section 5.4 involves further investigation to the issue of setting α that may lead the algorithm to work well generally.

5.3.2 Convergence Rate

Besides analysing its time complexity, another method used to evaluate the efficiency of the algorithm is to compute its convergence rate. The complexity of analysing the convergence rate of a population-based random optimisation algorithm is no less significant than that of computing the expected value or variance of $K(\varepsilon)$. Although some work has been done in the analysis of convergence rate [16–21], it is still an open problem for arbitrary objective functions. The approach considered in this monograph differs from those in existing literatures. Define the convergence rate, $c_n \in (0,1)$, at the nth step as

$$c_n = E\left[\frac{|f_n - f_*|}{|f_{n-1} - f_*|} \middle| f_{n-1} \right] = E\left[\frac{f_n - f_*}{f_{n-1} - f_*} \middle| f_{n-1} \right]. \qquad (5.100)$$

Therefore,

$$E\left[(f_n - f_*) \middle| f_{n-1} \right] = c_n (f_{n-1} - f_*). \qquad (5.101)$$

Note that the speed of convergence increases with smaller values of $c_n \in (0,1)$. Let $\varepsilon_n = E(f_n - f_*)$ be the expected error at any $n \geq 0$. If there exists a constant $c \in (0,1)$ called the expected convergence rate, then $\varepsilon_n = c^n \varepsilon_0$, for every $n \geq 0$. Taking logarithm on both sides leads to

$$n = \frac{\log_{10}(\varepsilon_n/\varepsilon_0)}{\log_{10}(c)} = -\frac{\Theta}{\log_{10}(c)}, \tag{5.102}$$

where $\Theta > 0$ denotes the order of magnitude the error is to be decreased. If Θ is fixed, then the time n required to reduce the error by Θ order of magnitude decreases as c decreases towards zero. Since the expected error after $K(\varepsilon)$ iterations is approximately ε, $\varepsilon_{K(\varepsilon)} = c^{K(\varepsilon)} \varepsilon_0 \approx \varepsilon$, from which $K(\varepsilon)$ can be approximated as

$$K(\varepsilon) \approx \frac{\log_{10}(\varepsilon/\varepsilon_0)}{\log_{10}(c)} = \frac{\log_{10}(\varepsilon/E[f_0 - f_*])}{\log_{10}(c)}$$

$$= \frac{\log_{10}\{\varepsilon/E[f_0 - f_*]\}}{\log(c)} = \frac{-\Theta_0}{\log(c)}, \tag{5.103}$$

where $\Theta_0 = \log_{10}((E[f_0] - f_*)/\varepsilon) > 0$. The following theorem states the relationship between c and c_n.

Theorem 5.18

Let $\bar{c} = \left(\prod_{i=1}^{n} \bar{c}_i\right)^{1/n}$ where $\bar{c}_i = E(c_i)$. If $\{c_n, n > 0\}$ and $\{f_n - f_*, n > 0\}$ are two negatively correlated (or positively correlated or uncorrelated) sequences of random variables, then $c < \bar{c}$ (or $c > \bar{c}$ or $c = \bar{c}$).

Proof: According to the properties of conditional expectations, one has

$$E[(f_n - f_*)] = E[E[(f_n - f_*)|f_{n-1}]], \quad n > 0. \tag{5.104}$$

If $\{c_n, n > 0\}$ and $\{f_n - f_*, n > 0\}$ are negatively correlated, it follows that

$$Cov(c_n, f_n - f_*) = E[c_n(f_n - f_*)] - E(c_n)E(f_n - f_*) < 0,$$

namely,

$$E[c_n(f_n - f_*)] < E(c_n)E(f_n - f_*), \quad n > 0. \tag{5.105}$$

By (5.104) and (5.105), one can deduce that

$$\varepsilon_n = E(f_n - f_*) = E\{E[(f_n - f_*)|f_{n-1}]\} = E[c_n \cdot (f_{n-1} - f_*)]$$

$$< E(c_n) \cdot E(f_{n-1} - f_*) = \bar{c}_n E(f_{n-1} - f_*) = \bar{c}_n E\{E[(f_{n-1} - f_*)|f_{n-2}]\}$$

$$= \bar{c}_n E[c_{n-1} \cdot (f_{n-2} - f_*)] < \bar{c}_n \bar{c}_{n-1} E(f_{n-2} - f_*) = \cdots < \bar{c}_n \bar{c}_{n-1} \bar{c}_{n-2} \cdots \bar{c}_1 E(f_0 - f_*)$$

$$= \left(\prod_{i=1}^{n} \bar{c}_i \right) \cdot E(f_0 - f_*) = \bar{c}^n \varepsilon_0, \tag{5.106}$$

which implies that $c = (\varepsilon_n/\varepsilon_0)^{1/n} < \bar{c}$.

If $\{c_n, n > 0\}$ and $\{f_n - f_*, n > 0\}$ are positively correlated,

$$Cov(c_n, f_n - f_*) = E[c_n(f_n - f_*)] - E(c_n)E(f_n - f_*) > 0,$$

which implies that $c > \bar{c}$.

If $\{c_n, n > 0\}$ and $\{f_n - f_*, n > 0\}$ are uncorrelated,

$$Cov(c_n, f_n - f_*) = E[c_n(f_n - f_*)] - E(c_n)E(f_n - f_*) = 0.$$

Replacing each sign of the inequality in (5.106) by an equal sign implies that $c = \bar{c}$.

This completes the proof of the theorem. ■

Since the sequence $\{f_n - f_*, n > 0\}$ decreases with n, the negative correlation between $\{c_n, n > 0\}$ and $\{f_n - f_*, n > 0\}$ implies that c_n increases or the speed of convergence decreases as f_n decreases. The convergence of f_n is said to be sublinear. When $\{c_n, n > 0\}$ and $\{f_n - f_*, n > 0\}$ are positively correlated, c_n decreases as f_n decreases, which means that convergence accelerates as f_n decreases. The convergence of f_n is said to be super-linear.

When $\{c_n, n > 0\}$ and $\{f_n - f_*, n > 0\}$ are uncorrelated, $c = \bar{c}$ for all $n > 0$, which implies that $c_n = c$. The convergence of f_n is said to be linear.

Linear convergence may occur in some idealised situations such as using pure adaptive search (PAS) [14] for the sphere function. Taking a two-dimensional (21) sphere function as an example, one has

$$E = [f_n - f_* \mid f_{n-1}] = E[f_n \mid f_{n-1}] = E\left[\|X_n\|^2 \|X_{n-1}\|^2\right].$$

Denote $r_{n-1}^2 = \|X_{n-1}\|^2$ and consider X_n to be distributed uniformly on $S_n = \left\{X : \|X\|^2 \leq r_{n-1}^2\right\}$, one obtains

$$E = [f_n - f_* \mid f_{n-1}] = \frac{1}{\pi r_{n-1}^2} \iint_{S_n} r^2 dX = \frac{1}{\pi r_{n-1}^2} \int_0^{2\pi} d\theta \int_0^{r_{n-1}} r^2 \cdot r dr$$

$$= \frac{r_{n-1}^2}{2} = \frac{\|X_{n-1}\|^2}{2} = \frac{f_{n-1} - 0}{2} = \frac{f_{n-1} - f_*}{2}.$$

It can be found that $c_n = 0.5$, for all $n > 0$, and $c = c_n = 0.5$. This implies that linear convergence is attained by using PAS.

Most practical stochastic algorithms, however, exhibit sublinear convergence in general. To evaluate an algorithm in terms of its convergence rate, the value of $\bar{c} = \left(\prod_{i=1}^{n} \bar{c}_i\right)^{1/n}$ may be computed and compared with that of other algorithms. It is also possible to compute the value of $|\bar{c} - c|$ in order to measure the 'linearity' of its convergence. However, determining the convergence rate analytically is no less difficult than the time complexity for most of the algorithms, including QPSO. In the remaining part of this section, attention is paid to the numerical determination of the convergence rate and the time complexity for certain types of problems.

5.3.3 Testing Complexity and Convergence Rate of QPSO

In this section, the sphere function $f(X) = X^T \cdot X$, which has its minimal value at zero, is used to examine empirically the convergence properties of QPSO. The sphere function is unimodal and is a special instance from the class of quadratic functions with positive definite Hessian matrix.

It is a common function used to test the convergence properties of a random search algorithm [5]. In the experiments performed, the initial search space used by the algorithms for the function is $[-10,10]^N$, where N is the dimension of the problem.

To evaluate the convergence of the algorithms, a fair time measurement is required. Usually, the number of iterations cannot be accepted as the time measure since the algorithms perform different amount of work in their inner loops with different population sizes. In the experiments, the number of fitness function (objective function) evaluations was used as a measure of time. The advantage of measuring complexity by counting the number of function evaluations is that there is a strong relationship between this measure and the processing time as the function complexity increases. Therefore, the subscript n in this section does not denote the number of iterations but the number of fitness function evaluations.

Two sets of experiments to evaluate the time complexity and convergence rate of the QPSO algorithm, respectively, were carried out. QPSO-Type-1, QPSO-Type-2 and PSO with constriction factor (PSO-Co) [3] were tested and the results were compared. The optimality region was set as

$$V(\varepsilon) = V(10^{-4}) = \{f : f \in V, f - f_* < 10^{-4}\},$$

and the number of function evaluations, $K(\varepsilon)$, when the tested algorithm first reached the region, is recorded. Each of the previous three algorithms was set to run 50 times on the sphere function with a given dimension. Statistics of the numerical results are collected for the mean number of function evaluations ($\bar{K}(\varepsilon)$), standard deviation of $K(\varepsilon)$ ($\sigma_{K(\varepsilon)}$) and standard error. Tables 5.2

TABLE 5.2   Statistics of the Time Complexity Using QPSO-Type-1

N	$\bar{K}(\varepsilon)$	$\sigma_{K(\varepsilon)}$	$\sigma_{K(\varepsilon)}/\sqrt{50}$	$\sigma_{K(\varepsilon)}/\left(N\sqrt{50}\right)$	$\bar{K}(\varepsilon)/N$
2	232.16	49.2471	6.9646	3.4823	116.0800
3	382.78	67.1183	9.4920	3.1640	127.5933
4	577.92	92.8789	13.1351	3.2838	144.4800
5	741.08	104.8719	14.8311	2.9662	148.2160
6	921.92	128.5678	18.1822	3.0304	153.6533
7	1124.70	115.4569	16.3281	2.3326	160.6714
8	1396.20	170.9966	24.1826	3.0228	174.5250
9	1586.36	159.0606	22.4946	2.4994	176.2622
10	1852.86	185.6496	26.2548	2.6255	185.2860

TABLE 5.3   Statistics of the Time Complexity Using QPSO-Type-2

N	$\bar{K}(\varepsilon)$	$\sigma_{K(\varepsilon)}$	$\sigma_{K(\varepsilon)}/\sqrt{50}$	$\sigma_{K(\varepsilon)}/\left(N\sqrt{50}\right)$	$\bar{K}(\varepsilon)/N$
2	306.26	69.4153	9.8168	4.9084	153.1300
3	455.06	71.1530	10.0626	3.3542	151.6867
4	619.38	60.0306	8.4896	2.1224	154.8450
5	748.92	69.8716	9.8813	1.9763	149.7840
6	883.92	94.4546	13.3579	2.2263	147.3200
7	1043.26	101.0267	14.2873	2.0410	149.0371
8	1171.68	102.8985	14.5520	1.8190	146.4600
9	1314.30	117.4271	16.6067	1.8452	146.0333
10	1477.20	112.9856	15.9786	1.5979	147.7200

TABLE 5.4   Statistics of the Time Complexity Using PSO

N	$\bar{K}(\varepsilon)$	$\sigma_{K(\varepsilon)}$	$\sigma_{K(\varepsilon)}/\sqrt{50}$	$\sigma_{K(\varepsilon)}/\left(N\sqrt{50}\right)$	$\bar{K}(\varepsilon)/N$
2	679.2	126.4995	96.0534	48.0267	339.6000
3	967.8	146.1330	136.8676	45.6225	322.6000
4	1235.3	137.2839	174.6978	43.6745	308.8250
5	1417.3	157.8286	200.4365	40.0873	283.4600
6	1686.0	174.2886	238.4364	39.7394	281.0000
7	1914.5	220.8512	270.7512	38.6787	273.5000
8	2083.3	186.5505	294.6231	36.8279	260.4125
9	2346.5	162.1101	331.8452	36.8717	260.7222
10	2525.2	200.4582	357.1172	35.7117	252.5200

through 5.4 list the results of QPSO-Type-1 with $\alpha = 1$, QPSO-Type-2 with $\alpha = 0.75$, and PSO with constriction factor $\chi = 0.7298$ and acceleration coefficients $c_1 = c_2 = 2.05$. Each algorithm used 20 particles.

Numerical results as shown in the last columns of Tables 5.3 and 5.4 indicate that there is a linear correlation between $\bar{K}(\varepsilon)$ and the dimension, that is, $\bar{K}(\varepsilon) = HN$. The constant H is related to the algorithm used. From Table 5.3, H appears to be approximately 150 for QPSO-Type-2. From Table 5.4, H appears to be approximately 300 for PSO. To investigate these properties, the statistics with other parameters are provided in Figures 5.4 through 5.6. Figure 5.4 displays the value of $\bar{K}(\varepsilon)/N$ which increases slowly as the dimension increases. This implies that the time complexity may increase non-linearly with the dimension. From Figures 5.5 and 5.6, it can be seen that the time complexities of QPSO-Type-2 and PSO increase

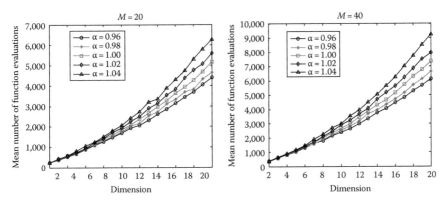

FIGURE 5.4 Results of the time complexity using QPSO-Type-1 for different values of α and population size (M).

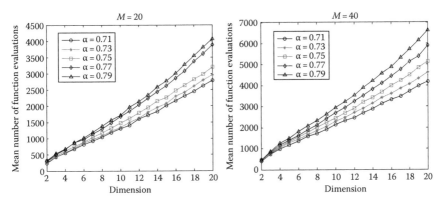

FIGURE 5.5 Results of the time complexity using QPSO-Type-2 for different values of α and population size (M).

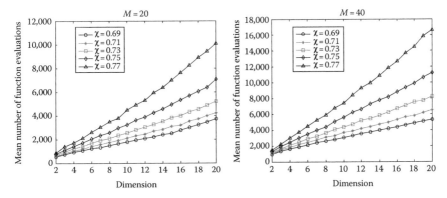

FIGURE 5.6 Results of the time complexity using PSO for different values of χ and population size (M).

almost linearly when the dimension varies in between 2 and 20. However, both QPSO-Type-1 and QPSO-Type-2 exhibit lower time complexities than PSO under the same set of parameters.

The linear correlation observed in the experiments may be explained as follows. Consider an idealised random search algorithm, say the somewhat adaptive search (SAS) [15], with the constant probability $\rho(0 < \rho \leq 1)$ of improving the objective function value. It is also known as 'ρ-adaptive search' since at each iteration, the probability that it behaves as PAS is ρ. Mathematically, the time complexity of ρ-adaptive search, $E[K_{\rho-SAS}(\varepsilon)]$, is given by

$$E[K_{\rho-SAS}(\varepsilon)] = \frac{1}{\rho} E[K_{PAS}(\varepsilon)], \qquad (5.107)$$

where $E[K_{PAS}(\varepsilon)]$ is the time complexity of PAS. It has been shown in [15] that $E[K_{PAS}(\varepsilon)]$ may be determined by

$$E[K_{PAS}(\varepsilon)] = \ln \frac{v(S)}{v(V_\varepsilon)}. \qquad (5.108)$$

One can deduce that

$$E[K_{\rho-SAS}(\varepsilon)] = \frac{1}{\rho} \ln \frac{v(S)}{v(V_\varepsilon)}, \qquad (5.109)$$

where $v(\cdot)$ is the Lebesgue measure. For the testing problem, V_ε, the optimal region of the sphere function, is an N-dimensional super-ball with radius $\sqrt{\varepsilon}$ with its volume given by

$$v(V_\varepsilon) = \left[\frac{\pi^{N/2}}{\Gamma((N/2)+1)} \right] \varepsilon^{N/2}, \qquad (5.110)$$

where $\Gamma(\cdot)$ is the gamma function. Assuming

$$\ln \Gamma\left(\frac{N}{2}+1\right) = O(N), \qquad (5.111a)$$

one obtains

$$E[K_{\rho-SAS}(\varepsilon)] = \frac{1}{\rho} \ln \frac{\delta^N}{v(V_\varepsilon)} = \frac{1}{\rho} \ln \left[\frac{\delta^N \cdot \Gamma((N/2)+1)}{(\pi\varepsilon)^{N/2}} \right]$$

$$= \frac{N}{\rho} \ln \left[\frac{\delta}{\sqrt{\pi\varepsilon}} \right] + \ln \Gamma \left(\frac{N}{2} + 1 \right) = O(N), \qquad (5.111b)$$

where δ is the length of the search space of each dimension. The earlier equation implies that ρ-adaptive search has linear time complexity and the constant H can be determined by ρ if the values of δ and ε are given.

For the algorithms used in the tests, the probability ρ generally varies with the number of function evaluations and dimension of the problem. Under certain parameter settings, the linear correlation between time complexity and dimension as produced by QPSO-Type-2 or PSO indicates that ρ is relatively stable when the number of function evaluations increases and the dimension varies within a certain interval. The value of ρ of QPSO-Type-1 seems to be less stable when the algorithm is running, leading the time complexity to increase non-linearly with the dimension. Nevertheless, since both types of QPSO may have larger ρ, the values of their respective H are smaller than that of PSO.

To test the convergence rate, the experiments consisted of 50 trial runs for every instance, with each run executing for a given maximum number of function evaluations, which were set to 200 times the dimensionality of the problem, that is, $n_{max} = 200N$. For each algorithm, two sets of experiments were performed, one with population size $M = 20$ and the other with $M = 40$. In order to compute the convergence rate, 30 independent samples of the particle positions were recorded before the next update of the position. For each sample position, the objective function value, f_n^k (k is the number of sampled position), was calculated and in turn was compared to that of the global best position f_{n-1}. If $f_n^k < f_{n-1}$, f_n^k was recorded; otherwise, it was replaced by f_{n-1}. After collecting 30 samples, the mean value, $E[f_n - f_* \mid f_{n-1}] = (1/30) \sum_{k=1}^{30} \left(f_n^k - f_* \right)$, was computed and the convergence rate c_n was then determined. Note that the sampling procedure did not affect the actual variables such as the current position of the particle, its personal best position, the global best position, its velocity (for PSO algorithm) and so forth. After sampling, the position of each particle was updated according to these variables.

For a given value of n, the value of $\bar{c}_n$ was obtained by calculating the arithmetic mean of all the c_n values resulted from those 50 runs. Hence, $\bar{c}$ can be obtained by

$$\bar{c} = \left(\prod_{n=1}^{n_{max}} \bar{c}^n \right)^{1/n_{max}}, \tag{5.112}$$

and the expected convergence rate c was worked out by

$$c = \left(\frac{\overline{f}_{n_{max}}}{\overline{f}_0} \right)^{1/n_{max}}. \tag{5.113}$$

The correlation coefficient between c_n and f_n, denoted as $\theta(c_n, f_n)$, was also computed. Tables 5.5 through 5.7 list the statistics generated by QPSO-Type-1 ($\alpha = 1.00$), QPSO-Type-2 ($\alpha = 0.75$) and PSO ($\chi = 0.7298$, $c_1 = c_2 = 2.05$).

TABLE 5.5   Statistics for the Convergence Rates Using QPSO-Type-1

N	n_{max}	$\overline{f}_0$	$\overline{f}_{n_{max}}$	$\bar{c}$	c	$\theta(c_n, f_n)$
$M = 20$						
2	400	6.3710	1.931×10^{-5}	0.9772	0.9687	−0.1306
3	600	19.1981	4.724×10^{-6}	0.9810	0.9750	−0.1089
4	800	32.2220	3.551×10^{-6}	0.9842	0.9802	−0.0893
5	1000	52.0046	6.632×10^{-6}	0.9869	0.9843	−0.0734
6	1200	74.4027	1.327×10^{-6}	0.9887	0.9871	−0.0830
7	1400	104.7848	1.326×10^{-6}	0.9902	0.9887	−0.0870
8	1600	119.6822	1.729×10^{-6}	0.9918	0.9902	−0.0745
9	1800	145.3857	7.288×10^{-6}	0.9926	0.9920	−0.0710
10	2000	166.8939	6.994×10^{-6}	0.9933	0.9927	−0.0581
$M = 40$						
2	400	2.8309	4.527×10^{-5}	0.9867	0.9728	−0.1513
3	600	11.1715	2.834×10^{-4}	0.9876	0.9825	−0.1341
4	800	23.8846	2.573×10^{-4}	0.9900	0.9858	−0.1333
5	1000	44.3261	6.890×10^{-4}	0.9912	0.9890	−0.1052
6	1200	63.5427	6.901×10^{-4}	0.9926	0.9905	−0.1120
7	1400	82.9243	0.0012	0.9936	0.9921	−0.0999
8	1600	103.6366	0.0027	0.9943	0.9934	−0.1009
9	1800	127.3277	0.0032	0.9950	0.9941	−0.0924
10	2000	139.2103	0.0046	0.9954	0.9949	−0.0848

TABLE 5.6   Statistics for the Convergence Rates Using QPSO-Type-2

N	n_{max}	$\bar{f}_0$	$\bar{f}_{n_{max}}$	$\bar{c}$	c	$\theta(c_n, f_n)$
$M = 20$						
2	400	6.2432	2.206×10^{-5}	0.9845	0.9691	−0.2165
3	600	14.0202	1.296×10^{-5}	0.9864	0.9771	−0.2134
4	800	33.3147	1.766×10^{-5}	0.9867	0.9821	−0.1644
5	1000	50.7010	7.177×10^{-6}	0.9881	0.9844	−0.1579
6	1200	68.6281	4.290×10^{-6}	0.9889	0.9863	−0.1405
7	1400	100.3869	2.688×10^{-6}	0.9899	0.9876	−0.1410
8	1600	132.1488	2.161×10^{-6}	0.9910	0.9889	−0.1244
9	1800	144.2586	1.595×10^{-6}	0.9915	0.9899	−0.1090
10	2000	172.0399	1.746×10^{-6}	0.9921	0.9908	−0.0782
$M = 40$						
2	400	3.5850	4.207×10^{-4}	0.9883	0.9776	−0.2589
3	600	12.6022	0.0012	0.9898	0.9847	−0.2304
4	800	25.9027	0.0019	0.9920	0.9882	−0.2289
5	1000	43.9053	0.0012	0.9927	0.9895	−0.1910
6	1200	62.3872	0.0018	0.9933	0.9913	−0.1529
7	1400	87.6376	0.0020	0.9939	0.9924	−0.1633
8	1600	100.1323	0.0023	0.9945	0.9933	−0.1700
9	1800	119.0672	0.0013	0.9951	0.9937	−0.1329
10	2000	154.5341	0.0017	0.9951	0.9943	−0.1274

The value of $\bar{c}$ for applying each algorithm to a problem is larger than c and the correlation coefficient becomes negative. This indicates that the algorithms show sublinear convergence. It can also be observed that the convergence rates of both types of QPSO are smaller than that of PSO for each problem with the same population size except when $N = 2$. This implies that QPSO may converge faster. A careful look at the three tables reveals that $\theta(c_n, f_n)$ increases as the dimension increases. The reason may be due to the fact that the improvement of the function value becomes relatively harder when the dimension is high, making the convergence rate c_n so close to 1 that it changes little as f_n decreases. Note also that when $M = 40$ for a given problem the convergence rates are larger than those when $M = 20$. It is difficult to conclude whether a larger population size leads to a higher convergence rate as the convergence rate also depends on other parameters. It is also difficult to conclude that smaller population size results in faster convergence for the problem with lower dimension, simply because the algorithm may encounter premature convergence when the dimension is higher. On the other hand, although the algorithm

TABLE 5.7   Statistics for the Convergence Rates Using PSO

N	n_{max}	$\bar{f_0}$	$\bar{f}_{n_{max}}$	$\bar{c}$	c	$\theta(c_n, f_n)$
$M = 20$						
2	400	6.2601	0.0050	0.9911	0.9823	−0.2151
3	600	18.2950	0.0061	0.9921	0.9867	−0.1664
4	800	36.3027	0.0053	0.9921	0.9890	−0.1333
5	1000	56.0108	0.0045	0.9932	0.9906	−0.1033
6	1200	75.1300	0.0042	0.9936	0.9919	−0.0812
7	1400	92.8699	0.0043	0.9940	0.9929	−0.0621
8	1600	125.3514	0.0036	0.9946	0.9935	−0.0731
9	1800	153.0013	0.0035	0.9949	0.9941	−0.0530
10	2000	170.3619	0.0034	0.9952	0.9946	−0.0446
$M = 40$						
2	400	2.9063	0.0252	0.9938	0.9882	−0.2362
3	600	10.1600	0.0636	0.9948	0.9916	−0.1726
4	800	22.8484	0.0830	0.9954	0.9930	−0.1478
5	1000	39.7976	0.1222	0.9954	0.9942	−0.1223
6	1200	60.8450	0.1102	0.9961	0.9948	−0.1088
7	1400	80.6101	0.1425	0.9964	0.9955	−0.0902
8	1600	98.5102	0.1624	0.9967	0.9960	−0.0608
9	1800	123.9556	0.1989	0.9969	0.9964	−0.0723
10	2000	135.5715	0.2071	0.9971	0.9968	−0.0636

with larger population size is not efficient for low-dimensional problems, the chance of premature convergence on high-dimensional problems is also smaller. It can be inferred that given a set of parameters, the convergence rate of the algorithm with smaller population sizes may exceed that of the same algorithm with larger population sizes when the dimension increases to exceed a certain number.

5.4 PARAMETER SELECTION AND PERFORMANCE COMPARISON

Parameter selection is the major concern when a stochastic optimisation algorithm is being employed to solve a given problem. It is noted that α is the most influential parameter on the convergence properties of QPSO apart from the population size. In Section 5.1, it is shown that it is necessary and sufficient to set $\alpha \le e^\gamma$ to prevent the individual particle from explosion and guarantee the convergence of the particle swarm. However, this does not mean that any value of α smaller than or equal to e^γ can lead to a satisfactory performance of QPSO in practical applications. This section intends to find out, through empirical studies, suitable control and selection of α so that QPSO may yield good performance in general. It also aims to compare the performance with other forms of PSO. Eight benchmark functions, f_1 to f_8, as presented in Chapter 2, are used here to test the algorithms.

5.4.1 Methods for Parameter Control

There are several control methods for the parameter α when QPSO is applied to practical problems. A simple approach is to set α as a fixed value when executing the algorithm. Another method is to decrease the value of α linearly during the course of the search process, that is,

$$\alpha = \frac{(\alpha_1 - \alpha_2) \times (n_{max} - n)}{n_{max}} + \alpha_2, \qquad (5.114)$$

where

α_1 and α_2 are the initial and final values of α, respectively
n is the current iteration number
n_{max} is the maximum number of allowable iterations

A variant of QPSO-Type-2 in which the mean best position C in (4.82) is replaced by the personal best position of a randomly selected particle in the swarm at each iteration. For convenience, QPSO-Type-2 with C is denoted as QPSO-Type-2-I and that with randomly selected personal best position as QPSO-Type-2-II.

5.4.2 Empirical Studies on Parameter Selection

The aforementioned two parameter controlling methods for α were tested by applying the algorithms QPSO-Type-1, QPSO-Type-2-I and QPSO-Type-2-II to three frequently used functions: Rosenbrock function (f_3), Rastrigin function (f_4) and Griewank's function (f_6). Rastrigin and Griewank functions are two difficult multimodal problems and Rosenbrock function is a unimodal problem. The expressions and bounds of the functions are shown in Chapter 2. The set of numerical tests consists of running each control method for α using each algorithm with 20 particles and 100 runs for each function. N was chosen as 30 for each of these functions. The initial position of each particle was determined randomly within the search domain. To determine the effectiveness of each algorithm with a control method, the best objective function value (i.e. the best fitness value) found after 3000 iterations was averaged over the same choice of α and the same benchmark function. The results obtained by the controlling methods were also compared across the three benchmark functions. The best α of each control method was selected by ranking the averaged best objective function values for each problem, summing the ranks, and taking the value that had the lowest summed rank, provided that the performance is acceptable (in the top half of the rankings) in all the tests for a particular setting of α.

The results for QPSO-Type-1 are presented in Table 5.8. When the fixed-value method was used, α was set to a range of values smaller than $e^\gamma \approx 1.781$ in each case. Results obtained for α outside the range [1.2, 0.85] were very poor and are not listed in the Table 5.8. It can be observed from Table 5.8 that very different objective function values were generated by using QPSO-Type-1 for different values of α. In particular, results for Rosenbrock function show that the function values obtained are very sensitive to the values of α. The best result occurs when $\alpha = 1.0$. When linearly varying α was used, α_1 and α_2 ($\alpha_1 > \alpha_2$) were selected from a series of different values less than $e^\gamma \approx 1.781$. Only acceptable results are listed in the table. It can be seen that the performance is sensitive to both α_1 and α_2. It is also found that decreasing α linearly from 1.0 to 0.9 leads to the best performance in general.

Table 5.9 records the results for QPSO-Type-2-I. For the fixed-value method, the results of the algorithm obtained outside the range [0.6, 1.2] are not listed because of their poor qualities. It can be observed from the results that the performance of QPSO-Type-2-I is less sensitive to α compared to QPSO-Type-1. The results seem fairly stable when α lies in the interval [0.8, 0.7]. The best results in this case occur when $\alpha = 0.75$. For the linearly varying method, the objective function values obtained by QPSO-Type-2-I using different ranges of α values are less different from each other as those obtained by QPSO-Type-1. It is also identified that varying α linearly from 1.0 to 0.5 could yield the best quality results.

The results for QPSO-Type-2-II are summarised in Table 5.10. It is clear from the results that the value of α, whether it is the fixed-value or time-varying method, should be set relatively small so that the algorithm is comparable in performance with the other types of QPSO. Results obtained with α outside the range [0.4, 0.8] were of poor quality and are not listed in Table 5.10. As shown in Table 5.10, α should be set in the range of 0.5–0.6 when the fixed-value method is used. The best results were obtained by setting $\alpha = 0.54$. On the other hand, the algorithm exhibits the best performance when α was decreasing linearly from 0.6 to 0.5 for the linearly varying method.

TABLE 5.8   Mean Best Fitness Values Obtained by QPSO-Type-1

α	Rosenbrock	Rastrigin	Griewank	α	Rosenbrock	Rastrigin	Griewank
Fixed α							
1.2	4.2826×10^4	76.3189	0.1332	**0.97**	145.8500	67.1396	0.0388
1.10	2.1139×10^4	**37.2340**	**0.0205**	**0.96**	288.8165	66.7416	0.0466
1.05	2.0526×10^4	43.6718	0.0213	**0.95**	98.3856	70.4229	0.0473
1.02	123.2954	47.0144	0.0217	**0.94**	71.9859	73.5172	0.0467
1.01	107.8675	54.8220	0.0266	**0.93**	351.1961	74.7908	0.1029
1.00	**69.9803**	53.7181	0.0267	**0.92**	117.1520	73.6466	0.0762
0.99	107.4633	59.3292	0.0316	**0.90**	74.5763	78.9895	0.1868
0.98	103.1208	60.1456	0.0430	**0.85**	235.1376	85.4868	0.4867
Linearly decreasing α							
1.2→0.8	1.2098×10^5	62.7922	0.0265	**1.0→0.8**	124.7534	61.2892	0.0278
1.2→0.7	525.6297	62.0354	0.0205	**1.0→0.7**	269.5334	67.1196	0.0347
1.2→0.6	494.7749	58.573	0.0239	**1.0→0.6**	170.8216	70.0846	0.0377
1.2→0.5	772.9795	65.142	0.0207	**1.0→0.5**	166.7593	68.6221	0.0357
1.1→0.9	157.9579	86.7267	0.0228	**0.9→0.8**	124.1540	82.2728	0.2623
1.1→0.8	553.5018	**56.2149**	0.0206	**0.9→0.7**	165.2973	80.7904	0.2896
1.1→0.7	158.7706	58.4138	**0.0181**	**0.9→0.6**	247.8022	80.7705	0.2867
1.1→0.6	441.2566	58.6426	0.0197	**0.9→0.5**	203.3426	84.2041	0.3143
1.1→0.5	237.9138	60.6624	0.0235	**0.8→0.7**	385.5187	83.785	1.1654
1.0→0.9	**110.8694**	57.8073	0.0229	**0.8→0.6**	1.3201×10^3	88.0189	1.1371

Note: Entries in bold are the best results for benchmarks with a certain parameter control method.

TABLE 5.9 Mean Best Fitness Values Obtained by QPSO-Type-2-I

α	Rosenbrock	Rastrigin	Griewank	α	Rosenbrock	Rastrigin	Griewank
Fixed α							
1.20	1.4692×10^9	280.8070	77.7691	0.75	**39.3163**	**32.6227**	0.0124
1.00	1.7487×10^5	209.9255	1.1187	0.74	52.4772	36.8626	0.0146
0.95	1.1765×10^3	192.551	0.6646	0.73	65.8057	35.0175	0.0170
0.90	90.1718	165.3004	0.2492	0.72	80.7611	41.4047	0.0311
0.85	66.5070	128.641	0.0456	0.71	79.6069	41.1116	0.0409
0.80	62.2914	75.907	0.0126	0.70	140.9301	44.6338	0.0610
0.78	53.4872	51.4421	**0.0095**	0.65	1.4055×10^4	52.9482	0.6207
0.76	56.6505	38.7811	0.0100	0.60	7.9429×10^5	64.1557	1.3434
Linearly decreasing α							
1.2→0.6	113.1691	37.2102	0.0113	1.0→0.4	178.3165	25.8537	0.0099
1.2→0.5	83.1475	27.0609	0.0100	1.0→0.3	126.3348	28.7571	0.0094
1.2→0.4	122.4921	27.4377	0.0116	0.9→0.6	86.3771	28.4534	0.0118
1.2→0.3	176.8883	30.7265	0.0088	0.9→0.5	124.9590	28.7131	0.0125
1.1→0.6	87.0748	34.2422	0.0089	0.9→0.4	109.3549	27.3782	0.0121
1.1→0.5	83.9509	26.7475	0.0121	0.9→0.3	112.0582	30.088	0.0099
1.1→0.4	95.8830	28.2161	0.0097	0.8→0.6	123.3768	29.3412	0.0097
1.1→0.3	120.9677	28.1815	**0.0079**	0.8→0.5	125.4396	30.4675	0.0102
1.0→0.6	**68.1485**	28.4897	0.0084	0.8→0.4	152.9611	33.908	0.0095
1.0→0.5	97.5625	**25.3874**	0.0087	0.8→0.3	133.5464	32.5239	0.0096

Note: Entries in bold are the best results for benchmarks with a certain parameter control method.

TABLE 5.10 Mean Best Fitness Values Obtained by QPSO-Type-2-II

α	Rosenbrock	Rastrigin	Griewank	α	Rosenbrock	Rastrigin	Griewank
Fixed α							
0.80	4.1108×10^6	256.6912	2.5629	**0.55**	92.6718	47.9632	0.0112
0.70	224.8298	210.8581	0.2911	**0.54**	81.7886	41.7568	0.0105
0.65	155.9584	184.3198	0.0378	**0.53**	89.9607	**40.769**	0.0128
0.60	83.2594	129.0775	0.0095	**0.52**	96.8675	44.996	0.0164
0.59	91.2091	109.9264	**0.0084**	**0.51**	89.9172	44.8129	0.0187
0.58	70.1766	92.6251	0.0114	**0.50**	92.9839	48.2355	0.0355
0.57	135.7775	72.4211	0.0129	**0.45**	3.8735×10^3	61.066	0.3608
0.56	**75.5104**	57.1841	0.0099	**0.40**	5.8156×10^6	68.5784	2.4911
Linearly decreasing α							
1.0→0.5	255.6318	174.985	0.0181	**0.8→0.3**	145.8116	40.0094	0.0143
1.0→0.4	371.6781	51.8245	0.0101	**0.8→0.2**	215.3517	40.2069	**0.0085**
1.0→0.3	667.5847	44.2418	0.0095	**0.7→0.5**	78.5191	94.3664	0.0109
1.0→0.2	594.4328	44.1004	0.0086	**0.7→0.4**	158.5018	**36.2573**	0.0115
0.9→0.5	154.9674	162.9026	0.0151	**0.7→0.3**	126.8576	38.0482	0.0105
0.9→0.4	180.5486	43.0173	0.0109	**0.7→0.2**	183.1478	40.6623	0.0099
0.9→0.3	252.4979	40.5222	0.0098	**0.6→0.5**	**78.1274**	44.0210	0.0095
0.9→0.2	553.9905	41.7187	0.0134	**0.6→0.4**	123.3330	37.0037	0.0116
0.8→0.5	111.3174	143.6441	0.0173	**0.6→0.3**	111.6767	43.3740	0.0105
0.8→0.4	111.0077	40.6233	0.0107	**0.6→0.2**	233.8916	40.7905	0.0103

Note: Entries in bold are the best results for benchmarks with a certain parameter control method.

5.4.3 Performance Comparison

To determine whether QPSO is as effective as other variants of PSO, including PSO with inertia weight (PSO-In) [22–24], and PSO with constriction factor (PSO-Co) [3,25], a series of experiments were conducted on the eight functions described in Chapter 2. Functions f_1 to f_3 are unimodal, while functions f_4 to f_8 are multimodal. Except that $N = 2$ for function f_7 (Schaffer's F6 function), the dimensionality of each function is 30. Each algorithm was run 100 times for each problem using 20 particles in searching the global best fitness value. At each run, the particles in the algorithms started in new and randomly generated positions which are uniformly distributed within the search scopes. Each run of each algorithm was executed 3000 iterations, and the best fitness value (objective function value) for each run was recorded.

For the three types of QPSO, both methods of controlling α were used and the parameters for each case were set as those suggested values that generated the best results in the previous experiments. For PSO-In, Shi and Eberhart showed that the PSO-In with linearly decreasing inertia weight performs better than the one with fixed inertia weight [23]. They varied the inertia weight linearly from 0.9 to 0.4 in the course of the run and fixed the acceleration coefficients (c_1 and c_2) at 2 in their empirical study [24]. For PSO-Co, Clerc and Kennedy found that the values of constriction factor χ and acceleration coefficients (c_1 and c_2) need to satisfy certain constraints in order for the particle's trajectory convergence without the restriction of velocities [3,25]. They recommended using a value of 4.1 for the sum of c_1 and c_2, which results in a value constriction factor $\chi = 0.7298$ and $c_1 = c_2 = 2.05$. Eberhart and Shi also used these values of the parameters when comparing the performance of PSO-Co with that of PSO-In [26]. These parametric values for PSO-In and PSO-Co were used in the experiments described here, though these might not be optimal.

The mean best fitness value and standard deviation out of 100 runs of each algorithm for each problem are presented in Tables 5.11 and 5.12. To investigate if the differences in the mean best fitness values between algorithms are significant, the mean values for each problem were analysed using a multiple comparison procedure, ANOVA (analysis of variance), with 0.05 as the level of significance. Unlike the honestly significant (THS) difference test used in [27], the procedure employed in here is known as the 'stepdown' procedure which takes into account that all but one of the comparisons are less different than the range. When doing all pairwise

TABLE 5.11   Mean and Standard Deviation of the Best Fitness Values after 100 Runs for f_1 to f_4

Algorithms	f_1 (Sphere Function)	f_2 (Schwefel's Problem 1.2)	f_3 (Rosenbrock Function)	f_4 (Rastrigin Function)
PSO-In (standard deviation)	3.5019×10^{-14} (4.8113×10^{-14})	419.2338 (241.8106)	201.1325 (423.1043)	39.8305 (11.5816)
PSO-Co	8.6167×10^{-38} (3.1475×10^{-37})	**0.0158** (0.0404)	57.9633 (91.3910)	73.5472 (19.4169)
QPSO-Type-1 ($\alpha = 1.00$)	2.2634×10^{-40} (9.6483×10^{-40})	15.0383 (10.3268)	69.9803 (114.1642)	51.7377 (23.3804)
QPSO-Type-1 ($\alpha = 1.00 \rightarrow 0.90$)	5.0922×10^{-48} (1.6942×10^{-47})	1.6964 (1.1189)	110.8694 (214.9120)	55.0409 (20.8591)
QPSO-Type-2-I ($\alpha = 0.75$)	$\mathbf{1.6029 \times 10^{-71}}$ (7.6101×10^{-71})	0.1369 0.1038	**39.3163** (35.9071)	31.0361 (12.7262)
QPSO-Type-2-I ($\alpha = 1.0 \rightarrow 0.5$)	2.9682×10^{-22} (1.6269×10^{-21})	227.8900 (128.3278)	97.5625 (171.1638)	**25.3874** (8.3176)
QPSO-Type2-II ($\alpha = 0.54$)	1.8152×10^{-69} (4.7739×10^{-69})	0.0523 (0.0541)	81.7886 (169.2082)	44.3119 (22.1026)
QPSO-Type2-II ($\alpha = 0.6 \rightarrow 0.5$)	4.7455×10^{-63} (1.2769×10^{-62})	0.3763 (0.3564)	78.1274 (147.3051)	44.0210 (26.4678)

Note: Entries in bold are the best results obtained for benchmarks.

TABLE 5.12   Mean and Standard Deviation of the Best Fitness Values after 100 Runs for f_5 to f_8

Algorithms	f_5 (Ackley Function)	f_6 (Greiwank Function)	f_7 (Schaffer's F6 Function)	f_8 (Schwefel Function)
PSO-In (standard deviation)	0.1096 (0.3790)	0.0139 (0.0151)	$\mathbf{1.9432 \times 10^{-4}}$ (0.0014)	4.0872×10^3 (743.9391)
PSO-Co	2.9731 (1.9449)	0.0479 (0.0741)	0.0035 (0.0047)	3.7826×10^3 (620.7381)
QPSO-Type-1 ($\alpha = 1.00$)	0.3791 (2.0506)	0.0219 (0.0291)	0.0058 (0.0048)	2.7570×10^3 (527.4648)
QPSO-Type-1 ($\alpha = 1.0 \rightarrow 0.90$)	0.3405 (1.2457)	0.0229 (0.0302)	0.0061 (0.0047)	2.8276×10^3 (629.0328)
QPSO-Type-2-I ($\alpha = 0.75$)	0.0799 (0.3216)	0.0129 (0.0150)	0.0033 (0.0046)	2.6859×10^3 (467.4396)
QPSO-Type-2-I ($\alpha = 1.0 \rightarrow 0.5$)	1.8491×10^{-12} (2.6806×10^{-12})	**0.0087** (0.0118)	0.0016 (0.0036)	3.4083×10^3 (900.6818)
QPSO-Type2-II ($\alpha = 0.54$)	0.0547 (0.3870)	0.0105 (0.0122)	0.0029 (0.0045)	$\mathbf{2.5635 \times 10^3}$ (480.8379)
QPSO-Type2-II ($\alpha = 0.6 \rightarrow 0.5$)	$\mathbf{7.8515 \times 10^{-15}}$ (3.3056×10^{-15})	0.0095 (0.0111)	0.0029 (0.0045)	2.6399×10^3 (574.0727)

Note: Entries in bold are the best results obtained for benchmarks.

comparisons, this approach is the best available if confidence intervals are not needed and sample sizes are equal [28].

A ranking system is used to identify the most effective algorithm for each problem. The algorithms that are not statistically different from each other would be given the same rank; those that are not statistically different from more than one other groups of algorithms are ranked with the best-performing of these groups. For each algorithm, the resulting rank and the total rank for each problem are shown in Table 5.13.

For the sphere function (f_1), both QPSO-Type-2 algorithms with the fixed-value method produced better results than the other methods. The results for Schwefel's Problem 1.2 (f_2) show that QPSO-Type-2-II with the fixed-value method and PSO-Co produced better results, but the performances of PSO-In and QPSO-Type-2-I with linearly decreasing α seem to be inferior to the other competitors. For benchmark problem (f_3), Rosenbrock function, QPSO-Type-2-I with the fixed-value method performed the best, and the performance of PSO-In seems to be inferior to the other algorithms. For Rastrigin's function (f_4), QPSO-Type-2-I with linearly decreasing α yielded the best result, and the results obtained by PSO-In and all other QPSO-based methods rank the second best. As far as the benchmark function (f_5) is concerned, both QPSO-Type II methods with time-varying α showed better performances for this problem than the others. The results for Griewank's function (f_6) suggest that both of QPSO-Type-2 algorithms with linearly varying α are able to find the solution for the function with better quality compared to the other methods.

TABLE 5.13 Ranking by Algorithms and Problems

Algorithms	f_1	f_2	f_3	f_4	f_5	f_6	f_7	f_8	Total Rank
In-PSO	7	8	7	7	=3	8	13	1	116
Co-PSO	=6	=1	6	8	=6	=1	=3	13	87
QPSO-Type-1 ($\alpha = 1.00$)	4	6	=5	=5	7	=1	=3	7	64
QPSO-Type-1 ($\alpha = 1.00 \rightarrow 0.90$)	3	5	4	=5	=6	=1	=3	1	46
QPSO-Type-2-I ($\alpha = 0.75$)	=6	=3	1	=2	=3	1	3	1	28
QPSO-Type-2-I ($\alpha = 1.00 \rightarrow 0.5$)	5	7	=5	=2	1	1	=3	1	41
QPSO-Type2-II ($\alpha = 0.54$)	1	=1	2	=2	3	1	=1	1	**25**
QPSO-Type2-II ($\alpha = 0.6 \rightarrow 0.5$)	2	=3	3	1	2	=1	=1	1	33

For Schaffer's F6 function (f_7), PSO-In showed the best performance in finding the global optimum. For Schwefel's function (f_8) QPSO-Type-2-II with the fixed-value method yielded the best results.

As shown by the total ranks listed in Table 5.13, the methods based on QPSO-Type-2 attain better overall performances than all of the other tested algorithms. It is also revealed by the total ranks that QPSO-Type-2-II performs slightly better than QPSO-Type-2-I. However, more detailed comparisons reveal that the QPSO-Type-2-I with the fixed-value method shows the most stable performance across all of the benchmark functions with the worst rank being 6 for f_7. It is also observed that for both versions of QPSO-Type-2, time-varying method did not show better overall performance than the fixed-value method, particularly for QPSO-Type-2-I.

The second best-performing algorithm is the QPSO-Type-1 algorithm as indicated by the total ranks. It can be seen that there is much difference between the performances of the two types of parameter control methods. Unlike QPSO-Type-2, QPSO-Type-1 appears to show better results by using the time-varying control method. PSO-Co is considered as the second best algorithm and is ranked fourth in the overall performance. It shows slightly better performance compared to the PSO-In algorithm, which is particularly good performance for benchmark function such as f_6.

5.5 SUMMARY

This chapter analyses the behaviour of the individual particle in QPSO, gives proofs of the global convergence of the algorithm, evaluates the performance of the algorithm in terms of computational complexity and convergence rate, provides the guideline of parameter selection for the algorithm and compares the performance between QPSO and several other forms of PSO.

Two types of particles corresponding to two search strategies of QPSO are analysed by using the theory of probability measures. Necessary and sufficient conditions for the position of a single particle to be convergent or probabilistic bounded are derived. The choice of the contraction–expansion coefficient, α, has important implication in the condition for convergence or probabilistic boundedness of the position of a particle. For a Type-1 particle, if $\alpha < e^\gamma$, the position of the particle converges to its local attractor in probability (or almost surely, in distribution); if $\alpha = e^\gamma$, the position is probabilistic bounded; otherwise, the position diverges. For a Type-2 particle, if $\alpha \leq e^\gamma$, the position is probabilistic bounded, or else it diverges.

The convergence of the QPSO algorithm is investigated from the perspectives of probability measures, Markov processes and PM spaces and shown to be a global convergent algorithm. In addition, convergence or boundedness of the individual particle which guarantees QPSO to converge to the global optimum with finite time complexity is explained. The effectiveness of the algorithm is also evaluated by considering time complexity for the sphere function. The convergence rate is also considered with numerical examples given in the chapter.

Two methods, one known as the fixed-value method and the other linearly varying method, of controlling α are examined. Several benchmark functions are used in the numerical studies. It can be concluded that QPSO, particularly QPSO-Type-2, is comparable with or even better than other forms of PSO in finding the optimal solutions of the tested benchmark functions.

The chapter concentrates on the search mechanism of each individual particle of QPSO. The theoretical part of the investigation exposed in this chapter forms the solid foundation for different applications of QPSO. It is still an open challenging question on the determination of the values of parameters of QPSO in order to improve the performance of the algorithm. It is hoped that this exposition has provided the reader with certain solid theoretical foundation and hopefully leads to further improvements of the QPSO algorithm.

REFERENCES

1. J. Sun, W. Fang, X. Wu, V. Palade, W. Xu. Quantum-behaved particle swarm optimization: Analysis of the individual particle's behaviour and parameter selection. *Evolutionary Computation*, in press.
2. A.N. Shiryayev. *Probability*. Springer-Verlag, Inc., New York, 1984.
3. M. Clerc, J. Kennedy. The particle swarm-explosion, stability and convergence in a multidimensional complex space. *IEEE Transactions on Evolutionary Computation*, 2002, 6(2): 58–73.
4. R. Courant. *Introduction to Calculus and Analysis*. Springer-Verlag, New York, 1989.
5. F.J. Solis, R.J.-B. Wets. Minimization by random search techniques. *Mathematics of Operations Research*, 1981, 6(1): 19–30.
6. W. Fang, J. Sun, W. Xu. Convergence analysis of quantum-behaved particle swarm optimization algorithm and study on its control parameter. *Acta Physica Sinica*, 2010, 59(6): 3686–3694.
7. D.-H. Shi, J.-P. Peng. A new framework for analyzing stochastic global optimi algorithms. *Journal of Shanghai University*, 1999, 6(3): 175–180.
8. B. Schweizer, A. Sklar. *Probabilistic Metric Spaces*. Dover Publications, New York, 1982.

9. S.-S. Chang. On some fixed point theorems in probabilistic metric space and its applications. *Probability Theory and Related Fields*, 1983, 63(4): 463–474.
10. Y. Yu, Z.-H. Zhou. A new approach to estimating the expected first hitting time of evolutionary algorithms. *Artificial Intelligence*, 2008, 172(15): 1809–1832.
11. J. He, X. Yao. Towards an analytic framework for analysing the computation time of evolutionary algorithms. *Artificial Intelligence*, 2003, 145(1–2): 59–97.
12. J. He, X. Yao. Drift analysis and average time complexity of evolutionary algorithms. *Artificial Intelligence*, 2001, 127(1): 57–85.
13. J. He, X. Yao. From an individual to a population: An analysis of the first hitting time of population-based evolutionary algorithms. *IEEE Transactions on Evolutionary Computation*, 2002, 6(5): 495–551.
14. B.Z. Zabinsky, R.L. Smith. Pure adaptive search in global optimization. *Mathematical Programming*, 1992, 53: 323–338.
15. W.P. Baritompa, B.-P. Zhang, R.H. Mlandineo, G.R. Wood, Z.B. Zambinsky. Towards pure adaptive search. *Journal of Global Optimization*, 1995, 7: 93–110.
16. M. Vose. Logarithmic convergence of random heuristic search. *Evolutionary Computation*, 1997, 4: 395–404.
17. X. Qi, F. Palmeiri. Theoretical analysis of evolutionary algorithms with infinite population size in continuous space, part I: Basic properties. *IEEE Transactions on Neural Networks*, 1994, 5: 102–119.
18. H.-G. Beyer. Toward a theory of evolution strategies: On the benefits of sex— The (μ/μ, λ) theory. *Evolutionary Computation*, 1995, 3: 81–111.
19. K. Leung, Q. Duan, Z. Xu, C.K. Wong. A new model of simulated evolutionary computation-convergence analysis and specification. *IEEE Transactions on Evolutionary Computation*, 2001, 5(1): 3–16.
20. D.R. Stark, J.C. Spall. Rate of convergence in evolutionary computation. In *Proceedings of the 2003 American Control Conferences*, Denver, CO, 2003, pp. 1932–1937.
21. G. Rudolph. Local convergence rates of simple evolutionary algorithms with Cauchy mutations. *IEEE Transactions on Evolutionary Computation*, 2007, 1(4): 249–258.
22. Y. Shi, R.C. Eberhart. A modified particle swarm optimizer. In *Proceedings of the 1998 IEEE International Conference on Evolutionary Computation*, Anchorage, AK, 1998, pp. 69–73.
23. Y. Shi, R.C. Eberhart. Parameter selection in particle swarm optimization. In *Proceedings of the Seventh Conference on Evolutionary Programming*, San Diego, CA, 1998, Vol. VII, pp. 591–600.
24. Y. Shi, R.C. Eberhart. Empirical study of particle swarm optimization. In *Proceedings of the 1999 Congress on Evolutionary Computation*, Washington, DC, 1999, Vol. 3, pp. 1945–1950.
25. M. Clerc. The swarm and the queen: Towards a deterministic and adaptive particle swarm optimization. In *Proceedings of the 1999 Congress on Evolutionary Computation*, Washington, DC, 1999, pp. 1951–1957.

26. R.C. Eberhart, Y. Shi. Comparing inertia weights and constriction factors in particle swarm optimization. In *Proceedings of the 2000 Congress on Evolutionary Computation*, San Diego, CA, 2000, Vol. 1, pp. 84–88.

27. T.J. Richer, T.M. Blackwell. The Levy particle swarm. In *Proceedings of the 2006 Congress on Evolutionary Computation*, 2006, British Columbia, Canada, pp. 808–815.

28. R.W. Day, G.P. Quinn. Comparisons of treatments after an analysis of variance in ecology. *Ecological Monographs*, 1989, 59: 433–463.

CHAPTER 6

Industrial Applications

THIS CHAPTER COVERS SEVERAL real-world applications of the QPSO algorithm, including inverse problems, optimal design of digital filters, economic dispatch (ED) problems, biological multiple sequence alignment (MSA) and image processing, showing details of implementation aspects so that the readers may be better equipped with suitable knowledge of applying the algorithm.

The material covered in each section is intended to be self-contained. Sufficient references and description of the problems are provided to the readers in order to make clear the work.

6.1 INVERSE PROBLEMS FOR PARTIAL DIFFERENTIAL EQUATIONS

6.1.1 Introduction

Inverse problems occur in many branches of sciences and mathematics where the value of certain model parameter(s) are required from observed data. Mathematically, an inverse problem is to find θ such that (at least approximately)

$$d = G(\theta), \tag{6.1}$$

where G is an operator describing the explicit relationship between the dataset d and the parameter set θ of the physical system. Inverse problems are typically ill-posed as opposed to the well-posed problems in which physical situations are modelled using known model parameters or material properties.

The subject of inverse problems has enjoyed a remarkable growth in the past few decades. This subject is too broad and it is difficult to provide a complete account of all recent developments here. In this chapter, attention is focused on the inverse problems for partial differential equations and dynamical systems. In particular, this section addresses solutions of the former by using the QPSO algorithm with inverse heat conduction problem (IHCP) in the form of a case study.

Heat conduction through a medium is governed by the thermophysical properties, such as the thermal conductivity, of the medium. The quantity of these properties has a significant influence on the analysis of the temperature distribution and heat flow rate when the material is heated and on the analysis of thermal instability problems. An efficient, robust and accurate method for estimating the thermal conductivity is useful especially in the material design industry. One application of IHCP in engineering and sciences is to predict the thermal conductivity from temperature measurements obtained at finite number of interior locations of the physical body under consideration.

The determination of the thermal conductivity from a measured temperature profile is a coefficient inverse problem of heat conduction [1,2]. In many practical engineering problems, the thermal conductivity is temperature dependent and the heat conduction equation is a non-linear equation which is usually difficult to solve. Many methods have been developed to determine the temperature-dependent thermal conductivity of materials. Huang used the conjugate gradient method (CGM) with an adjoint equation to search for the thermal conductivity [3,4]. Terrol applied a least squares method to estimate thermal conductivity functions [5]. Kim formulated the problem to find the solution through the direct integral method [6], and Yeung et al. predicted the thermal conductivity by solving the discretised linear equations from the whole domain temperature measurements [7–9]. In [10], a sensitivity equation was used to estimate the parameters in a known functional form of the thermal conductivity. However, most of the methods involve either steady-state problems or parameter estimation with a priori functional form of the thermal conductivity. In the open literature, only the CGM with an adjoint equation was used to estimate unknown thermal conductivities in IHCPs without a priori knowledge of the functional form. However, the initial guess of the unknown quantities must be chosen carefully to guarantee the convergence of the method.

In the remaining part of this section, the QPSO algorithm is used to predict temperature-dependent thermal conductivity without a priori information on its functional form—function estimation in inverse problems. Details of such implementation of the algorithm are discussed. In general, the inverse problem is converted to an optimisation problem which involves using the least squares method to evaluate the positions of the particles. Tikhonov regularisation is used with QPSO in order to stabilise the ill-posed inverse problem. Comparisons of QPSO with PSO, GA and CGM are also presented.

6.1.2 IHCPs

Consider a typical inverse problem of estimating the temperature-dependent thermal conductivity function of a two-dimensional (2D) slab with thickness L and infinite length (Figure 6.1).

For time $t > 0$, the boundary surfaces at $x = 0$ and $x = L$ are subject to the prescribed time-varying heat fluxes $q_1(t)$ and $q_2(t)$, respectively. As the slab is of infinite length, the mathematical model governing the heat conduction process may be reduced to a one-dimensional (1D) problem as in the following:

$$
\begin{cases}
\rho c \dfrac{\partial T}{\partial t} = \dfrac{\partial}{\partial x}\left[K(T)\dfrac{\partial T}{\partial x} \right] + g(x,t) & 0 < x < L,\ t > 0, \\[3mm]
-K(T)\dfrac{\partial T}{\partial t}\bigg|_{x=0} = q_1(t) & t > 0, \\[3mm]
-K(T)\dfrac{\partial T}{\partial t}\bigg|_{x=L} = q_2(t) & t > 0, \\[3mm]
T(x,0) = T_0 & 0 < x < L.
\end{cases}
\tag{6.2}
$$

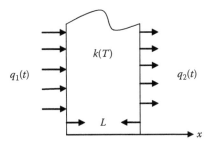

FIGURE 6.1   A simplified one-dimensional slab.

For simplicity, the physical properties may be taken as $\rho c = L = 1$. In real engineering problem, the thermal conductivity $K(T)$ is to be predicted from temperature measurements obtained from the sensors located inside of the slab. Here simulated temperature, $T(x, t)$, is generated by solving the direct problem and is used as the measured temperature. The direct problem involved in Equation 6.2 is non-linear since the thermal conductivity is a function of temperature, that is, $K(T)$. An iterative technique is needed to solve the discretised direct problem resulting from applying an implicit finite difference method to Equation 6.2 as given in the following:

$$\rho c \frac{T_i^{j+1} - T_i^j}{\Delta t} = \frac{K_{i+1}^{j+1} - K_{i-1}^{j+1}}{2\Delta x} \frac{T_{i+1}^{j+1} - T_{i-1}^{j+1}}{2\Delta x} + K_i^{j+1} \frac{T_{i+1}^{j+1} - 2T_i^{j+1} + T_{i-1}^{j+1}}{\Delta x^2}$$
$$+ g\big((i-1)\Delta x, (j-1)\Delta t\big), \tag{6.3}$$

where

T_i^j and K_i^j are the temperature and thermal conductivity, respectively, at the jth time step along the ith grid point
$i = 1, 2, \ldots, N_x - 1$
Δx is the mesh size
Δt is the temporal step size

Here N_x is the number of grid points along x axis such that $\Delta x = L/N_x - 1$. Along the boundaries, a second-order finite difference approximation is used leading to

$$-K_0^j \frac{-3T_0^j + 4T_1^j - T_2^j}{2\Delta x} = q_1(t_j), \tag{6.4}$$

$$-K_{N_x}^j \frac{-3T_{N_x}^j + 4T_{N_x-1}^j - T_{N_x-2}^j}{2\Delta x} = q_2(t_j), \tag{6.5}$$

where $t_j = (j - 1)\Delta t$. In the case of an inverse problem, the thermal conductivity $K(x, t)$ at any specific time and position (x, t) is to be estimated so that the temperature history computed from the mathematical model matches the measured temperature history. The problem may be solved by minimising an objective function defined as the sum of the squared differences between the measured temperatures and the computed temperatures, that is,

$$J[K(T)] = \|T(K) - Y\|^2, \tag{6.6}$$

where Y is the measured temperature. The average error, E, between the exact and predicted values for thermal conductivity $K(x, t)$ may be defined as

$$E = \frac{1}{N_x \times N_t} \sqrt{\sum_{j=1}^{N_t} \sum_{i=1}^{N_x} \left(K_i^j - \tilde{K}_i^j \right)^2} \tag{6.7}$$

where

N_t is the number of time steps
N_x is the number of spatial grid points
K_i^j is the predicted thermal conductivity at the jth time step along the ith grid point
$\tilde{K}_i^j$ is the exact thermal conductivity at the jth time step along the ith grid point

6.1.3 QPSO for IHCPs

In this section, the QPSO method is considered for solving the IHCP of estimating the thermal conductivity. The thermal conductivity $K(x, t)$ is discretised spatially at the time $t_j = (j - 1)\Delta t$ as $K(x, t_j) = \{ K(x_0, t_j), K(x_1, t_j), \ldots, K(x_i, t_j), \ldots, K(x_{N_x}, t_j) \}$ in the numerical computation and the optimisation process. The dimension, N, of the particle positions equals to the number of spatial grid points N_x used in the discretisation of $K(x, t_j)$. The problem is to find such thermal conductivity $K(x, t)$, for which the temperature history computed from the mathematical model matches the measured temperature history.

Consider the optimisation problem of finding a minimum of the objective function $J[K(T)]$. For this optimisation problem, at each iteration of the QPSO algorithm, a population of particles needs to be maintained, that is,

$$X_n = \{ X_{1,n}, X_{2,n}, \ldots, X_{i,n}, \ldots, X_{M,n} \}, \tag{6.8}$$

where $X_{i,n}$ is the current position of particle i,

$$X_{i,n} = \left(X_{i,n}^1, X_{i,n}^2, \ldots, X_{i,n}^j, \ldots, X_{i,n}^N \right). \tag{6.9}$$

Equation 6.9 represents a feasible solution of $K(x,t_j) = \left\{ K(x_0,t_j), K(x_1,t_j), \ldots, K(x_i,t_j), \ldots, K(x_{N_x},t_j) \right\}$. Substituting $X_{i,n}$ into the constraints described by Equation 6.2, the temperatures $T(x, t_j)$ can be computed. Each feasible solution $X_{i,n}$ is evaluated by computing the fitness function (6.6).

At each iteration, the positions of the particles are updated according to Equations 4.81 through 4.83. This iterative process is repeated until a predefined number of generations is reached or the solution converges.

Inverse problems are usually ill-posed and the solutions are very sensitive to random noise generated from measurements by using sensors. Therefore, special techniques for its solution are needed in order to satisfy the stability condition. One approach to reduce such instabilities is to use the procedure known as the Tikhonov regularisation, which modifies the least squares norm by the addition of a regularisation term as follows:

$$J\left[K\left(x,t_j \right) \right] = \sum_{i=1}^{N_x} \left(T_i^j - Y_i^j \right)^2 + \lambda \sum_{i=1}^{N_x} \left(K_i^j \right)^2 \tag{6.10}$$

where λ is the regularisation parameter and the second term on the right-hand side is the whole-domain zeroth-order regularisation term [2]. The values chosen for the regularisation parameter λ affects the stability of the solution. As $\lambda \to 0$, the solution exhibits oscillatory behaviour and becomes unstable. On the other hand, with large values of λ the solution is damped and deviates from the exact result. Tikhonov suggested that λ should be selected according to the discrepancy principle—the minimum value of the objective function is equal to the sum of the squares of the errors due to the measurements. It is also possible to use the L-shape curve method [11,12] in order to find the best value of λ. The Tikhonov first-order regularisation procedure involves the minimisation of the following modified least squares norm:

$$J\left[K\left(x,t_j \right) \right] = \sum_{i=1}^{N_x} \left(T_i^j - Y_i^j \right)^2 + \lambda \sum_{i=1}^{N_x-1} \left(K_{i+1}^j - K_i^j \right)^2. \tag{6.11}$$

The procedure of QPSO for inverse problems is described as follows in Figure 6.2.

```
for j = 1:N_t
        Initialize particles with random positions X_i = {X_{1,0},X_{2,0},...,
        X_{3,0},...,X_{M,0}};
        P_0 = X_0; n = 0; α = 1;
        Find the global best position G_0;
        while n < n_max
                Compute the mean best position by Equation 4.81;
                Update contraction-expansion coefficient according
                to Equation 5.114
                Compute p by Equation 4.83;
                Update the positions of all particles according to
                Equation 4.88;
                for i = 1: M
                        If J(X_{i,n}) < J(P_{i,n-1}) then P_{i,n} = X_{i,n};
                        else P_{i,n} = P_{i,n-1}
                        Find G^n;
                end for
                n = n + 1;
        end while
                K(x, t_j) = G_{n_max};
end for
```

FIGURE 6.2   The procedure of QPSO for IHCPs.

6.1.4 Numerical Experiments and Results

To illustrate the validity and stability of the QPSO method in predicting temperature-dependent thermal conductivity $K(T)$, one example with known exact analytic functional form is considered in this section.

Consider the linear temperature-dependent thermal conductivity

$$K(T) = a_0 + a_1T, \tag{6.12}$$

where the constants a_0 and a_1 were both taken to be 0.5 in the numerical experiments. This slab material has initial temperature $T_0 = \sin(x)$ and heat source $g(x, t) = -0.5 \cdot \exp(-t)(\sin x - \exp(-t)\cos(2x))$. When $t > 0$, the left and right boundaries are subjected to time-dependent heat fluxes $q_1(t)$ and $q_2(t)$, respectively. Analytic solution of the form $T(x, t) = \sin(x)\exp(-t)$, $0 < x < 1.0, 0 < t \leq 1.0$ was used to obtain the simulated temperature measurements. The spatial mesh size was chosen as $\Delta x = 0.05$ and the temporal step size was taken as $\Delta t = 0.05$.

QPSO, PSO, GA and CGM were tested using the example. The parameters in the QPSO method were set, respectively, as $M = 20$, $n_{max} = 1000$, $N = 21$, the contraction–expansion coefficient α was controlled to be decreased linearly from 1.0 to 0.5. Those in the PSO algorithm were set as

$M = 20$, $n_{max} = 1000$, $N = 21$, the inertia weight w was chosen to decrease linearly from 0.9 to 0.4, $c_1 = c_2 = 2$. The GA used in the experiment was configured as $M = 50$, $N = 21$, $n_{max} = 1000$, crossover probability $p_c = 0.8$, mutation probability $p_m = 0.2$.

Normally distributed uncorrelated errors with zero mean and constant standard deviation were added to the exact solution in order to study the performance of the algorithm for random errors in measurements. The simulated inexact measurement Y can be expressed as

$$Y = Y_{exact} + \sigma\varepsilon, \qquad (6.13)$$

where

Y_{exact} is the solution of the direct problem with the exact value of $K(T)$
σ is the standard deviation of the measurements
ε is a random number which lies within a specified confidence bound

In the present test, 99% confidence bound is used, that is, $-2.57 < \varepsilon < 2.576$.

Note that in generating the simulated measurement temperature Y, the exact $K(T)$ is used in the direct problem. The problem is non-linear and the iterative technique described previously is needed in the solution process. However, in the inverse estimation, the thermal conductivity exists in the form of $K(x, t)$, and the problem becomes linear.

The predicted $K(x, t)$ obtained by using the four methods with exact measurements, $\sigma = 0.0$, are shown in Figure 6.3. The total objective function values over N_t runs and the average error generated by each method are listed in Table 6.1. These results indicate good performance of the QPSO algorithm for the problem. The average errors resulted from using the initial approximations of K, $K^0(x, t) = 0.5$ and $K^0(x, t) = 0.8$, are 4.4445×10^{-6} and 4.9954×10^{-5}, respectively. This implies that the performance of CGM depends strongly on its initial guess.

The predicted thermal conductivity using inexact temperature measurements generated by using Equation 6.13 are shown in Figure 6.4. Here the initial approximation of K was chosen as 0.5. The total objective function values and the average errors generated by each method are listed in Table 6.2. Note that relative good results are obtained by using QPSO. This illustrates the robustness of the QPSO method for IHCPs.

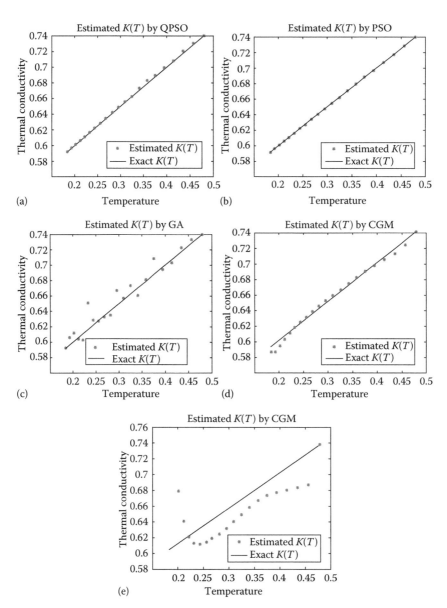

FIGURE 6.3 (a) QPSO, (b) PSO, (c) GA, (d) CGM with $K^0(x, t) = 0.5$, and (e) CGM with $K^0(x, t) = 0.8$ show the predicted thermal conductivity at $x = 0.5$ using the temperature exact measurement with $\sigma = 0.0$.

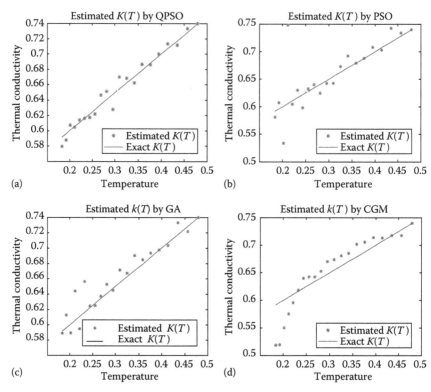

FIGURE 6.4 (a) QPSO, (b) PSO, (c) GA, and (d) CGM with $K^0(x, t)$ show the predicted thermal conductivity at $x = 0.5$ taking $\sigma = 0.005$.

TABLE 6.1 Objective Function Values and Average Errors at $x = 0.5$ Taking $\sigma = 0.0$

Algorithms	Objective Function Value	Average Error
QPSO	2.01×10^{-6}	1.45×10^{-4}
PSO	2.23×10^{-6}	1.48×10^{-4}
GA	1.18×10^{-4}	8.45×10^{-4}
CGM ($K^0(x, t) = 0.5$)	1×10^{-3}	4.31×10^{-4}
CGM ($K^0(x, t) = 0.8$)	2.15×10^{-2}	3.80×10^{-3}

TABLE 6.2 Total Objective Function Values and Average Errors at $x = 0.5$ and $\sigma = 0.005$

Algorithms	Objective Function Value	Average Error
QPSO	7.75×10^{-4}	1.73×10^{-3}
PSO	6.96×10^{-3}	4.62×10^{-3}
GA	2.44×10^{-3}	1.86×10^{-3}
CGM ($K^0(x, t) = 0.5$)	5.08×10^{-2}	1.60×10^{-3}

6.1.5 Summary

Numerical predictions of temperature-dependent thermal conductivity without a priori information of its functional form in IHCPs are presented. Tikhonov regularisation together with QPSO is used to handle ill-posedness and to maintain the stability in the inverse solution process. Numerical results demonstrate the efficiency, stability and viability of the QPSO method for solving IHCPs. In addition, the stochastic algorithm avoids the complicated gradient computation as in gradient-based method and guarantees the convergence to the global optimum.

6.2 INVERSE PROBLEMS FOR NON-LINEAR DYNAMICAL SYSTEMS

6.2.1 Introduction

In general, a physical system may be described by systems of ordinary equations (ODEs) or by differential algebraic equations. A system of ODEs with suitable initial conditions is known as a dynamical system which is very often non-linear and even chaotic in nature. Solution of parameter estimation problems, that is, inverse problems, plays a key role in the construction of non-linear dynamic models. During the past few years, many different techniques have been proposed to tackle the inverse problems of non-linear dynamical systems, especially those of chaotic systems.

Annan and Hargreaves introduced a method to identify the parameters of chaotic systems using Kalman filter [13]. Parlitz [14] and Chen et al. [15] employed the synchronisation method to estimate the parameters of chaotic systems based on the theory of Lyapunov function, and Li et al. [16] further investigated into the properties of this method. Guan et al. introduced the concept of state observer to identify the unknown parameters [17]. Lü and Zhang [18] and Lie et al. [19] used Guan's method to control chaotic systems. Gu et al. applied linear associate memory to the estimation of the unknown parameters [20]. Alvarez et al. discussed how to estimate parameter from the two-valued symbolic sequences generated by iterations of quadratic map when its initial value is known [21]. Wu et al. studied the two-valued symbolic sequences using Grey codes [22]. In addition, work using modern heuristic methods for the identification of chaotic systems has been carried out in several occasions. For example, Wang proposed a parameter estimation method based on genetic algorithms (GAs) through the construction of a suitable fitness function [23]. Gao et al. invented a method for the estimation problem using particle swarm optimisation (PSO) [24]. Li et al. proposed the chaotic ant swarm (CAS) algorithm to estimate the parameters of chaotic systems [25].

This section examines the QPSO algorithm for the solutions of inverse problems in non-linear and chaotic systems. QPSO, along with PSO and GA, is used to estimate the parameters of three non-linear dynamical systems, including two chaotic ones. The performances of these three algorithms are presented.

6.2.2 Identification of Chaotic Systems by QPSO

Let

$$\dot{\vec{x}} = G(\vec{x}, \theta), \tag{6.14}$$

be a continuous non-linear dynamical system, where $\vec{x} = (x_1, x_2, \ldots, x_N) \in R^N$ is the state vector of the chaotic system, $\dot{\vec{x}}$ is the temporal derivative of $\vec{x}$ and $\theta = (\theta_1, \theta_2, \ldots, \theta_D)$ is the unknown parameter vector of the chaotic system, where D is the number of the parameters.

In order to evaluate the performance of the QPSO algorithm for parameter estimation of the non-linear dynamic system (Equation 6.14), pseudo-experimental data are generated by substituting the actual value of parameters $\theta = (\theta_1, \theta_2, \ldots, \theta_D)$ into Equation 6.14 and a fourth-order Runge–Kutta method is applied to the resulting equation. Selecting a point at random as the initial state at time $t = 0$ and assuming the dynamical system evolves from this initial state until the time $t = N_t \Delta t$, one can obtain $\vec{x}(t) = (x_1(t), x_2(t), \ldots, x_N(t))$ at time $t = \Delta t, 2\Delta t, \ldots, N_t \Delta t$, where the temporal step size is chosen as $\Delta t = 0.01$ and N_t is set to be 30 in the simulation. Then the pseudo-experimental data $\vec{x}(t) = (x_1(t), x_2(t), \ldots, x_N(t))$ is used as the experimental data for the subsequent simulation.

The current position vector of particle i in QPSO at the nth iteration, $\hat{\theta}_i^n = \left(\hat{\theta}_{i,1}^n, \hat{\theta}_{i,2}^n, \ldots, \hat{\theta}_{i,D}^n \right)$, represents an estimation of θ. Given the experimental data of the measurable state vector $\vec{x}(t) = (x_1(t), x_2(t), \ldots, x_N(t))$, the objective function or fitness function

$$f\left(\hat{\theta}_i^n \right) = \sum_{t=0}^{N_t} \left[\left(x_1(t) - x_{i,1}^n(t) \right)^2 + \cdots + \left(x_N(t) - x_{i,N}^n(t) \right)^2 \right], \tag{6.15}$$

where $t = 0, 1, \ldots, N_t$, is defined. The goal of estimating the parameters of the dynamical system (Equation 6.14) is to find a suitable value of $\hat{\theta}_i^n$ such that the fitness function (6.15) is globally minimised. To obtain $x_i^n(t) = \left(x_{i,1}^n(t), x_{i,2}^n(t), \ldots, x_{i,N}^n(t) \right)$, the estimation $\hat{\theta}_i^n$ is substituted into

Equation 6.14 and a fourth-order Runge–Kutta method is used to integrate the resulting equation.

6.2.3 Simulation Results

The QPSO algorithm, GA and PSO were tested in this performance study. The parameters of the algorithms were configured as follows. For PSO, the inertia weight w was set to be 0.729, and the acceleration coefficients c_1 and c_2 were set to be 1.49. For QPSO, the value of α was chosen to vary linearly from 1.0 to 0.5. For GA, the crossover probability was taken as 0.8 and the mutation probability was chosen to increase linearly from 0.001 to 0.6 during the evolution. The population size of each algorithm was taken as 30. These parameters were selected according to those used in Section 2.4 and Section 4.5 and/or the authors' experience with preliminary numerical experiments. The Brusselator, the Lorenz system and the Chen chaotic system described in the following were selected to test the performance of the algorithms. Each algorithm was executed 30 times for each example, with each run executing different number of iterations for the different system.

6.2.3.1 Brusselator

The Brusselator is a system of equations that model a hypothetical chemical reaction [26] as described in the following:

$$\begin{cases} \dot{x} = \theta_1 - \theta_2 x + x^2 y - x, \\ \dot{y} = \theta_2 x - x^2 y, \end{cases} \tag{6.16}$$

where
    x and y are the state variables
    θ_1 and θ_2 are unknown positive constant parameters

It can be shown by using the Poincaré–Bendixson theorem [27] that if $\theta_2 > 1 + \theta_1^2$ and $\theta_1 > 0$, the system has a periodic orbit. In the numerical simulation, the two parameters are chosen as $\theta_1 = 1$ and $\theta_2 = 3$. The system evolves as shown in Figure 6.5.

In the numerical experiments, each single run of the algorithm took 50 iterations. The statistics of the best fitness value were obtained, its standard deviation and the best value of identified parameters θ_1 and θ_2 are listed in Table 6.3. Numerical results show that the mean and the best parameters identified by QPSO exhibit better accuracy than those identified by PSO and GA. Figure 6.6 depicts the relative estimation

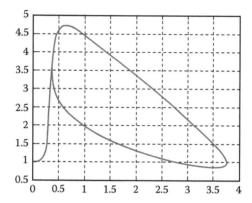

FIGURE 6.5 Evolution process of Brusselator when $\theta_1 = 1$ and $\theta_2 = 3$.

TABLE 6.3 Statistics of the Three Algorithms for Brusselator

Algorithms	Mean of the Best Fitness	Standard Deviation of the Best Fitness	The Best Values Obtained of Identified Parameters	
			$\theta_1 = 1$	$\theta_2 = 3$
QPSO	1.501×10^{-12}	3.4056×10^{-12}	0.99999998	2.99999992
PSO	2.6634×10^{-7}	4.609×10^{-7}	0.99997748	3.00006389
GA	8.4732×10^{-3}	2.7147×10^{-2}	0.99987791	2.99975583

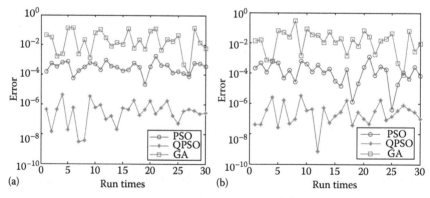

FIGURE 6.6 The relative estimation error of (a) θ_1 and (b) θ_2.

errors from 30 independent runs of each algorithm. The relative esti-
mation errors for the parameters are defined by $|\hat{\theta}_1 - 1|/1$ and $|\hat{\theta}_2 - 3|/3$,
respectively, where $\hat{\theta}_1$ and $\hat{\theta}_2$ are the identified values of the parameters. It can be seen that the estimation errors averaged over the 30 runs
of QPSO are the smallest amongst all the algorithms. Figure 6.7 shows
that QPSO converges to the optimal solution more rapidly than PSO
and GA.

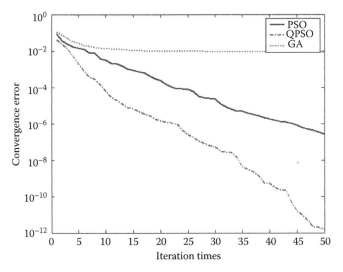

FIGURE 6.7   Convergence history of different algorithms.

6.2.3.2 Lorenz System

The second example system to be considered here is the well-known Lorenz system [28] defined by

$$\begin{cases} \dot{x}_1 = \theta_1(x_2 - x_1), \\ \dot{x}_2 = (\theta_3 - x_3)x_1 - x_2, \\ \dot{x}_3 = x_1 x_2 - \theta_2 x_3, \end{cases} \tag{6.17}$$

where

x_1, x_2 and x_3 are the state variables

θ_1, θ_2 and θ_3 are the unknown parameters to be determined

The system is in the chaotic state when $\theta_1 = 10$, $\theta_2 = 8/3$ and $\theta_3 = 28$. Figure 6.8 shows the evolution process of the Lorenz system using the set of parameters.

To identify the parameters of the aforementioned chaotic system, each single run of an algorithm was executed 50 iterations. The statistics of the best fitness value obtained, its standard deviation and the best value of the identified parameters are listed in Table 6.4. It is shown that the best value of the parameters identified by QPSO are more accurate than those identified by PSO and GA. Figure 6.9 depicts the relative estimation errors from 30 runs of each algorithm, with each single run executing

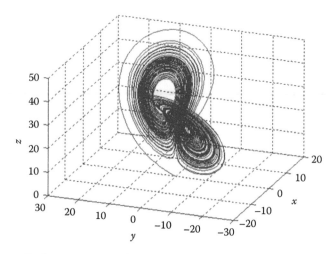

FIGURE 6.8   Evolution process of Lorenz system when $\theta_1 = 10$, $\theta_2 = 8/3$ and $\theta_3 = 28$.

TABLE 6.4   Statistics of the Three Algorithms for the Lorenz System

Algorithms	Mean of Best Fitness	Standard Deviation of Best Fitness	The Best Value Obtained of Identified Parameters		
			$\theta_1 = 10$	$\theta_2 = 8/3$	$\theta_3 = 28$
QPSO	9.381616×10^{-10}	2.187490×10^{-9}	9.999999	2.666667	28.000000
PSO	0.0019379	0.0045299	9.999104	2.666868	28.000317
GA	0.215385	0.298406	9.995112	2.665362	28.000010

50 iterations. Here the relative estimation errors for the three parameters are defined by $|\hat{\theta}_1 - 10|/10$, $|\hat{\theta}_2 - 8/3|/(8/3)$ and $|\hat{\theta}_3 - 28|/28$, respectively, where $\hat{\theta}_1$, $\hat{\theta}_2$ and $\hat{\theta}_3$ are the identified values of the parameters. Numerical results show that the errors obtained by using QPSO are the smallest on average over the 30 runs. It is also demonstrated in Figure 6.10 that QPSO exhibits faster convergence than PSO and GA.

6.2.3.3 Chen System

The third example system is the Chen system [29,30] described by

$$\begin{cases} \dot{x}_1 = \theta_1(x_2 - x_1), \\ \dot{x}_2 = (\theta_3 - \theta_1)x_1 + \theta_3 x_2 - x_1 x_3, \\ \dot{x}_3 = x_1 x_2 - \theta_2 x_3, \end{cases} \quad (6.18)$$

where
  x_1, x_2 and x_3 are the state variables
  θ_1, θ_2 and θ_3 are unknown parameters

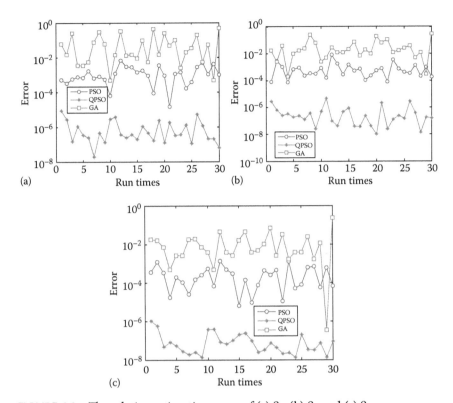

(a)

(b)

(c)

FIGURE 6.9 The relative estimation error of (a) θ_1, (b) θ_2 and (c) θ_3.

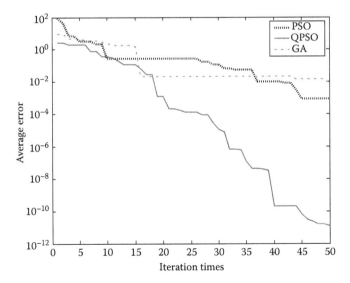

FIGURE 6.10 The convergence history of the fitness value for the Lorenz system.

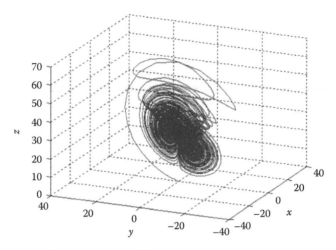

FIGURE 6.11 Evolution process of Chen system when $\theta_1 = 35$, $\theta_2 = 3$ and $\theta_3 = 28$.

The system is in the chaotic state when $\theta_1 = 35$, $\theta_2 = 3$ and $\theta_3 = 28$. Figure 6.11 shows the evolution process of the Chen system using the set of parameters.

To identify the parameters of the aforementioned chaotic system, each of the three algorithms was executed 200 iterations in each single run. It seems that the parameters of the Chen system are more difficult to identify than the Lorenz system. The statistics in Table 6.5 indicates that QPSO is able to identify the parameters of the system more accurately. Figure 6.12 depicts the relative estimation errors over 30 runs of each algorithm, with each single run executing 200 iterations. Here the relative estimation errors for the three parameters are defined by $|\hat{\theta}_1 - 35|/35$, $|\hat{\theta}_2 - 3|/3$ and $|\hat{\theta}_3 - 28|/28$, where $\hat{\theta}_1$, $\hat{\theta}_2$ and $\hat{\theta}_3$ are the identified values of the parameters. Numerical results show that the errors obtained by using QPSO are the smallest on average over 30 runs from Figure 6.12 and that QPSO has faster convergence speed than PSO and GA, as shown in Figure 6.13.

TABLE 6.5 Statistics of the Three Algorithms for the Chen System

Algorithms	Mean of Best Fitness	Standard Deviation of Best Fitness	The Best Value Obtained of Identified Parameters		
			$\theta_1 = 35$	$\theta_2 = 3$	$\theta_3 = 28$
QPSO	1.4904	4.3142	35.061386	3.001224	28.029025
PSO	1.7176	2.1645	34.783951	3.000978	27.895262
GA	133.61	189.99	31.142857	2.937378	26.197802

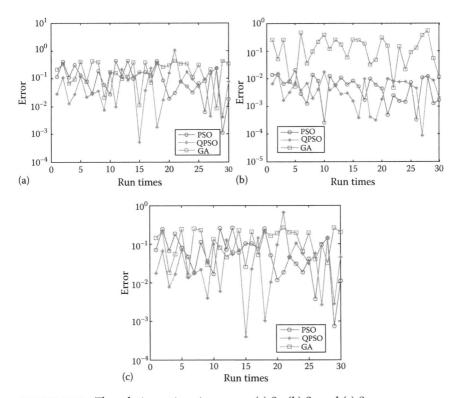

FIGURE 6.12 The relative estimation errors (a) θ_1, (b) θ_2 and (c) θ_3.

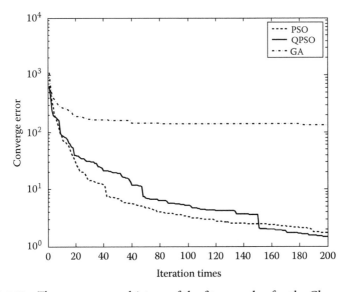

FIGURE 6.13 The convergence history of the fitness value for the Chen system.

6.2.4 Summary

In this section, the QPSO algorithm is used to predict the parameters of three non-linear dynamical systems. The results of the performed experiments show that for the given parameter configurations and the maximum number of iterations for the tested algorithms, QPSO identifies the parameters of the Brusselator, Lorenz system and Chen system accurately and rapidly when compared with PSO and GA.

6.3 OPTIMAL DESIGN OF DIGITAL FILTERS

6.3.1 Introduction

In signal processing, the digital filters are broadly classified into two main categories, namely, finite impulse response (FIR) filters and infinite impulse response (IIR) filters [31,32]. An FIR filter has finite duration of its impulse response. The output of such a filter is calculated solely from the current and previous input values. This type of filter is said to be non-recursive. On the other hand, an IIR filter has its impulse response (theoretically) continues forever in time. They are also termed as recursive filters. The current output of such a filter depends on previous output values as well as on the current input value. FIR filters are generally easier to implement, as they are non-recursive and always stable (by definition). Several methods have been proposed to solve the optimal design problem of them. An obvious choice consists of using integer-programming techniques [33] or conducting a local univariate or bivariate search near the optimal coefficients [34]. Work has been done in applying the simulated annealing (SA) approach [35] or evolutionary computation techniques such as GAs [36,37], PSO [38] and immune algorithm [39]. On the other hand, the key practical benefit of using IIR filters is that they are much more efficient than FIR systems, which require higher order transfer functions (i.e. more multiplications) [40]. Overall, when properly designed, IIR filters show more accurate response than those obtained by FIR filters [32]. In digital IIR filter design, there are principally two approaches: transformation approach and optimisation approach. In the former approach, the analogue filter must be designed first before it is transformed into the digital IIR filter. In the latter approach, with suitability criteria, various optimisation methods may be used to obtain the optimal filter performances to some extent, where the p-error, mean square error (MSE) and ripple magnitudes (tolerances) of both passband and stopband are usually used as criteria to measure the performances of the designed digital IIR filters. In fact, the digital IIR filter design is essentially a multi-parameter and multi-criteria optimisation problem with multiple local optima. It is necessary to develop efficient optimisation algorithms to deal with the digital IIR filter design problems. However, conventional methods based on gradient search can easily be held at local minima of the error surface. Modern global optimisation algorithms such as the SA [41–43], GA [44–49], ant colony optimisation (ACO) [50], tabu search (TS) algorithm [51], differential evolution (DE) [52] and immune algorithm [53] began to be investigated. Among these algorithms, GA is the one that has been used a lot more than the others for the IIR filter design.

2D IIR filters play an important role in multidimensional digital signal processing (MDSP) and has many applications in the area of denoise of digital images, biomedical imaging and digital mammography, x-ray image enhancement, seismic data processing, etc. [54–56]. During the last three decades, 2D filter design has received growing attention by researchers and practitioners. The most popular design methods for 2D IIR filters are based either on an appropriate transformation of a 1D filter [55,56] or on appropriate optimisation techniques such as linear programming, the Remez exchange algorithm and non-linear programming based on gradient methods, direct search methods, Newton and Gauss–Newton methods, Fletcher–Powell and the conjugate gradient [55–63]. However, most of the existing methods may result in an unstable filter [54,56–63]. Thus, many techniques have been adopted to resolve these instability problems, but the outcome is likely to be a system that has a very small stability margin and hence not of much practical importance [46,64].

Many modern heuristic methods have been employed for 2D IIR design problems, such as GA [45,65], neural network (NN) [64], PSO [66] and the computer language GENETICA [66]. These techniques can be used to obtain better solutions than those discussed in the previous paragraph. The focus of this section is to select a 2D IIR filter design using an optimisation approach because it is more challenging to implement and has higher degrees of precision and practical efficiency. In the present study, the design task of 2D recursive filters is formulated as a constrained optimisation problem. The stability criterion is presented as constraints to the minimisation problem.

6.3.2 Problem Formulation

The design task of 2D recursive filters amounts to finding a transfer function $H(z_1, z_2)$ as described in Equation 6.19 such that the function $M(\omega_1, \omega_2) = H(e^{-j\omega_1}, e^{-j\omega_2})$ approximates the desired amplitude response $M_d(\omega_1, \omega_2)$, where $\omega_1, \omega_2 \in [-\pi, \pi]$ are the frequencies and $z_1 = e^{-j\omega_1}$, $z_2 = e^{-j\omega_2}$. For design purposes, the function $M(\omega_1, \omega_2)$ is equivalent to a class of non-symmetric half-plane (NSHP) filters to which the 2D transfer function $H(z_1, z_2)$ is given by

$$H(z_1, z_2) = H_0 \frac{\sum_{i=0}^{K} \sum_{j=0}^{K} a_{ij} z_1^i z_2^j}{\prod_{k=1}^{K} (1 + q_k z_1 + r_k z_2 + s_k z_1 \cdot z_2)}, \quad a_{00} = 1, \quad (6.19)$$

where K is the dimension of the filter which is 2 for 2D filters. The variables z_1 and z_2 can be interpreted as complex indeterminants in the discrete Laplace transform (z-transformation).

It is a general practice to normalise all a_{ij}s with respect to the value of a_{00} leading to $a_{00} = 1$. The design task for 2D filter can be reduced to finding the transfer function $H(z_1, z_2)$ such that the frequency response $H(e^{-j\omega_1}, e^{-j\omega_2})$ approximates the desired amplitude response $M_d(\omega_1, \omega_2)$ as close as possible. The approximation can be achieved by minimising [44,63–66]:

$$J = J(a_{ij}, q_k, r_k, s_k, H_0) = \sum_{k_1=0}^{K_1} \sum_{k_2=0}^{K_2} [|M(\omega_1, \omega_2)| - M_d(\omega_1, \omega_2)]^p, \quad (6.20)$$

where

$$M(\omega_1, \omega_2) = H(z_1, z_2)\big|_{z_1 = e^{-j\omega_1}, z_2 = e^{-j\omega_2}} \quad (6.21)$$

$\omega_1 = (\pi/K_1)k_1$, $\omega_2 = (\pi/K_2)k_2$
p is a positive even integer ($p = 2$, or 4, 8)

Equation 6.20 can be rewritten as

$$J = \sum_{k_1=0}^{K_1} \sum_{k_2=0}^{K_2} \left[\left|M\left(\frac{\pi k_1}{K_1}, \frac{\pi k_2}{K_2}\right)\right| - M_d\left(\frac{\pi k_1}{K_1}, \frac{\pi k_2}{K_2}\right)\right]^p. \quad (6.22)$$

Here the prime objective is to reduce the difference between the desired and actual amplitude responses of the filter at $K_1 \times K_2$ points. For bounded

input bounded output (BIBO) stability, the prime requirement is that the z-plane poles of the filter transfer function should lie within the unit circle. Since the denominator contains only first-degree factors, the stability conditions [45,54,56] hold,

$$|q_k + r_k| - 1 < s_k < 1 - |q_k - r_k|, \quad k = 1, 2, \ldots, K. \tag{6.23}$$

Thus, the design of 2D recursive filters is equivalent to the following constrained minimisation problem:

$$Minimize \ J = \sum_{k_1=0}^{K_1} \sum_{k_2=0}^{K_2} \left[\left| M\left(\frac{\pi k_1}{K_1}, \frac{\pi k_2}{K_2} \right) \right| - M_d\left(\frac{\pi k_1}{K_1}, \frac{\pi k_2}{K_2} \right) \right]^p, \tag{6.24a}$$

$$\text{s.t.} \quad \begin{aligned} &|q_k + r_k| - 1 < s_k & k = 1, 2, \ldots, K \\ &s_k < 1 - |q_k - r_k| & k = 1, 2, \ldots, K, \end{aligned} \tag{6.24b}$$

where

$p = 2$, or 4, 8

K_1 and K_2 are positive integers

6.3.3 A 2D Filter Design Using QPSO

6.3.3.1 Design Example

Assuming $K = 2$, the transfer function in Equation 6.19 can be rewritten as

$$H(z_1, z_2)$$

$$= H_0 \frac{a_{00} + a_{01} z_2 + a_{02} z_2^2 + a_{10} z_1 + a_{20} z_1^2 + a_{11} z_1 z_2 + a_{12} z_1 z_2^2 + a_{21} z_1^2 z_2 + a_{22} z_1^2 z_2^2}{(1 + q_1 z_1 + r_1 z_2 + s_1 z_1 z_2)(1 + q_2 z_1 + r_2 z_2 + s_2 z_1 z_2)}. \tag{6.25}$$

Substituting z_1 and z_2 into Equation 6.21, $M(\omega_1, \omega_2)$ becomes

$$M(\omega_1, \omega_2)$$

$$= H_0 \left[\begin{array}{c} \dfrac{a_{00} + a_{01} f_{01} + a_{02} f_{02} + a_{10} f_{10} + a_{20} f_{20} + a_{11} f_{11} + a_{12} f_{12} + a_{21} f_{21} + a_{22} f_{22}}{D} \\ - \dfrac{j[a_{00} + a_{01} g_{01} + a_{02} g_{02} + a_{10} g_{10} + a_{20} g_{20} + a_{11} g_{11} + a_{12} g_{12} + a_{21} g_{21} + a_{22} g_{22}]}{D} \end{array} \right] \tag{6.26}$$

with $f_{ij} = \cos(i\omega_1 + j\omega_2)$, $g_{ij} = \sin(i\omega_1 + j\omega_2)$ $(i, j = 0,1,2)$ and

$$D = [(1 + q_1 f_{10} + r_1 f_{01} + s_1 f_{11}) - j(q_1 g_{10} + r_1 g_{01} + s_1 g_{11})]$$
$$\cdot [(1 + q_2 f_{10} + r_2 f_{01} + s_2 f_{11}) - j(q_2 g_{10} + r_2 g_{01} + s_2 g_{11})] \qquad (6.27)$$

Hence, the compact form of $M(\omega_1, \omega_2)$ is

$$M(\omega_1, \omega_2) = H_0 \frac{A_R - jA_I}{(B_{1R} - jB_{1I})(B_{2R} - B_{2I})}, \qquad (6.28)$$

where

$$A_R = a_{00} + a_{01} f_{01} + a_{02} f_{02} + a_{10} f_{10} + a_{20} f_{20} + a_{11} f_{11} + a_{12} f_{12} + a_{21} f_{21} + a_{22} f_{22},$$
$$A_I = a_{00} + a_{01} g_{01} + a_{02} g_{02} + a_{10} g_{10} + a_{20} g_{20} + a_{11} g_{11} + a_{12} g_{12},$$
$$B_{1R} = 1 + q_1 f_{10} + r_1 f_{01} + s_1 f_{11}, \quad B_{1I} = q_1 g_{10} + r_1 g_{01} + s_1 g_{11}, \qquad (6.29)$$
$$B_{2R} = 1 + q_2 f_{10} + r_2 f_{01} + s_2 f_{11}, \quad B_{2I} = q_2 g_{10} + r_2 g_{01} + s_2 g_{11},$$

The magnitude of $M(\omega_1, \omega_2)$ is given by

$$|M(\omega_1, \omega_2)| = H_0 \sqrt{\frac{\left(A_R^2 + A_I^2\right)}{\left(B_{1R}^2 + B_{1I}^2\right)\left(B_{2R}^2 + B_{2I}^2\right)}}. \qquad (6.30)$$

The same design problem as that considered in [45,64,65,67] was tested, with the user specification for the desired circular symmetric low-pass filter response is given by

$$M_d(\omega_1, \omega_2) = \begin{cases} 1 & \sqrt{\omega_1^2 + \omega_2^2} \leq 0.08\pi, \\ 0.5 & 0.08\pi < \sqrt{\omega_1^2 + \omega_2^2} \leq 0.12\pi, \\ 0 & \text{otherwise} \end{cases} \qquad (6.31)$$

A continuous differentiable form of the constraints can be obtained from Equation 6.23 in the form

$$-(1+s_k) < (q_k + r_k) < (1+s_k), \quad -(1-s_k) < (q_k - r_k) < (1-s_k)$$

$$(1+s_k) > 0, \quad (1-s_k) > 0 \tag{6.32}$$

Choosing the values $p = 2$, 4 and 8, and $K_1 = 50$ and $K_2 = 50$, the corresponding constrained optimisation problem (6.24a and b) becomes

$$\text{Minimize } J = \sum_{k_1=0}^{50} \sum_{k_2=0}^{50} \left[\left| M\left(\frac{\pi k_1}{50}, \frac{\pi k_2}{50} \right) \right| - M_d\left(\frac{\pi k_1}{50}, \frac{\pi k_2}{50} \right) \right]^2, \tag{6.33}$$

subject to the constraints imposed by Equation 6.32 with $k = 1, 2$.

6.3.3.2 Representation of the Particle and Constraint Handling

In order to apply the QPSO algorithm to the design problem formulated in Equation 6.33, each trial solution is represented as a particle in a multidimensional space. Since a_{00} is always set to 1 as in Equation 6.19, the dimensionality of the present problems is 15, and each particle has 15 positional coordinates represented by the vector

$$x = (a_{01}, a_{02}, a_{10}, a_{11}, a_{12}, a_{20}, a_{21}, a_{22}, q_1, q_2, s_1, s_2, r_1, r_2, H_0)^T. \tag{6.34}$$

Here x contains real number components and represents the position of particles in the algorithm.

To handle the constraints, methods described in [45,64–67] are followed: (1) any feasible solution is preferred to any infeasible solution; (2) between feasible solutions, the one with a better objective function value is preferred; (3) between two infeasible solutions, the one with a smaller violation of the constraint is preferred. As far as the constraints presented in Equation 6.32, a population of around 200 particles with randomly initialised positional coordinates is to be used. M (population size) particles are selected out of the population with spatial coordinates satisfying the constraints in Equation 6.32. If more than M particles are initially found to satisfy the constraints, the selection takes into account the initial fitness value of these particles. Suppose a particle has its trajectory through the search space yields the lowest fitness value found so far, its position would be memorised as the global best position by all other particles in the swarm only if it satisfies the constraints.

6.3.4 Experimental Results and Discussions

A set of numerical tests were carried out by using three population-based optimisation algorithms, namely, QPSO, the PSO with inertia weight w and the binary encoded GA suggested by Mastorakis et al. [45], for the design of circular symmetric zero-phase low-pass filter according to the user specification summarised in Equation 6.31. The results from NN method [64] and the computer language GENETICA [67] were used for performance comparison in the test.

6.3.4.1 Parameter Configurations

As suggested in [45,64], the initial approximations of the parameters of the vector in Equation 6.34 were chosen randomly from the interval (−3, 3). In the cases of the QPSO algorithm, the contraction–expansion coefficient was chosen to vary linearly from 1.0 to 0.5. The inertia weight in the cases of PSO was decreased linearly from 0.9 to 0.4. The number of particles in these experiments was chosen to be 40 and 20 in the PSO and QPSO, respectively. Table 6.6 shows the parameter configurations used in

TABLE 6.6   Parameter Configurations for the Competitor Algorithms

Algorithm	Parameter	Value
GA	Population size M	250
	Number of bits	32
	Mutation probability P_m	0.05
	Part of genetic material interchanged during cross-over N_p	12
	Maximum number of children from each pair of parents N_c	10
PSO1	Population size M	40
	Inertia weight w	$0.9 \rightarrow 0.4$
	c_1, c_2	2
	V_{max}	3
PSO2	Population size M	20
	Inertia weight w	$0.9 \rightarrow 0.4$
	c_1, c_2	2
	V_{max}	3
QPSO1	Population size M	40
	CE coefficient α	$1.0 \rightarrow 0.5$
QPSO2	Population size M	20
	CE coefficient α	$1.0 \rightarrow 0.5$

all numerical experiments, and PSO1 and QPSO1 are used to denote cases with 40 particles, and PSO2 and QPSO2 are used to denote cases with 20 particles.

6.3.4.2 Simulation Results

Two aspects of the algorithms were investigated, first their accuracy and second their speed of convergence. In order to compare the speed of convergence of different algorithms, a suitable measurement of computational time is needed. The number of iterations cannot be used as a reference of computational time as the algorithms perform different amount of works in their inner loops and have different population sizes. Here, the number of fitness function evaluations (FEs) was used as a measure of the computational time. The advantage of measuring complexity by counting the FEs is that the processor time can be accurately reflected.

To study the accuracy of the algorithms, each of the three algorithms involved in the test was run separately for 100 times, with each single run executing 40,000 FEs. The best values of the parameters and the mean best J value along with its standard deviation obtained from 100 runs of each algorithm are reported in Tables 6.7 and 6.8. Table 6.7

TABLE 6.7   The Best Values of Filter Coefficients with Exponent $p = 2$ after 40,000 FEs Generated by the Competitor Algorithms

	NN	GA	GENETICA	PSO1	PSO2	QPSO1	QPSO2
a_{01}	1.8922	1.8162	−0.2247	0.9301	1.3357	−0.6317	−0.83657
a_{02}	−1.2154	−1.1060	2.5248	1.5378	−2.7052	−2.2518	−1.862
a_{10}	0.0387	0.0712	−0.3498	0.3317	−2.656	−0.5204	−1.6195
a_{11}	−2.5298	−2.5132	−2.0915	−2.9988	0.93592	1.3449	2.994
a_{12}	0.3879	0.4279	0.0317	2.0503	0.36258	0.9652	0.07868
a_{20}	0.6115	0.5926	2.4656	1.8594	−1.0322	−1.7862	−1.7383
a_{21}	−1.4619	−1.3690	0.1652	1.9410	1.0865	−0.0122	0.54907
a_{22}	2.5206	2.4326	0.7713	0.7672	−0.81088	−0.2554	−0.67628
q_1	−0.8707	−0.8662	−0.9316	0.2083	−0.06855	−0.7764	−0.53496
q_2	−0.8729	−0.8907	−0.0249	−0.9145	−0.8659	−0.5187	−0.82565
r_1	−0.8705	−0.8531	−0.9309	0.1955	−0.55581	−0.8888	−0.34733
r_2	−0.8732	−0.8388	−0.0326	−0.9249	−0.08978	−0.0154	−0.87021
s_1	0.7756	0.7346	0.8862	−0.2056	−0.27798	0.6953	−0.01655
s_2	0.7799	0.8025	−0.8082	0.8488	0.06095	−0.3215	0.72768
H_0	0.0010	0.0009	0.0010	−0.0020	−0.0061	−0.0024	0.0018

TABLE 6.8   Mean Value and Standard Deviations of J Values with Exponent $p = 2, 4$ and 8 after 40,000 FEs Generated by the Competitor Algorithms

Value of J_p	J_2	J_4	J_8
GA	18.9614 ± 4.7974	2.4621 ± 1.2028	0.1025 ± 0.0754
PSO1	16.4010 ± 3.1299	1.1766 ± 0.5315	0.0258 ± 0.0207
PSO2	16.9224 ± 2.4654	1.2371 ± 0.2608	0.0206 ± 0.0102
QPSO1	13.9381 ± 1.5987	0.9856 ± 0.1135	0.0111 ± 0.0032
QPSO2	13.9032 ± 1.9924	0.9899 ± 0.1319	0.0104 ± 0.0025

TABLE 6.9   Unpaired t-Tests Applied to the Data of the Values of J with Exponent $p = 2$ in Table 6.8

	QPSO2–QPSO1	QPSO1–PSO1	PSO1–PSO2	PSO1–GA
Standard error	0.2555	0.3515	0.3984	0.5728
T value	−0.1366	−7.0068	−1.3087	−6.4267
95% confidence interval	(−0.5357, 0.4659)	(−0.8799, 0.4979)	(−1.7740, −3.1518)	(−3.6831, −1.4377)
P value	0.4443	$<10^{-7}$	0.1515	$<10^{-7}$
Significance	Not significant	Extremely significant	Not significant	Extremely significant

also lists the filter parameters yielded by NN [64] and GENETICA [67] for comparison. The notation J_p has been used to denote the three sets of experiments performed where the value of J is obtained with exponent $p = 2, 4$ and 8. Table 6.9 summarises the results of the unpaired test on the J_2 value (standard error of the difference of the two means, 95% confidence interval of the difference, the t value, and the two-tailed P value) between results of competitor algorithms in Table 6.9. For all cases listed in Table 6.9, the example size is 100.

Figures 6.14 through 6.21 show the ideal low-pass filter frequency response and the frequency response of the filter designed using the competitor algorithms. All these figures are drawn from the results obtained by using the minimisation algorithm for J_2 with 40,000 FE as well as from the results yielded by NN [64] and GENETICA [67]. Figure 6.22 shows the convergence history of the algorithms towards the optima of the search space with the number of FE.

Figures 6.15 through 6.21 reveal that the QPSO (including those with 20 and 40 particles) algorithms proposed by the authors yield better

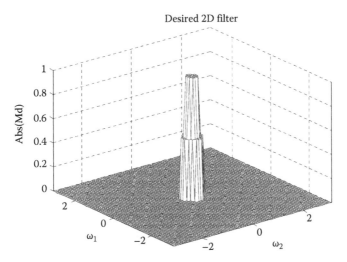

FIGURE 6.14 Desired amplitude response $|M_d(\omega_1, \omega_2)|$ of the 2D filter defined by Equation 6.31.

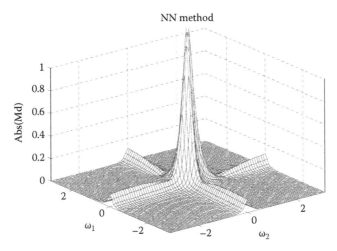

FIGURE 6.15 Amplitude response $|M(\omega_1, \omega_2)|$ of the 2D filter using NN method in [63].

approximations of the desired response as compared to the work presented in [45] or in [64] and yield a considerably good filter response with the same number of FEs. The ripples in the stop band of Figures 6.20 and 6.21 are lesser as compared to Figures 6.14 through 6.19. As evident from the results of the t-test applied to the data of Table 6.9, the QPSO method shows its superiority compared to its nearest competitor in a statistically

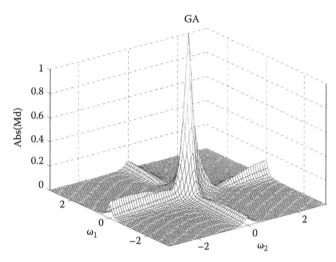

FIGURE 6.16   Amplitude response $|M(\omega_1, \omega_2)|$ of the 2D filter using GA in [44].

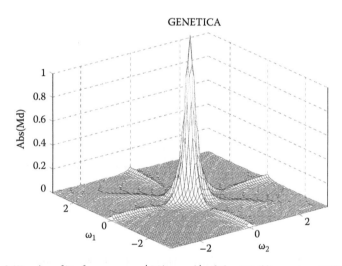

FIGURE 6.17   Amplitude response $|M(\omega_1, \omega_2)|$ of the 2D filter using GENETICA language in [66].

significant manner. The comparison also reflects the fact that GENETICA could generate better filter coefficients than the GA, NN and PSO methods and comparable results with the QPSO. The earlier comparison shows that with different number of particles but equal number of FEs, the performance of the PSO or the QPSO do not show significant statistical differences.

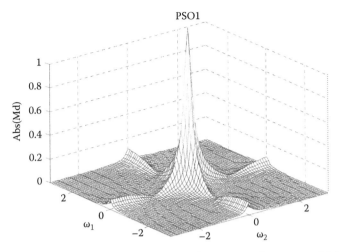

FIGURE 6.18   Amplitude response $|M(\omega_1, \omega_2)|$ of the 2D filter using PSO1.

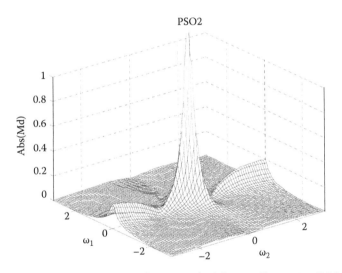

FIGURE 6.19   Amplitude response $|M(\omega_1, \omega_2)|$ of the 2D filter using PSO2.

Figure 6.22 shows the convergence history of the GA, PSO (with 40 particles) and QPSO (with 40 particles). It is obvious that the QPSO algorithm has better convergence property than the PSO and GA for 2D digital filter design.

6.3.5 Summary

In this section, the QPSO algorithm is applied to the real-world problem of designing 2D zero-phase recursive filters. The simulation results show that

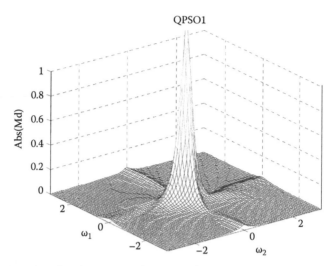

FIGURE 6.20 Amplitude response $|M(\omega_1, \omega_2)|$ of the 2D filter using QPSO1.

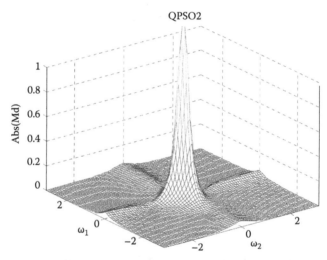

FIGURE 6.21 Amplitude response $|M(\omega_1, \omega_2)|$ of the 2D filter using QPSO2.

the filter obtained by QPSO has good stability margins (the stability criterion is incorporated as constraints to the minimisation task). Comparison of numerical results obtained by using methods such as GA [45], NN [64] and PSO [67] concluded that the QPSO method used here yields a better design. The QPSO algorithm also demonstrates its superiority compared to PSO and GA in a statistical significant manner for the design of 2D IIR digital filters.

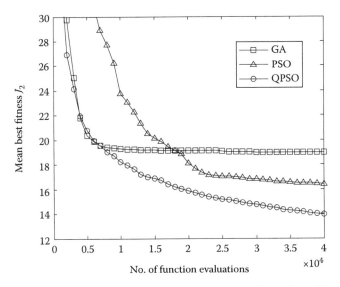

FIGURE 6.22 Performance of different optimisation algorithms for J_2 taking 40,000 FEs.

6.4 ED PROBLEMS

6.4.1 Introduction

The power ED is one of the most important problems in power system operation. Its objective is to minimise the total generation cost of the generating units while satisfying various constraints of the units and the system. It is usually a non-linear optimisation problem because of its non-linear characteristics, including discontinuous prohibited zones, unit power limits, ramp rate limits and cost functions [68].

In classical ED problems, the cost function of each generator is approximated by a quadratic function and the optimisation problem is solved by various mathematical programming methods, including the lambda-iteration method, the base point and participation factors method, the interior point method, dynamic programming (DP) and the gradient method [68–72]. However, none of these traditional approaches is able to provide an optimal solution, because they are local search techniques and usually stuck at a local optimum.

In the past decade, a wide variety of heuristic optimisation methods such as GA [73–75], PSO [76–78], DE [79,80], evolutionary programming

(EP) [81–83], TS [84], NN [85,86] and artificial immune (AI) [87] have been applied to solve the ED problem. Bakirtzis et al. presented a GA method and an enhanced GA to solve the ED problem [72]. According to their work, the results obtained are better than the DP method. Chen and Chang developed a lambda-based GA approach for solving the ED problem, and the method is faster and more robust than the well-known lambda-iteration method for large-scale systems [74]. Chiang suggested an improved GA with multiplier updating to solve the ED problem with valve-point effects and multiple fuels [75]. Gaing proposed a PSO method for solving the ED problem in power systems, and the simulation results show that the PSO method is indeed capable of obtaining better quality solutions than GA method in ED problems [76]. Park et al. designed a dynamic search-space reduction strategy to accelerate the optimisation process in the PSO method to solve the ED problem [77]. Coelho and Mariani combined the DE method with the generator of chaos sequences and sequential quadratic programming (SQP) technique to optimise the performance of solving ED problems [80], and their proposed method outperforms other state-of-the-art algorithms in solving load dispatch problems with the valve-point effect.

This section explores the applicability of the QPSO algorithm in solving the ED problem written in the form of a constrained optimisation problem with a penalty function. The feasibility and the global convergence characteristics of the algorithm are demonstrated by two power systems and are compared to those obtained by GA, PSO and DE.

6.4.2 Problem Description

The ED problem in power systems is a non-linear optimisation problem that determines the power output level of an online generator results in a system state of the least cost. Practically, while the scheduled combination units at each specific period of operation are listed, the ED planning must perform the optimal generation dispatch among the operating units to satisfy the system load demand, spinning reserve capacity and practical operation constraints of generating units that include the ramp rate limit and the prohibited operating zones [88].

6.4.2.1 Formulation of the ED Problem

The objective of the classical ED problem is to minimise the total system fuel cost over some appropriate period (1 h typically) while satisfying various constraints. The problem can be defined as the following constrained optimisation problem.

$$\text{Minimize } F_{cost} = \sum_{j=1}^{N_g} F_j(P_j) \tag{6.35}$$

subject to

$$\sum_{j=1}^{N_g} P_j = P_D + P_L, \tag{6.36}$$

$$P_j^{min} < P_j < P_j^{max} \quad (j = 1,\ldots, N_g), \tag{6.37}$$

where
  $F_j(P_j)$ is the cost function of the jth generating unit (in \$/h)
  P_j is the real output of the jth generating unit (in MW)
  N_g is the total number of generating units in this power system

The equality constraint in Equation 6.36 means that the total system generation includes load demand of the system and the transmission losses. While the total generation cost is being minimised, the total generation $\sum_{j=1}^{N_g} P_j$ should be equal to the total system demand P_D (in MW) plus the transmission network loss P_L (in MW).

The inequality constraint in Equation 6.37 requires that the generation of each unit to lie between its minimum $\left(P_j^{min}\right)$ and maximum $\left(P_j^{max}\right)$ production limits which are directly related to the design of the machine.

The cost function of each generating unit is related to the actual power injected to the system. It is typically modelled by the smooth quadratic function

$$F_j(P_j) = a_j + b_j P_j + c_j P_j^2, \tag{6.38}$$

where a_j, b_j and c_j are the cost coefficients of the jth generating unit.

6.4.2.2 System Transmission Losses

The most popular approach for finding an approximate value of the losses is by means of Kron's loss formula [68]. It represents the losses as a function of the output level of the system generating units:

$$P_L = \sum_{j=1}^{N_g} \sum_{k=1}^{N_g} P_j B_{jk} P_k + \sum_{j=1}^{N_g} P_j B_{j0} + B_{00}, \tag{6.39}$$

where B_{jk}, B_{j0} and B_{00} are known as the loss coefficients or B coefficients. Using the matrix notation, the loss formula can be written as

$$P_L = P^T B P + B_0 P + B_{00}. \tag{6.40}$$

6.4.2.3 Ramp Rate Limits

A number of studies have focused on the economical aspects of the problem under the assumption that the unit generation output can be adjusted instantaneously. Even though this assumption simplifies the problem, it does not reflect the actual operating processes of the generating unit.

Practically, the operating range of all online units is restricted by their ramp rate limits. According to [74], the inequality constraints due to the ramp limits are given:

1. If the output increases

$$P_j - P_j^0 \le U_j. \tag{6.41}$$

2. If the output decreases

$$P_j^0 - P_j \le D_j, \tag{6.42}$$

where
P_j^0 (in MW) is the previous output power
U_j (in MW/h) is the up-ramp limit of the jth generator
D_j (in MW/h) is the down-ramp limit of the jth generator

6.4.2.4 Prohibited Operating Zone

Due to operations of steam valves or vibration in shaft bearings, the system contains certain operating zone. In the actual power system, the unit loading must avoid the prohibited zones. The feasible operating zones of the jth unit can be described as follows:

$$P_j^{min} \leq P_j \leq P_{j,1}^l$$

$$P_{j,k-1}^u \leq P_j \leq P_{j,k}^l, \quad k = 2,3,...,n_j, \tag{6.43}$$

$$P_{j,n_j}^u \leq P_j \leq P_j^{max}$$

where

$P_{j,k}^l$ and $P_{j,k}^u$ are the lower and upper bound of the kth prohibited zone of the jth unit

n_j is the number of prohibited zones of the jth unit

Combining (6.41) through (6.43) and (6.35) through (6.37), the constrained optimisation problem is modified as

$$\text{Minimize } F_{cost} = \sum_{j=1}^{N_g} F_j(P_j) \tag{6.44}$$

subject to

$$\sum_{j=1}^{N_g} P_j = P_D + P_L \tag{6.45}$$

$$\max\left(P_j^{min}, P_j^0 - D_j\right) \leq P_j \leq \min\left(P_j^{max}, P_j^0 + U_j\right) \tag{6.46}$$

$$P_j^{min} \leq P_j \leq P_{j,1}^l,$$

$$P_{j,k-1}^u \leq P_j \leq P_{j,k}^l, \quad k = 2,3,...,n_j \tag{6.47}$$

$$P_{j,n_j}^u \leq P_j \leq P_j^{max}.$$

6.4.3 Solving ED Problems by the QPSO Algorithm

Before applying the QPSO algorithm to a constrained ED problem, the following definitions are required.

6.4.3.1 Representation of Individual Particle

In this book, the power output generated by each unit is considered as a component and many components comprise the spatial position of a particle. Each particle within the population represents a candidate solution of the ED problem. For example, if there are N_g units that are being operated to provide power to loads, P_{gi} denoting the position of the ith particle can be defined as follows.

$$P_{gi} = \left[P_{i,1}, P_{i,2}, \ldots, P_{i,j}, \ldots, P_{i,N_g} \right], \quad i = 1, 2, \ldots, M, \tag{6.48}$$

where
  M means the population size
  the subscript j is used to denote the jth generator
  $P_{i,j}$ is the jth component of the ith particle, representing an estimation
    of the power output generated by the jth generation unit

6.4.3.2 Constraints Handling and Objective Function
A penalty function is used to handle the equality constraint in Equation 6.45. Adding the penalty function, the objective function gives

$$\text{Minimize } F = \sum_{j=1}^{N_g} F_j(P_j) + K \left| \sum_{j=1}^{N_g} P_j - P_D - P_L \right|, \tag{6.49}$$

where
  K is a positive number
  $|\cdot|$ is the absolute value of the mathematical expression, $\displaystyle\sum_{j=1}^{N_g} P_j - P_D - P_L$

In order to limit each particle of the population within the feasible range, before the objective function value of the particle is estimated, the power output generated by each unit must be checked against the constraints in (6.46) and (6.47). If one particle satisfies these constraints, it is a feasible particle and is evaluated by the objective function described in (6.49). If the particle fails to satisfy the constraints in (6.46), the power output generated by each unit is adjusted by randomly selecting its value from the feasible interval and checking against constraints (6.47). If (6.47) is violated, the objective function value of the particle is penalised with a very large positive constant.

6.4.4 Case Studies and Discussion
To assess the efficiency of the QPSO algorithm in solving ED problems, two different power systems were tested. In these numerical tests, the ramp rate limits and prohibited zones of units were taken into account in practical application. The QPSO method was compared with the existing

well-known heuristic methods for ED problems, such as binary-coded GA [74], DE [79] and PSO [76]. For each example system, the same objective function is used across all methods. In order that the roulette selection can be executed, a fitness value is defined for GA as

$$fitness = 2F^{max} - F, \tag{6.50}$$

where F^{max} is the largest objective function value among the population at a certain generation.

6.4.4.1 Parameter Configurations

The number of objective FEs for which every optimisation algorithm executed was set to be 20,000. Two groups of experiments were executed by using each of the four optimisation methods for each example system, one group with population size $M = 100$ and maximum number of iterations $n_{max} = 200$, the other with population size $M = 20$ and maximum number of iterations $n_{max} = 1000$. There were 100 independent runs performed by each of the optimisation methods with the given values of M and n_{max} for each sample system. The positive number K in the objective function was set as $K = 100 \times n$, where n is the number of iterations. The parameter configurations of all approaches used are listed as follows:

- GA: The parameters of GA were set as those in [74], that is, the cross-over probability $p_c = 0.8$ and the mutation probability $p_m = 0.1$.

- DE: A constant mutation factor of $f_m = 0.4$ and a crossover rate of CR = 0.8 [79].

- PSO: w decreasing linearly from 0.9 to 0.4; $c_1 = c_2 = 2$; $V_{max,j} = -V_{min,j} = \left(P_j^{max} - P_j^{min}\right)/2$ as in [76].

- QPSO: The contraction–expansion coefficient decreasing linearly from the initial value $\alpha_0 = 1.0$ to the final value $\alpha_1 = 0.5$.

6.4.4.2 Simulation Results

The performance of the tested algorithms may be described in terms of the solution quality and robustness with respect to the mean costs and standard deviations obtained from 100 runs of each optimisation algorithm. In addition, the statistics collected for the convergence properties of the algorithms were averaged over 100 runs and compared through visualising the dynamic changes of the objective function value during the runs.

Example 6.1

This system consists of 6 thermal units, 26 buses and 46 transmission lines [89]. The load demand is 1263 MW. The characteristics of the six thermal units are given in Tables 6.10 and 6.11. In normal operation of the system, the loss coefficients B with the 100 MW base capacity are as listed in Table 6.12.

The results obtained for Example 6.1 are given in Table 6.13, which shows that the QPSO with $M = 100$ and $n_{max} = 200$ is able to obtain the best solution, the lowest standard deviation and the best mean value of costs amongst the tested optimisation algorithms. Table 6.13 reveals that when $M = 100$ and $n_{max} = 200$, QPSO is better than the other methods; when $M = 20$ and $n_{max} = 1000$, QPSO also outperforms the others.

The solution found by using the QPSO algorithm has a relatively low standard deviation, which demonstrates the robustness of the method (see Table 6.13). In addition, from Table 6.13, the best solution with a minimum cost of $15,442.76/h was found by QPSO with

TABLE 6.10   Generator Unit Capacity and Coefficients of the Six-Unit System

Unit	P_j^{min}	P_j^{max}	a_j	b_j	c_j
1	100	500	240	7.0	0.0070
2	50	200	200	10.0	0.0095
3	80	300	220	8.5	0.0090
4	50	150	200	11.0	0.0090
5	50	220	220	10.5	0.0080
6	50	120	190	12.0	0.0075

TABLE 6.11   Generator Unit Ramp Rate Limits and Prohibited Zones of the Six-Unit System

Unit	P_j^0	U_j	D_j	Prohibited Zones
1	440	80	120	[210, 240] [350, 380]
2	170	50	90	[90, 110] [140, 160]
3	200	65	100	[150, 170] [210, 240]
4	150	50	90	[80, 90] [110,120]
5	190	50	90	[90, 110] [140, 150]
6	110	50	90	[75, 85] [100, 105]

TABLE 6.12 Loss Coefficients B of the Six-Unit System

B_{jk}	1	2	3	4	5	6
1	0.0017	0.0012	0.0007	−0.0001	−0.0005	−0.0002
2	0.0012	0.0014	0.0009	0.0001	−0.0006	−0.0001
3	0.0007	0.0009	0.0031	0	−0.001	−0.0006
4	−0.0001	0.0001	0	0.0024	−0.0006	−0.0008
5	−0.0005	−0.0006	−0.001	−0.0006	0.0129	−0.0002
6	−0.0002	−0.0001	−0.0006	−0.0008	−0.0002	0.0150
B_{0k}	−0.0004	−0.0001	0.0007	0.0001	0.0002	−0.0007
B_{00}	0.056					

TABLE 6.13 Optimised Results Taking 100 Runs of Various Optimisation Methods for the Six-Unit System (the Algorithm Marked with '*' Performs Best)

	Minimum Cost	Mean Cost	Standard Deviation	Maximum Cost
$M = 100$, $n_{max} = 200$				
GA	15,445.60	15,453.83	9.73	15,491.48
DE	15,444.95	15,448.90	6.99	15,472.07
PSO	15,444.78	15,456.70	7.92	15,483.97
*QPSO	**15,442.76**	**15,445.02**	**2.28**	**15,455.29**
$M = 20$, $n_{max} = 1000$				
GA	15,446.48	15,453.73	10.71	15,493.20
DE	15,442.98	15,455.24	13.74	15,489.90
PSO	15,443.84	15,456.83	13.37	15,529.61
QPSO	**15,442.89**	**15,450.13**	**9.35**	**15,478.19**

Note: Bold entry is the best result for each setting of M and n_{max}.

$M = 100$ and $n_{max} = 200$ for the 100 runs. The corresponding solution vector P_j, $j = 1, 2,\ldots, 6$ is given in Table 6.14.

Figure 6.23 shows the convergence history of the tested optimisation algorithms for the ED problem of the system. In general, the QPSO algorithm has better convergence properties than the other approaches.

Example 6.2

This example involves a system with 15 thermal units whose characteristics are given in Tables 6.15 and 6.16 [90]. The load demand of the system is 2630 MW. The B loss coefficients are listed in Equation 6.51. Table 6.17 summarises the minimum, mean and

TABLE 6.14   The Best Solution Taking 100 Runs
Obtained for the Six-Unit System Using QPSO
with $M = 100$ and $n_{max} = 200$

Power Output		Power Output	
P_1 (MW)	445.2541	P_4 (MW)	141.0687
P_2 (MW)	172.7916	P_5 (MW)	163.8578
P_3 (MW)	263.5285	P_6 (MW)	88.8558
Total power output (MW)		1,275.3565	
Power loss (MW)		12.3598	
Total generation cost ($/h)		15,442.76	

maximum values of the costs and the standard deviation achieved by various optimisation algorithms for 100 runs.

From Table 6.17, the QPSO algorithm shows the best fuel cost, the best mean cost and the lowest standard deviation for the ED problem of the 15-unit system when $M = 100$ and $n_{max} = 200$. In addition, the QPSO algorithm also shows the best maximum cost among all the tested optimisation algorithms. When $M = 20$ and $n_{max} = 1000$, QPSO exhibits the best quality solution of the ED problem.

The best solution of the ED problem of this system is achieved by the QPSO algorithm with $M = 20$ and $n_{max} = 1000$. The result obtained for solution vector $P_j, j = 1, 2,..., 15$ with a minimum cost of \$32,652.34/h is presented in Table 6.18. Figure 6.24 also shows that the QPSO technique has better convergence properties than the other optimisation methods in the test.

$B_{ik} =$

$$
\begin{bmatrix}
0.0014 & 0.0012 & 0.0007 & -0.0001 & -0.0003 & -0.0001 & -0.0001 & -0.0001 & -0.0003 & 0.0005 & -0.0003 & -0.0002 & 0.0004 & 0.0003 & -0.0001 \\
0.0012 & 0.0015 & 0.0013 & 0.0000 & -0.0005 & -0.0002 & 0.0000 & 0.0001 & -0.0002 & -0.0004 & -0.0004 & -0.0000 & 0.0004 & 0.0010 & -0.0002 \\
0.0007 & 0.0013 & 0.0076 & -0.0001 & -0.0013 & -0.0009 & -0.0001 & 0.0000 & -0.0008 & -0.0012 & -0.0017 & -0.0000 & -0.0026 & 0.0111 & -0.0028 \\
-0.0001 & 0.0000 & -0.0001 & 0.0034 & -0.0007 & -0.0004 & 0.0011 & 0.0050 & 0.0029 & 0.0032 & -0.0011 & -0.0000 & 0.0001 & 0.0001 & -0.0026 \\
-0.0003 & -0.0005 & -0.0013 & -0.0007 & 0.0090 & 0.0014 & -0.0003 & -0.0012 & -0.0010 & -0.0013 & 0.0007 & -0.0002 & -0.0002 & -0.0024 & -0.0003 \\
-0.0001 & -0.0002 & -0.0009 & -0.0004 & 0.0014 & 0.0016 & -0.0000 & -0.0006 & -0.0005 & -0.0008 & 0.0011 & -0.0001 & -0.0002 & -0.0017 & 0.0003 \\
-0.0001 & 0.0000 & -0.0001 & 0.0011 & -0.0003 & -0.0000 & 0.0015 & 0.0017 & 0.0015 & 0.0009 & -0.0005 & 0.0007 & -0.0000 & -0.0002 & -0.0008 \\
-0.0001 & 0.0001 & 0.0000 & 0.0050 & -0.0012 & -0.0006 & 0.0017 & 0.0168 & 0.0082 & 0.00079 & -0.0023 & -0.0036 & 0.0001 & 0.0005 & -0.0078 \\
-0.0003 & -0.0002 & -0.0008 & 0.0029 & -0.0010 & -0.0005 & 0.0015 & 0.0082 & 0.0129 & 0.0116 & -0.0021 & -0.0025 & 0.0007 & -0.0012 & -0.0072 \\
-0.0005 & -0.0004 & -0.0012 & 0.0032 & -0.0013 & -0.0008 & 0.0009 & 0.0079 & -0.0116 & 0.0200 & -0.0027 & -0.0034 & 0.0009 & -0.0011 & -0.0088 \\
-0.0003 & -0.0004 & -0.0017 & -0.0011 & 0.0007 & 0.0011 & -0.0005 & -0.0023 & -0.0021 & -0.0027 & 0.0140 & 0.0001 & 0.0004 & -0.0038 & 0.0168 \\
-0.0002 & -0.0000 & -0.0000 & -0.0000 & -0.0002 & -0.0001 & 0.0007 & -0.0036 & -0.0025 & -0.0034 & 0.0001 & 0.0054 & -0.0001 & -0.0004 & 0.0028 \\
0.0004 & 0.0004 & -0.0026 & 0.0001 & -0.0002 & -0.0002 & -0.0000 & 0.0001 & 0.0007 & -0.0009 & 0.0004 & -0.0001 & 0.0103 & -0.0101 & 0.0028 \\
0.0003 & 0.0010 & 0.0111 & 0.0001 & -0.0024 & -0.0017 & -0.0002 & 0.0005 & -0.0012 & -0.0011 & -0.0038 & -0.0004 & -0.0101 & 0.0578 & -0.0094 \\
-0.0001 & -0.0002 & -0.0028 & -0.0026 & -0.0003 & 0.0003 & -0.0008 & -0.0078 & -0.0072 & -0.0088 & 0.0168 & 0.0028 & 0.0028 & -0.0094 & 0.1283
\end{bmatrix}
$$

$B_{0k} = \begin{bmatrix} -0.0001 & -0.0002 & 0.0028 & -0.0001 & 0.0001 & -0.0003 & -0.0002 & -0.0002 & 0.0006 & 0.0039 & -0.0017 & -0.0000 & -0.0032 & 0.0067 & -0.0064 \end{bmatrix}$,

$B_{00} = 0.055$.

$$(6.51)$$

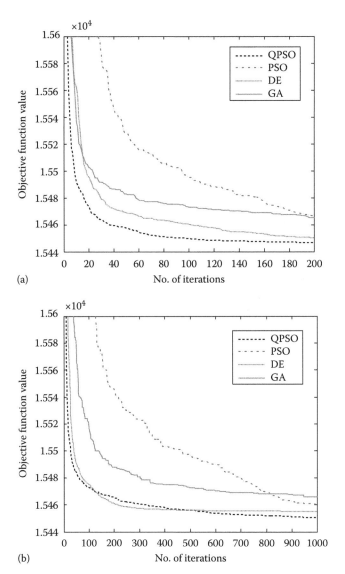

FIGURE 6.23 Convergence properties of the tested optimisation methods for the six-unit system for the cases (a) $M = 100$ and $n_{max} = 200$ and (b) $M = 20$ and $n_{max} = 1000$.

6.4.5 Summary

The QPSO method, along with other optimisation techniques, such as GA, PSO and DE, is tested for the ED problems on three different power systems. Non-linear characteristics of the generator such as ramp rate limits, valve-point zones and non-smooth functions are also taken account of

TABLE 6.15   Generator Unit Capacity
and Coefficients of the 15-Unit System

Unit	P_j^{min}	P_j^{max}	a_j	b_j	c_j
1	150	455	671	10.1	0.000299
2	150	455	574	10.2	0.000183
3	20	130	374	8.8	0.001126
4	20	130	374	8.8	0.001126
5	150	470	461	10.4	0.000205
6	135	460	630	10.1	0.000301
7	135	465	548	9.8	0.000364
8	60	300	227	11.2	0.000338
9	25	162	173	11.2	0.000807
10	25	160	175	10.7	0.001203
11	20	80	186	10.2	0.003586
12	20	80	230	9.9	0.005513
13	25	85	225	13.1	0.000371
14	15	55	309	12.1	0.001929
15	15	55	323	12.4	0.004447

TABLE 6.16   Generator Unit Ramp Rate Limits
and Prohibited Zones of the 15-Unit System

Unit	P_j^0	U_j	D_j	Prohibited Zones
1	400	80	120	
2	300	80	120	[185, 225] [305, 335] [420, 450]
3	105	130	130	
4	100	130	130	
5	90	80	120	[180, 200] [305, 335] [390, 420]
6	400	80	120	[230, 255] [365, 395] [430, 455]
7	350	80	120	
8	95	65	100	
9	105	60	100	
10	110	60	100	
11	60	80	80	
12	40	80	80	[30, 40] [55, 65]
13	30	80	80	
14	20	55	55	
15	20	55	55	

TABLE 6.17   Optimised Results Obtained by Taking 100 Runs
of Each of the Optimisation Methods for the 15-Unit System
(the Algorithm Marked with "" Performs Best)

	Minimum	Mean	Standard Deviation	Maximum
$M = 100, G_{max} = 200$				
GA	32,939.52	33,049.36	100.13	33,231.62
DE	32,818.58	32,801.24	115.09	33,478.87
PSO	32,715.10	32,933.46	121.87	33,450.01
QPSO	32,711.59	32,775.512	48.37	32,963.15
$M = 20, n_{max} = 1000$				
GA	32,905.36	33,121.25	88.91	33,273.17
DE	32,678.64	32,865.45	219.32	33,213.32
PSO	32,735.70	32,967.49	102.05	33,297.22
*QPSO	**32,652.34**	**32,749.86**	**68.47**	**32,956.64**

Note: Bold entry is the best result for each setting of M and n_{max}.

TABLE 6.18   The Best Solution Obtained by Taking 100 Runs for the 15-Unit System
Using QPSO with $m = 20$ and $n_{max} = 1000$

Power Output

$P_1 \sim P_5$ (MW)	454.8093	379.9742	129.8458	129.9152	169.0867
$P_6 \sim P_{10}$ (MW)	459.6428	429.9559	454.8093	49.7381	160.0000
$P_{11} \sim P_{15}$ (MW)	79.8596	79.3174	26.1522	18.7287	15.0645
Total power output (MW)		2,655.3650	Power loss (MW)		25.3696
Total generation cost ($/h)			**32,652.34**		

for practical generation operation in these tests. In order to obtain solutions that satisfy the equality constraint, a penalty function is added to the cost function, and the sum of them is used as the objective function. The inequality constraints are handled by penalty method and position adjustment method. The simulation results of the case studies reveal that the QPSO algorithm shows superior features, such as high-quality solutions, robustness and good convergence properties.

6.5 MSA

6.5.1 Introduction

MSA of nucleotides or amino acids is one of the most important and challenging problems in bioinformatics. It is an extension of the pairwise alignment method to incorporate more than two sequences at a time. MSA methods aim to align all of the sequences in a given query set.

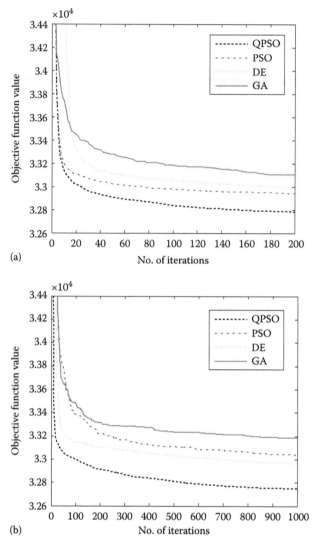

FIGURE 6.24 Convergence properties of the tested optimisation methods for the 15-unit system for the cases (a) $M = 100$ and $n_{max} = 200$ and (b) $M = 20$ and $n_{max} = 1000$.

The resulting aligned sequences are often used to construct phylogenetic trees, to find protein families, to predict secondary and tertiary structures of new sequences and to demonstrate the homology between new sequences and existing families [91]. MSA is computationally difficult to produce, and most formulations of the problem lead to NP-complete combinatorial optimisation problems [92]. Nevertheless, the utility of these alignments in bioinformatics has led to the development of a variety of methods suitable for aligning three or more sequences.

The technique of DP may theoretically be applied to any number of sequences. Such technique is usually computationally expensive in both time and memory and hence it is rarely used for more than three or four sequences in its most basic form. One method to tackle this problem is to use the so-called progressive alignment strategies [93,94]. The progressive alignment strategy involves the repetition of the steps described as follows. First, two sequences are chosen from the given multiple sequences. Second, the two sequences are aligned by the DP algorithm. Third, the two sequences are replaced with the resulting pairwise alignment. This strategy allows us to align a large number of given sequences in a practical amount of computational time, but the resulting alignment is not necessarily optimal. This is due to the fact that only local or partial information produced by each pairwise alignment would enter the result. One well-known MSA program implementing progressive alignments is ClustalW (CW) [94].

An alternative to progressive alignment methods is to use stochastic optimisation method, such as SA [95,96] or evolutionary algorithms (EAs) [97–99]. An objective function is involved in optimisation methods which measures multiple alignment quality. In order to find an optimal alignment, several steps are involved of updating the alignment to improve the objective function value. In the case of EAs, a population of alignments is evolved in a quasi-evolutionary manner, and the fitness of the population is gradually improved as measured by the objective function.

Another efficient and popular approach for MSA is based on probabilistic models such as hidden Markov models (HMMs) [100,101]. HMMs have been applied to MSA and shown to be efficient tools for MSA problems [102]. HMMs are used to create an operation sequence of gap insertion and deletion instructions to align the sequences. Generally, an HMM topology used for the MSA problem requires roughly as many states as the average length of the sequences in the problem. Therefore, one issue of the HMM approach is that there is no known deterministic algorithm that can guarantee to yield an optimally trained HMM within reasonable computational time. One way to tackle this problem is to use an approximation algorithm based on statistics and re-estimation, such as the Baum–Welch (BW) algorithm which is known as forward–backward algorithm [103]. The gradient methods [100] have also been used to optimise the parameters of an HMM. However, these methods are local search techniques that usually result in suboptimally trained HMM. Another way is to estimate the parameters of an HMM by stochastic optimisation algorithms, such as SA [104] and EAs [105,106]. SA or EAs ensure a higher chance of reaching a global optimum by starting with single or multiple random search points and updating the candidate solutions randomly. However, SA and EAs encounter certain shortcomings such as lack of local search ability, premature convergence and slow convergence speed. ACO and PSO are two typical paradigms of this kind of stochastic optimisation algorithms and have been shown to be efficient tools for solving the MSA problem [107,108].

This section presents the performance of the QPSO algorithm in training the HMM for the solutions of the MSA problem. Three sets of benchmark data are tested by using the HMMs trained with the QPSO, PSO and BW algorithms and performance comparisons are studied.

6.5.2 HMM for MSA

6.5.2.1 Topology of HMM for MSA

The HMM structure used in this study is the standard topology for the MSA problem originally suggested by Krogh et al. [101]. Figure 6.25 shows an example of a simple topology of the HMM as a directed graph. The HMM consists of a set of q states, $(S_1, S_2,..., S_q)$, which are divided into three groups: match (M), insert (I) and delete (D). In addition, there are two special states: the begin state and the end state. States are connected to each other by the transition probability a_{ij}, where $a_{ij} \geq 0$ $(1 \leq i, j \leq q)$ and $\sum_{j=1}^{q} a_{ij} = 1 (1 \leq i \leq q)$. A match or insert state, S_j, emits an observable symbol, v_k, from an output alphabet Σ with a probability $b_j(k)$ where $b_j(k) \geq 0$ $(1 \leq j \leq q, 1 \leq k \leq m)$ and $\sum_{k=1}^{m} b_j(k) = 1 (1 \leq j \leq q)$. Here, m is the number of observable symbols. Delete states, begin states and end states do not emit observable symbols and are called silent states.

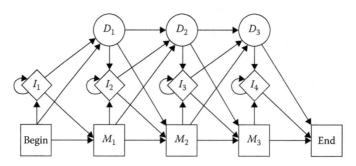

FIGURE 6.25   An example of HMM of length 3 for MSA problem.

Starting from the begin state to the end state, the HMM generates sequences (strings of observable symbols) by making non-deterministic walks that randomly go from one state to another according to the transitions. Each work yields a path, $\pi = (\pi_1, \pi_2,..., \pi_p)$, consisting of the visited states and a sequence consisting of emitted observable states on the path. When the HMM is applied to MSA, the sequence of observable symbols is given in the form of an unaligned sequence. The goal of MSA is thus to find a path π which generates the best alignment. It is possible to use the *forward* and *Viterbi* algorithm to determine $P(o|\lambda)$, the probability of a given symbol sequence (o) generated by an HMM (λ), and derive the path π with maximal probability of the generating the sequence (o) [109].

6.5.2.2 Training HMM for MSA

For a given sequence (o) and an HMM (λ), the goal of training the HMM is to determine the parameters (transition and emission probabilities) of λ such that $P(o|\lambda)$ is maximised. This task is usually tackled by using either the BW technique, which is based on statistical re-estimation formulae [109], or by stochastic search methods such as SA [104] or EAs [99,105,106]. Before training, the length of the HMM should be determined. A commonly used estimate is the average length of the sequences to be aligned. After training, the length may be chosen by using a better model with a heuristic method known as *model surgery* [101].

The quality of the HMM needs to be evaluated during the training. Generally, a log-odds score is used for this purpose, which is based on a log-likelihood score [109] given by

$$\text{Log-odds}\,(O, \lambda) = \frac{1}{N} \sum_{i=1}^{N} \log_2 \frac{P(O_i \mid \lambda)}{P(O_i \mid \lambda_N)}, \qquad (6.52)$$

where
$O = (O_1, O_2,..., O_N)$ is the set of unaligned sequences
λ is the trained HMM
λ_N is a null-hypothesis model

Here, a random model is chosen as the null-hypothesis model.

The final step after the training of the HMM is to interpret the learned sequences as an MSA. The HMM model from the training phase is considered to be a profile for the set of sequences. Thus the unaligned sequence can be aligned by such a profile HMM (λ).

6.5.3 QPSO-Trained HMMs for MSA

6.5.3.1 Training HMMs by QPSO

This section concerns the application of the QPSO algorithm for the training of the HMM for MSA. When QPSO algorithm is used to train HMMs, the length of the HMM is kept constant during the training. Only the parameters, transition and emission probabilities of the HMM, need to be optimised. The candidate solution for an HMM is represented as the position of a particle with real encoding of l transitions and m emission probabilities. Thus, the dimension of the search space for the HMM training is $l + m$.

A copy of the population is created at each iteration of the QPSO method. All particles in this copy are normalised such that the constraints on the transition and emission probabilities mentioned in Section 6.5.2 are satisfied. Each particle in the copy population X' is evaluated by either using the log-odds in Equation 6.52 or evaluating the sum-of-pairs (SOP) score described in the next sub section.

The procedure of HMM training with QPSO is outlined as follows in Figure 6.26.

6.5.3.2 MSA with the Trained HMM

After training the HMM with QPSO, the output of the global best positions of the particles consists of the optimised parameters, transition and emission probabilities of the HMM. The trained HMM can be considered as a profile for the set of sequences. Thus multiple sequences based on the HMM

Step 0: Initialize the populations of the current positions, X, of the particles and the personal best positions P, and normalize P.

Step 1: For $n = 1$ to n_{max} (maximum number of iterations), execute the following steps;

Step 2: Copy populations of particles' current positions X to X to X';

Step 3: Calculate the mean best position C among the particles according to (4.81);

Step 4: Select the value of α;

Step 5: Normalize the population X';

Step 6: For each particle, compute its objective function at its position in X' by $f(X'_{i,n})$. If $f(X'_{i,n}) > f(P_{i,n})$, then $P_{i,n} = X'_{i,n}$;

Step 7: Select the current global best position G_n;

Step 8: For each dimension of each particle, obtain the stochastic point $p^j_{i,n}$ by Equation 4.83;

Step 9: Update each component of the current position in population X by Equation 4.82 and return to Step 1;

FIGURE 6.26 The procedure of HMM training with QPSO.

using the Viterbi algorithm [109] can be aligned. Finally, the resulting alignment of multiple sequences is evaluated according to the SOP score [108].

6.5.3.3 Scoring of MSA

There are two different methods for scoring the testing alignment obtained by the numerical experiments. The first one does not rely on any form of prior knowledge about the structure of resulting alignment, while the second one requires a reference alignment for comparison. The reference alignment is referred to as a manually refined alignment that is believed to be of high quality. Both scoring methods are based on the widely used SOP scoring function.

For the experiments without prior knowledge regarding the structure of the resulting alignments, the standard SOP scoring given in the following may be used:

$$\text{Sum-of-pairs (SOP)} = \sum_{i=1}^{n-1} \sum_{j=i+1}^{n} D(l_i, l_j), \tag{6.53}$$

where
l_i is the aligned sequence i
D is a distance metric

For example, in this study, the widely accepted *BLOSUM62* replacement matrix in [110] is used as the distance metric for amino acid. To prevent the accumulation of many gaps in an alignment, an affine gap cost is deduced from the SOP function leading to

$$\text{Gap cost} = \text{GOP} + n \times \text{GEP}, \tag{6.54}$$

where
GOP is a fixed penalty for opening a gap
GEP is the penalty for extending the gap
n is the number of gap symbols in the gap

With the gap cost calculated for each gap in each of the aligned sequences, the sum of these costs is then deducted from the SOP score.

In the experiments when the reference alignment is available, a modified SOP is employed to evaluate the alignment [111]. Given the tested alignment of N sequences consisting of M columns, the ith column of the alignment may be denoted by $A_{i1}, A_{i2}, \ldots, A_{iN}$. For each pair of residues A_{ij} and A_{ik}, p_{ijk} is defined as follows: $p_{ijk} = 1$ if the residues A_{ij} and A_{ik} from the test alignment are aligned with each other in the reference alignment, otherwise $p_{ijk} = 0$. Define the column score by

$$S_i = \sum_{j=1, j\neq k}^{N} \sum_{k=1}^{N} p_{ijk}.$$ (6.55)

The SOP score for the test alignment is then given by

$$\text{Sum-of-pairs (SOP)} = \sum_{i=1}^{N} \left(\frac{S_i}{\sum_{i=1}^{M_r} S_{ri}} \right),$$ (6.56)

where
  M_r is the number of columns in the reference alignment
  S_{ri} is the score S_i for the ith column in the reference alignment

6.5.4 Numerical Experiments
6.5.4.1 Benchmark Datasets
In order to evaluate the performance of the trained HMMs for MSA, QPSO, PSO and the BW algorithm were used to train the HMMs for the MSA problems. Three benchmark datasets, including simulated nucleotide dataset, four protein families and amino acid dataset from the benchmark alignment database, were used in the test. The performance of the previous three methods was compared with the CW program.

6.5.4.1.1 Simulated Nucleotide Dataset Nucleotide sequences (20 sequences with each having approximately 300 characters) generated by using the program Rose [112] (using the model proposed by Juke and Cantor [113] and taking the mean substitution rate = 0.013 and the insert/delete probability = 0.03) are listed as follows. A randomly generated root sequence (of length 500) was evolved on a random tree to yield sequences of 'low' or 'high' mean divergences, that is, with an average number of substitutions per site of 0.5 or 1.0, respectively. Furthermore, the insertion/deletion length distribution was set to 'short' (frequencies of gaps of length 1–3 = 0.8, 0.1, 0.1) or 'long' (frequencies of gaps of length 1–7 = 0.3, 0.2, 0.1, 0.1, 0.1, 0.1, 0.1). Hence, the four 'mean divergence-mean gap length' conditions tested here are 'low-short' 'low-long', 'high-short' and 'high-long'. Fifty random sequences were generated under each of the four combinations of settings, and the

TABLE 6.19   The Four Protein Families

Family	N	LESQ (min, max)	LESQ	T
G5	277	79 (67, 88)	79	150
CagY_M	549	31 (24, 35)	31	150
Interferon	375	164 (23, 200)	164	150
Biopterin_H	343	170 (13, 359)	170	150

N, number of sequences; LSEQ, length of sequences; and T, size of training set.

average performance was recorded. For this dataset, the SOP given by Equation 6.53 was employed to score the resulting alignments.

6.5.4.1.2 Four Protein Families   The second dataset consists of four large sets of unaligned sequences for the four protein families, namely, G5, CagY_M, Interferon and Biopterin_H, that were extracted from the *Pfam* database located at http://pfam.jouy.inra.fr/browse.shtml [114]. The average, minimum, maximum length, number of sequences and size of training set of each family are listed in Table 6.19. All of these protein families have previously been used in MSA studies using HMMs [104] and GAs [99].

In order to check over-fitting of the HMM model, the four protein families were divided into training and validation sets. The size of each training set was taken as 150 as shown in Table 6.19. The validation sets were the original datasets excluding the training sets. For this dataset, the SOP given by Equation 6.53 was used as the score function.

6.5.4.1.3 Amino Acid Dataset   The third group of dataset was extracted from BAliBASE (benchmark alignment database) database located at http://www.cs.nmsu.edu/~jinghe/CS516BIOINFO/Fall05/BaliBASE/ align_index.html. It contains several manually refined MSAs specifically designed for the evaluation and comparison of MSA methods. Twelve sequence sets were selected from the first reference set from BAliBASE database and are listed in Table 6.20. The first column in the table lists the names of selected sequence sets, the second column lists the number of sequences in each set, the third column includes the average, minimum and maximum lengths of the sequences in each set and the fourth column lists the identity of the sequences. Since the reference alignments for these datasets are available, the SOP score in Equation 6.56 is used to determine the quality of the resulting test alignments.

TABLE 6.20 The 12 Benchmark Datasets from BAliBASE

Name	N	LSEQ (Min, Max)	Identity (%)
lidy	5	(49, 58)	<25
451c	5	(70, 87)	20–40
1krn	5	(66, 82)	>35
kinase	5	(263, 276)	<25
1pii	4	(247, 259)	20–40
5ptp	5	(222, 245)	>35
1ajsA	4	(258, 387)	<25
glg	5	(438, 486)	20–40
1tag	5	(806, 928)	>35

6.5.4.2 Experimental Settings

6.5.4.2.1 Parameter Settings of the Algorithms   The parameters of PSO and QPSO were chosen as follows:

PSO: population size = 20; w decreases linearly from 0.9 to 0.4; $c_1 = c_2 = 2.0$; $V_{max} = 1.0$.

QPSO: population size = 20; α decreases linearly from 1.0 to 0.5.

The initial positions of the particles in both QPSO and PSO were chosen to be uniformly distributed on (0,1).

6.5.4.2.2 Other Experimental Configurations   For nucleic acids, the 'swap' substitution score table from CW 1.81 version was used as the distance metric in the SOP score function, and a GOP of 15 and a GEP of 7 were used as penalties. For amino acid, the *BLOSUM62* replacement matrix in [110] was used as the distance metric in the SOP score function, and GOP and GEP were set to be 11 and 2, respectively. These parameter values were configured as in [108].

For each set of sequences in the three datasets, the experiments for HMM training were studied by the four training algorithms with log-odds score and SOP score as the objective function. Except the BW algorithm, which is a deterministic algorithm providing a fixed initial HMM for all training experiments, the experiment of each stochastic algorithm was repeated 20 times with each run executing 1000 iterations.

6.5.4.3 Experimental Results

6.5.4.3.1 Results for Simulated Nucleotide Dataset   Tables 6.21 and 6.22 summarise the experimental results for HMM training using simulated nucleotide dataset. Table 6.21 lists the results of experiments showing the

TABLE 6.21   Results of the HMM Log-Odds Scores and the Average Execution
Time over 20 Runs of Each Experiment for Nucleotide Sequences

Nucleotide	Algorithms	Average Log-Odds Score (Standard Deviation)	Normalised Score	Average Execution Time (h)
Low-short	BW	394.6	−0.957	**0.0404**
	PSO	418.4 (7.35)	−0.081	4.3503
	QPSO	**448.8 (9.68)**	**1.038**	4.3435
Low-long	BW	226.5	−0.4714	**0.0378**
	PSO	215.8 (8.26)	−0.6772	4.2912
	QPSO	**310.7 (4.68)**	**1.1486**	4.3277
High-short	BW	186.3	−0.6998	**0.0345**
	PSO	195.6 (2.31)	−0.4456	5.3179
	QPSO	**253.8 (3.45)**	**1.1453**	5.4497
High-long	BW	280.8	−0.2929	**0.0371**
	PSO	256.4 (8.52)	−0.8209	5.3424
	QPSO	**345.8 (9.73)**	**1.1137**	5.3689

Note: Bold entry is the best result for each sequence set.

TABLE 6.22   Results of the SOP Scores and the Average Execution Time over
20 Runs of Each Experiment for the Final Alignments of Nucleotide Sequences

Nucleotide	Algorithms	Average SOP Score (Standard Deviation)	Normalised Score	Average Execution Time (h)
Low-short	CW	2514.9	0.6257	—
	BW	2024.7	−1.0612	**0.0430**
	PSO	2154.3 (25.36)	−0.6152	5.9012
	QPSO	**2638.4 (20.364)**	**1.0507**	5.9824
Low-long	CW	8760.7	−0.249	—
	BW	8623.9	−0.46	0.0419
	PSO	8430.8 (15. 0154)	−0.7579	5.9170
	QPSO	**9873.3 (18.462)**	**1.467**	5.9249
High-short	CW	4316.5	0.2358	—
	BW	4270.2	−1.0998	0.0477
	PSO	4294.7 (13.5643)	−0.3931	5.8703
	QPSO	**4351.9 (13.375)**	**1.2571**	5.8937
High-long	CW	7781.7	0.6663	—
	BW	7418.2	−1.129	0.0469
	PSO	7538.5 (22.8645)	−0.5349	6.5177
	QPSO	**7848.8 (29.6881)**	**0.9977**	6.5305

Note: Bold entry is the best result for each sequence set.

log-odds score used in the quality measure for the HMMs. It can be seen that the QPSO algorithm is able to generate the HMMs with a better average log-odds scores than the HMMs trained with BW and PSO. Table 6.22 shows the average best SOP scores for alignments produced by the HMMs of the training sets. The SOP scores for alignments generated by CW are also listed in Table 6.22. From the results, it is possible to conclude that the QPSO algorithm achieved the best SOP scores amongst all the methods tested. Alignments with CW yield better scores than those produced by the HMM trained with PSO and BW.

Figure 6.27 illustrates typical examples of the convergence of the average best log-odds scores and SOP scores of all the algorithms during the runs. It can be seen that the BW algorithm converges very fast to the local optima. PSO exhibits the fastest convergence speed compared with QPSO but may encounter premature convergence.

6.5.4.3.2 Results for the Four Protein Families   Tables 6.23 through 6.25 list the experimental results for HMM training and validation sets of four protein families. Tables 6.23 and 6.24 show the results of the experiments listing the log-odds scores used as the fitness value. These results demonstrate that the QPSO algorithm has better log-odds scores for HMM training than PSO and BW. The PSO algorithm does not possess better log-odds scores than BW for some sequence sets such as family G5, interferon, validation sets of CagY_M and validation sets of interferon. Table 6.25 summarises the results for using the SOP scores as the fitness values. It can be seen that the QPSO algorithm yields the best scores. Figure 6.28 shows that the QPSO algorithm exhibits better convergence properties than their competitors.

6.5.4.3.3 Results for Amino Acid Dataset   Table 6.26 shows the results of the experiments conducted on amino acid dataset, with the log-odds scores being used as the fitness function. The QPSO algorithm trains the HMMs that show better average log-odds scores than the HMMs trained by using BW and PSO.

Table 6.27 shows the experimental results performed on the amino acid dataset with the SOP scores being used as the fitness values. Apart from the results for the HMM trained with the three methods, it also shows the scores achieved by the well-known and widely used CW. It can be easily seen that the results of QPSO are better than those of PSO and BW, but not as good as those of CW on most of the sequence sets.

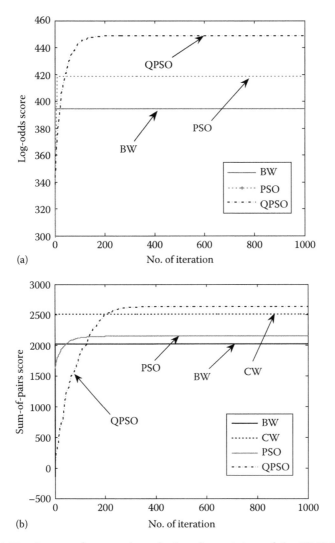

FIGURE 6.27   Average fitness values during the training of the HMM for low-short nucleotide sequences: (a) log-odds scores and (b) SOP scores.

However, the results are still remarkable considering that CW is a highly specialised tool for MSA, which, for instance, estimates the evolutionary distance between all sequences in the set to create an evolutionary tree before it aligns the sequences (using iterative pairwise alignment). Simple comparison shows that the QPSO algorithm produces better scores than BW, PSO as well as SA reported in [108] for all the sequence sets. This is a remarkable result showing the advantage of using QPSO in the training of HMM.

TABLE 6.23   Results of the HMM Log-Odds Scores and the Average Execution Time over 20 Runs of Each Experiment for the Training Sets of the Four Protein Families

Nucleotide	Algorithms	Average Log-Odds Score (Standard Deviation)	Normalised Score	Average Execution Time (h)
G5	BW	103.146	−0.5492	**0.0595**
	PSO	101.4500 (1.89)	−0.605	8.6404
	QPSO	**154.94 (1.06)**	**1.1543**	8.9469
CagY_M	BW	11.178	−1.0143	**0.0357**
	PSO	20.090 (1.14)	0.0292	7.0527
	QPSO	**28.255 (0.92)**	**0.9851**	7.5136
Interferon	BW	158.314	−0.0819	**0.0714**
	PSO	141.736 (2.41)	−0.9565	12.2936
	QPSO	**179.549 (1.43)**	**1.0384**	12.5230
Biopterin_H	BW	162.431	−1.1547	**0.0754**
	PSO	179.521 (4.38)	0.5759	14.3029
	QPSO	**179.549 (1.43)**	**0.5788**	14.4857

Note: Bold entry is the best result for each sequence set.

TABLE 6.24   Results of the HMM Log-Odds Scores and the Average Execution Time over 20 Runs of Each Experiment for the Validation Sets of the Four Protein Families

Nucleotide	Algorithms	Average Log-Odds Score (Standard Deviation)	Normalised Score	Average Execution Time (h)
G5	BW	78.357	−1.1389	**0.0529**
	PSO	141.4500 (2.42)	0.4044	6.9691
	QPSO	**154.94 (1.12)**	**0.7344**	7.0767
CagY_M	BW	12.832	−1.1545	**0.1729**
	PSO	20.090 (2.13)	0.5578	10.4372
	QPSO	**20.255 (1.08)**	**0.5967**	10.4611
Interferon	BW	102.652	−0.5014	**0.3196**
	PSO	95.736 (1.54)	−0.6501	17.9688
	QPSO	**179.549 (1.28)**	**1.1515**	18.1971
Biopterin_H	BW	171.281	0.0278	**0.3254**
	PSO	162.292 (2.38)	−1.0136	19.9303
	QPSO	**179.549 (1.28)**	**0.9858**	20.268

Note: Bold entry is the best result for each sequence set.

TABLE 6.25   Results of the SOP Scores (Divided by 1000) and the
Average Execution Time over 20 Runs of Each Experiment for the Final
Alignments of the Training Sets of the Four Protein Families

Nucleotide	Algorithms	Average SOP Score (Standard Deviation)	Normalised Score	Average Execution Time (h)
G5	CW	189	−0.3296	—
	BW	192	−0.1978	0.0737
	PSO	176 (113.1)	−0.901	13.1803
	QPSO	**229 (63.7)**	**1.4284**	13.2774
CagY_M	CW	−142	−0.9291	—
	BW	−138	−0.6893	0.0602
	PSO	−120 (131.5)	0.3896	9.8439
	QPSO	**−106 (103.37)**	**1.2288**	9.8653
Interferon	CW	3226	−0.8897	—
	BW	3294	−0.7309	0.0873
	PSO	3772 (101.4)	0.3853	15.7066
	QPSO	**4136 (98.5)**	**1.2353**	15.7198
Biopterin_H	CW	4015	−0.4612	—
	BW	4113	0.1038	0.0945
	PSO	3924 (59.27)	−0.9859	18.2017
	QPSO	**4328 (92.39)**	**1.3433**	18.6513

Note: Bold entry is the best result for each sequence set.

Figure 6.29 shows that the QPSO algorithm has the best convergence properties amongst all the training algorithms. Moreover, the reference alignments of the lidy sequences from BAliBASE database and the best resulting alignments obtained by HMMs trained with BW, PSO and QPSO are shown in Figures 6.30 through 6.33.

6.5.4.3.4 Further Evaluation   In order to study the overall performance amongst all the tested methods, the normalised score

$$\text{Normalized score (NS)} = \frac{(S_i - \bar{S})}{\sigma_S}, \tag{6.57}$$

is introduced, where
S_i is the score
$\bar{S}$ is the mean of the scores
σ_S is the standard deviation of the scores

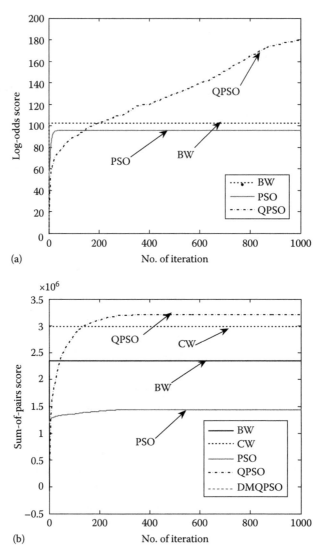

FIGURE 6.28 Average fitness values during the training of the HMM for interferon protein sequences from *Pfam* database: (a) log-odds scores and (b) SOP scores.

The average scores over all the tested algorithms listed in Tables 6.21 through 6.27 are normalised and are listed in the corresponding tables. The normalised log-odds scores, the normalised SOP scores and the total normalised scores are averaged over all the tested sequence sets and are listed in Table 6.28.

It can be seen that the normalised scores produced by the QPSO-trained HMM are better than the other competitors. This indicates that the HMM

TABLE 6.26   Results of the HMM Log-Odds Scores and the Average Execution Time over 20 Runs of Each Experiment for the BAliBASE Test Sets

Nucleotide	Algorithms	Average SOP Score (Standard Deviation)	Normalised Score	Average Execution Time (h)
1idy	BW	42.0576	−1.061	**0.0011**
	PSO	59.7932 (0.4202)	0.1359	0.0355
	QPSO	**71.4864 (0.7630)**	**0.9251**	0.0378
451c	BW	68.3522	−1.0306	**0.0016**
	PSO	89.1605 (0.5126)	0.0643	0.0426
	QPSO	**106.3024 (0.6950)**	**0.9663**	0.0440
1krn	BW	69.0222	−0.9068	**0.0013**
	PSO	81.9846 (0.9212)	−0.1657	0.0402
	QPSO	**103.6417 (0.8287)**	**1.0725**	0.0413
kinase	BW	214.9693	−0.555	0.0074
	PSO	211.2745 (0.8959)	−0.5995	0.7359
	QPSO	**356.8937 (0.6798)**	**1.1544**	0.7513
1pii	BW	213.0459	−1.0353	**0.0061**
	PSO	277.0576 (0.9306)	0.0747	0.6736
	QPSO	**328.1439 (0.8513)**	**0.9606**	0.6925
5ptp	BW	266.5928	−0.8246	**0.0058**
	PSO	311.5647 (0.9254)	−0.2877	0.6474
	QPSO	**428.8537 (0.6151)**	**1.1123**	0.6612
1ajsA	BW	326.4896	−0.8875	**0.0089**
	PSO	381.6639 (0.8374)	−0.1959	0.7655
	QPSO	483.7352 (0.5643)	1.0835	0.7994
glg	BW	380.7306	−0.6982	**0.0180**
	PSO	395.1211 (0.9201)	−0.4474	5.7376
	QPSO	**486.5318 (0.5950)**	**1.1456**	5.7533
1tag	BW	729.3726	−0.7873	**0.0415**
	PSO	763.7521 (1.0002)	−0.3378	8.9221
	QPSO	**875.6509 (0.7841)**	**1.1252**	9.1275

Note: Bold entry is the best result for each sequence set.

trained with QPSO had the best overall performance on the MSA problems for all the sequence sets. CW outperformed the PSO-trained HMM. However the total normalised score produced by CW is not as good as the QPSO-trained HMM. This implies that a good training algorithm is crucial for the HMM in order to fully expose its advantages in solving MSA problems.

A further performance comparison of the training algorithms was made by comparing the computational costs of the algorithms for the

TABLE 6.27   Results of the SOP Scores and the Average Execution
Time over 20 Runs of Each Experiment for the BAliBASE Test Sets

Nucleotide	Algorithms	SOP Score	Normalised Score	Average Execution Time (h)
1idy	CW	0.705	0.5338	—
	BW	0.5132	−1.0431	**0.0021**
	PSO	0.5658	−0.6106	0.0412
	QPSO	**0.7763**	**1.1199**	0.0439
451c	CW	**0.7190**	**1.4301**	—
	BW	0.3989	−0.8488	**0.0025**
	PSO	0.4519	−0.4715	0.0518
	QPSO	0.5027	−0.1098	0.0540
1krn	CW	**1.000**	1.0464	—
	BW	0.8182	−0.6949	**0.0023**
	PSO	0.7863	−1.0005	0.0472
	QPSO	0.9585	0.6489	0.0491
kinase	CW	**0.7360**	**1.1635**	—
	BW	0.2268	−0.9913	**0.0087**
	PSO	0.3061	−0.6557	0.8055
	QPSO	0.5753	0.4835	0.8318
1pii	CW	**0.8640**	**1.1717**	—
	BW	0.1647	−0.9898	**0.0073**
	PSO	0.2738	−0.6526	0.7369
	QPSO	0.6372	0.4707	0.7514
5ptp	CW	**0.9660**	**1.1487**	—
	BW	0.6053	−1.054	**0.0062**
	PSO	0.6831	−0.5789	0.7183
	QPSO	0.8572	0.4843	0.7372
1ajsA	CW	0.571	0.7971	—
	BW	0.2864	−0.9797	**0.0097**
	PSO	0.3245	−0.7418	1.2831
	QPSO	**0.5914**	**0.9244**	1.3207
Glg	CW	**0.9410**	1.0698	—
	BW	0.5691	−1.1144	**0.0246**
	PSO	0.6684	−0.5312	6.5017
	QPSO	0.8569	0.5758	6.6381
1tag	CW	**0.9630**	**1.3432**	—
	BW	0.6453	−0.9168	**0.0457**
	PSO	0.6931	−0.5767	15.7178
	QPSO	0.7953	0.1503	15.807

Note: Bold entry is the best result for each sequence set.

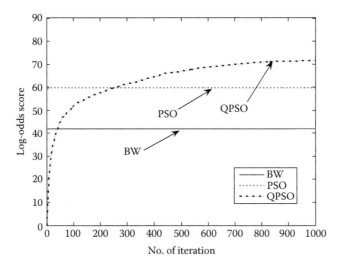

FIGURE 6.29   Average log-odds scores during the training of the HMM for the lidy sequences from BAliBase database.

```
1idy  MEVKKTSWTEEEDRILYQAHKRLG-NRWAEIAKLLP------GRTDNAIKNHWNSTMRRKV
1hstA ---SHPTYS-EMIAAAIRAEKSRGGSSRQSIQKYIKSHYKVGHNADLQIKLSIRRLLAAGV
1tc3C ---RGSALS-DTERAQLDVMKLLN-VSLHEMSRKIS-------RSRHCIRVYLKDPVSYGT
1aoy  --MRSSAKQEELVKAFKALLKEEKFSSQGEIVAALQEQGFDN-INQSKVSRMLTKFGAVRT
1jhgA TPDEREALG-TRVRIIEELLRGE--MSQRELKNELG-------AGIATITRGSNSLKAAPV
```

FIGURE 6.30   Lidy reference alignments.

```
1idy  ----------------------MEVKKTS---------WTEED--RILYQ------------------AHK-------------RLGN-RW-AEIAKLLPGRTDNAIKNHWNSTMRRKV
1hstA ----------------------SHPTYSEMIAAAIRAEKSRGGSSRQSIQKYIKSHYKVGHNADLQIKLSI------------RRLLAAGV---------------------
1tc3C ----------------------RGSALS---------DTERAQ---LDVM-----------KLLNVSLHEMSRKISRSRHCIRVYLK-DPVS-------YGT----------------
1aoy  MRSSAKQEELVKAFKALLKEEKFSSQGEIV-------AALQEQGF--DNINQ------------------SK----------VSRMLTK-FG-A-------VRT----------------
1jhgA ----------------TPDEREALG-TRV--RIIEELLRGEMSQ-----RE---------LKNELGAGIATI---------TRGSNSLKAAPV----------------
```

FIGURE 6.31   The best resulting alignments generated by the HMM trained with BW for the lidy sequences.

```
1idy  MEVKK---T-SWT-E-EED--RI-LYQA-HKRLG----------NRWA----EIAKLLPGRTDNAIKNHWNSTMRRKV
1hstA S-HPT---Y-S-E-MIAAA---IRAEKS-R-G-GSSRQSIQKYIKSHY--------KVGHNADLQIKLSIRRLLAAGV
1tc3C R-GSALSDTER-AQL-DVM---KLLNVSLH-E-M----SRKISRSRHC-----------------IRVYLKDPVSYGT
1aoy  M--RS---SA-K-Q-E-ELV---K-AFKALLKE-E-------KFSSQGEIVAALQEQGFDNINQSKVSRMLTKFGAVRT
1jhgA T--PD---E-R-E-A-LGTRVRI-IEELLRGE-M---SQRELKNELGA-------------GIATITRGSNSLKAAPV
```

FIGURE 6.32   The best resulting alignments generated by the HMM trained with PSO for the lidy sequences.

```
1idy   --MEVKKTSWTEE--EDRI---LYQAHKR--LG------NRWAEIAKLLPGRTDNAIKNHWNSTMRRKV
1hstA  --SHP--T-Y-SE--M-IA--AAIRAEKS-RGGSSRQSIQKYIKSHYKVGHNADLQIKLSIRRLLAAGV
1tc3C  --RGS--A-L-SD-TE-RA---QLDVMKL--LN-------VSLHEMSRKISRSRHCIRVYLKDPVSYGT
1aoy   MRSSA--K-Q-EE-LV-KAFKALLKEEKFSSQG-----EIVAALQEQGFDNINQSKVSRMLTKFGAVRT
1jhgA  --TPD--E-R--EALGTRV---RIIEELL--RG------EMSQRELKNELGAGIATITRGSNSLKAAPV
```

FIGURE 6.33 The best resulting alignments generated by the HMM trained with QPSO for the lidy sequences.

TABLE 6.28 The Average and Total Normalised Scores for All Sets of Tested Sequences

Algorithms	Average Normalised Log-Odds Score	Average Normalised SOP Score	Total Normalised Score
CW	—	0.4926	0.4926
BW	−0.7512	−0.8175	−0.7808
PSO	−0.2591	−0.5431	−0.3862
QPSO	1.0103	0.8680	0.9442

HMM training. The CW program is a progressive alignment method that does not involve HMM training and can achieve MSA within 1 min, so its computational consumption was not recorded for comparison. The average execution times over 20 runs of each HMM training experiment are listed in Tables 6.21 through 6.26. It is obvious that the BW algorithm requires the least computational time since it is a local search technique. On average, QPSO was slightly more time-consuming in HMM training than PSO. However, it is worthwhile to spend more computational time to obtain the significant performance advantages of QPSO over its competitors.

6.5.5 Summary

In this section, the QPSO algorithm, together with PSO and BW, was tested for training the HMMs on three benchmark datasets for MSA problems. From the experimental results of training the HMMs for MSA, it can be observed that with the log-odds scores as the objective functions (or fitness values), QPSO was able to produce the HMMs that has better average log-odds scores than the HMMs trained with BW and PSO for all of the three training sets. It is also observed that with SOP as the objective function, the QPSO algorithm yields better HMMs than its competitors. On the third dataset, the resulting alignments obtained by the QPSO-trained HMMs do not have better SOP scores on some sequence sets compared to those obtained by the CW program. In order to make an overall performance evaluation, an average of the normalised scores of the tested methods was obtained. It is found that the QPSO-trained HMM produces the best average normalised scores. Moreover, execution times of the algorithms show that the QPSO algorithm requires slightly more CPU time. However, considering its significant advantages in performance over its competitors, it can be concluded that the QPSO algorithm is an efficient approach for the training of HMMs for MSA problems.

6.6 IMAGE SEGMENTATION

6.6.1 Introduction

The goal of image segmentation is to extract meaningful objects from an input image. It is useful to discriminate objects from objects that have distinct grey levels or different colours. In recent years, many segmentation methods have been developed, including methods based on fuzzy C means (FCMs) and its variants, mean shift filters and non-linear diffusion [115–117]. Among all the existing segmentation techniques, the thresholding technique is one of the most popular one due to its simplicity, robustness and accuracy [118,119].

There are two categories of thresholding techniques. The first category involves analyzing the profile characteristics of the image histogram in order to determine the optimal thresholds. Hertz and Schafer [120] considered a multi-thresholding technique where a thinned edge field, obtained from the grey-level image, is compared with the edge field derived from the binarised image. Russ [121] noted that experts in microscopy subjectively adjust the thresholding level to a point where the edges and the shape of the object become stabilised. The second category involves optimising an objective function in order to determine the optimal thresholds, including maximisation of posterior entropy to measure homogeneity of segmented classes [122,123], minimisation of Bayesian error [124,125], maximisation of the measure of the separability on the basis of between-class variance [123], etc. The OTSU method, named after Nobuyuki Otsu, was proved to be one of the best thresholding methods for the uniformity and shape measures [119,126,127]. However the problem becomes more and more complex when one tries to achieve segmentation with greater details by employing multilevel thresholding. In recent years, several EAs have been introduced into the field of image segmentation for their fast computing ability. Tao et al. [128] used ACO method to obtain the optimal combination of the fuzzy parameters. Cao et al. [129] proposed a GA with a learning operator (GA-L) to accelerate the multilevel thresholding and to improve the optimal stability. PSO seems to be comparable to many other population-based stochastic optimisation methods and even shows its performance in searching ability for many hard optimisation problems with faster and stable convergence rates [130]. It has been shown to be very effective optimum tools in thresholding [131,132]. However, Ratnaweera et al. [133] suggested that the lack of population diversity in PSO algorithm is understood to cause premature convergence.

In this section, the QPSO algorithm is used for multilevel thresholding problem based on OTSU measure in image segmentation. Performance evaluation among QPSO, PSO and other well-known methods are presented.

6.6.2 OTSU Criterion–Based Measure

The OTSU criterion, as proposed by Otsu, has been frequently used in determining whether the optimal thresholding method is able to provide histogram-based image segmentation with satisfactory desired characteristics [134,135]. It was originally developed for bilevel thresholding and can be easily evolved into multilevel thresholding directly. Consider an image containing N pixels of grey levels from 0 to L. Let $h(i)$ represent the number of pixels with the ith grey level and P_i be the probability of a pixel with the ith grey level. Then

$$P_i = \frac{h(i)}{N}, \quad N = \sum_{i=0}^{L} h(i). \tag{6.58}$$

Suppose there are $M - 1$ thresholds, $\{t_1, t_2, \ldots, t_{M-1}\}$, which divide the original image into M classes: C_1 for $[0, \ldots, t_1]$, C_2 for $[t_1 + 1, \ldots, t_2], \ldots, C_M$ for $[t_{M-1}, \ldots, L]$. The optimal thresholds, $\{t_1^*, t_2^*, \ldots, t_{M-1}^*\}$, chosen by the OTSU method can be described as follows:

$$\{t_1^*, t_2^*, \ldots, t_{M-1}^*\} = \arg\max\{\sigma_B^2(t_1, t_2, \ldots, t_{M-1})\}, \quad 0 \le t_1 \le t_2 \le \cdots \le t_{M-1} \le L, \tag{6.59}$$

where

$$\sigma_B^2 = \sum_{k=1}^{M} \omega_k \left(u_k - u_{t_k^*}\right)^2, \tag{6.60}$$

with $\omega_k = \sum_{i \in C_k} P_i, \quad u_k = \sum_{i \in C_k} \frac{iP_i}{\omega_k}, \quad k = 1, 2, \ldots, M$.

In the OTSU method, the best thresholds, $\left\{t_1^*, t_2^*, \ldots, t_{M-1}^*\right\}$, are chosen in such a way as to maximise Equation 6.60 by using an exhaustive searching method. The OTSU method has been proven to be an efficient method in image segmentation for bilevel thresholding. However, when this method is extended to multilevel thresholding, the computation time grows exponentially with the number of thresholds. In particular, the computational complexity of an exhaustive search is $O(L^{M-1})$ which certainly limits the OTSU method for real applications. Thus, a new optimal multilevel thresholding algorithm for OTSU is needed. Here the QPSO algorithm is adopted for solving multilevel thresholding of OTSU. Each particle in the QPSO algorithm represents the potential solution of the optimal thresholds, and Equation 6.60 is used as the fitness function. The difference between the traditional OTSU method and the QPSO-based OTSU approach is that the former finds the solution by testing the thresholds in the range of the grey level using an exhaustive method. On the other hand, the latter obtains the solution by the operations of QPSO and, therefore, will shorten computation time, particularly in multilevel thresholding. The flowchart of QPSO for the OTSU method is shown in Figure 6.34.

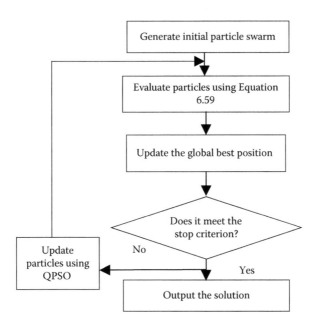

FIGURE 6.34 Flowchart of the QPSO algorithm for the OTSU method.

6.6.3 Performance Evaluation

In order to evaluate the performance of the QPSO-based OTSU approach, a wide variety of images were used to test the algorithmic implementation. The results obtained by using the QPSO-based scheme are compared with other popular algorithms developed in the literature so far, including the ACO algorithm [128], the GA algorithm with learning method [129] and the PSO algorithm [131]. In the experiments, the parameters of these algorithms were set as those used in the original papers to achieve the best results. A Core 2 Duo 1.86 GHz PC was used to perform the numerical experiments using MATLAB®, and the computational time of each algorithm on the platform was counted. The population size was chosen to be 20. Each algorithm was repeated 30 times. Furthermore, for fair comparison and to satisfy the time requirement of segmentation, the run time of each algorithm was set as 2 s. Benchmark images, namely, LENA, BANBOO, PEPPER and HUNTER (each of 512×512 pixels original sizes), were used in the experiments and the comparison studies.

Figure 6.35 presents the four original images. Equation 6.60 is used as an objective function in the comparison of performances. Table 6.29 presents the values of the objective function values attained (with $M - 1 = 2, 3, 4$) and the corresponding optimal thresholds obtained by the OTSU methods using the QPSO algorithm and each of the other algorithms named earlier. When the thresholding number is $M - 1 = 4$, there is no correlating values listed in the experiments because the CPU time is too long. The objective function values and the corresponding optimal thresholds obtained by the various algorithms (with $M - 1 = 2, 3, 4, 5$) are shown in Tables 6.30 and 6.31. The larger values in Table 6.30 mean that the algorithm may obtain better results. It can be seen that in almost all cases the QPSO method outperforms other algorithms.

To analyse the stability of the proposed PSO-based algorithm, all of the global best positions in each of the converged runs are checked to see whether the same value is arrived by computing

$$std = \sqrt{\frac{\sum_{i=1}^{n} (\sigma - \tilde{\sigma})^2}{n}}, \tag{6.61}$$

where
  std is the standard deviation of σ
  n is the run time of each of the algorithms
  σ is the best fitness value the algorithm recorded in each trial
  $\tilde{\sigma}$ represents the average value of σ

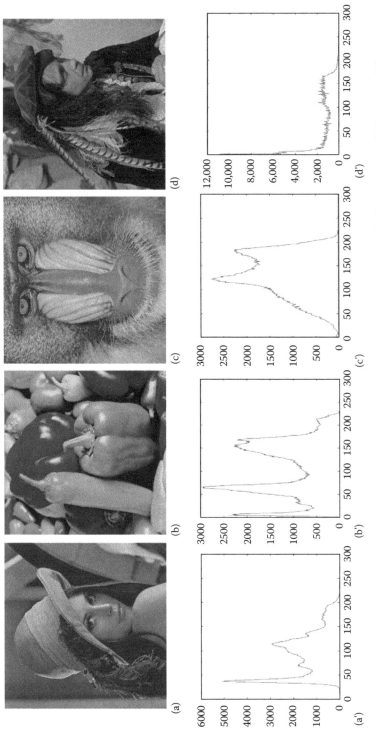

FIGURE 6.35 The test images: (a) LENA, (b) PEPPER, (c) BANBOO and (d) HUNTER, and the histograms of (a') LENA, (b') PEPPER, (c') BANBOO and (d') HUNTER.

TABLE 6.29 Objective Values and Thresholds by the OTSU Method

Image	M – 1 = 2		M – 1 = 3		M – 1 = 4	
	Objective Values	Optimal Thresholds	Objective Values	Optimal Thresholds	Objective Values	Optimal Thresholds
LENNA	9,344.9023	134, 165	11,333.5126	121, 151, 176	12,557.7434	111, 140, 158, 180
PEPPER	9,351.5285	134, 176	11,268.7031	113, 158, 184	12,525.4762	103, 140, 167, 189
BANBOO	9,042.8115	131, 171	11,051.0609	120, 149, 177	12,304.7347	112, 135, 160, 181
HUNTER	5,449.0945	102, 146	6,426.0889	86, 129, 155	6,972.5	69, 112, 137, 158
Computation time (s)	5.1267		253.575		16,452.133	

TABLE 6.30   Comparison between the Objective Function Values of the QPSO Algorithm and the Other Algorithms

Images	M − 1	Objective Values			
		QPSO	PSO	GA-L	ACO
LENNA	2	9,344.8932	9,344.8932	9,340.5147	9,344.6823
	3	11,333.2451	11,331.8123	11,282.8318	11,323.0786
	4	12,542.4476	12,523.6333	12,409.0375	12,491.098
PEPPER	2	9,351.5285	9,351.4688	9,348.4121	9,351.2942
	3	11,267.8193	11,266.6595	11,220.8713	11,260.2463
	4	12,510.8578	12,502.9363	12,341.4587	12,466.2279
BANBOO	2	9,042.8044	9,042.7083	9,038.1796	9,042.226
	3	11,050.8472	11,049.587	10,976.3514	11,044.6595
	4	12,296.9642	12,280.2584	12,100.1788	12,229.489
HUNTER	2	5,449.0759	5,449.0847	5,444.5908	5,448.6717
	3	6,423.1534	6,421.0423	6,383.0461	6,414.6335
	4	6,955.5496	6,948.1049	6,871.2479	6,924.7067

TABLE 6.31   The Thresholds Derived by the QPSO Algorithm and the Other Algorithms

Images	M − 1	Optimal Thresholds			
		QPSO	PSO	GA-L	ACO
LENNA	2	134, 165	134, 165	133, 165	134, 164
	3	121, 151, 176	121, 151, 177	114, 148, 174	125, 152, 177
	4	110, 139, 157, 180	109, 140, 156, 179	97, 130, 154, 179	111, 142, 162, 190
PEPPER	2	134, 176	134, 176	136, 177	133, 176
	3	112, 158, 184	114, 158, 184	107, 156, 183	115, 159, 184
	4	105, 142, 167, 188	103, 138, 165, 188	107, 154, 173, 192	108, 148, 171, 189
BANBOO	2	131, 171	131, 171	129, 170	130, 170
	3	120, 149, 178	120, 150, 178	114, 145, 174	121, 149, 177
	4	111, 135, 159, 181	111, 135, 161, 180	99, 132, 155, 177	118, 141, 162, 181
HUNTER	2	102, 146	102, 146	99, 146	102, 146
	3	82, 129, 155	89, 129, 155	88, 124, 154	79, 129, 155
	4	65, 109, 137, 158	62, 104, 135, 158	74, 124, 146, 162	79, 120, 143, 163

TABLE 6.32   Standard Deviation Obtained by Different Algorithms

Images	$M-1$	Standard Deviation			
		CQPSO	PSO	GA-L	ACO
LENNA	2	0.0912	0.0912	6.8173	0.626
	3	0.3505	3.0055	39.7188	40.0669
	4	8.0898	33.3878	75.7674	81.5615
PEPPER	2	1.6371×10^{-11}	0.4738	3.3477	0.8681
	3	4.1766	3.2586	31.7815	19.8069
	4	16.4163	22.1596	84.7434	97.0416
BANBOO	2	0.0702	0.3015	5.3859	1.6034
	3	0.9878	5.8266	59.2328	13.7832
	4	12.3659	28.5912	104.077	193.175
HUNTER	2	0.1299	0.0968	3.9608	1.4757
	3	4.9088	6.4211	19.9374	15.8994
	4	11.5647	17.3434	47.2423	63.3187

The standard deviation for 30 runs of each of the algorithms is listed in Table 6.32. Note that larger values of *std* lead to unstable numerical results.

From the results shown in Tables 6.30 and 6.31 and Figure 6.36, one can observe that QPSO and PSO show better segmentation performance in comparison with ACO and GA-L. The results obtained by QPSO are the best and very close or the same as those obtained by the OTSU method as shown in Table 6.29. The average CPU times recorded by using OTSU method were 4.975, 254.534 and 16,430.346 s when $M-1$ equals 2, 3 and 4, respectively. When $M-1 = 5$, the computational time by using QTSU was too long to be recorded, but the QPSO method generated the approximate solution in only 2 s. It is possible to conclude that QPSO is more efficient than the OTSU method. This may be explained as follows. QPSO inherits the advantage of fast convergence from PSO but is also able to resolve efficiently the premature convergence. From Table 6.31, which shows the stability of different algorithms, it is observed that the stability for each algorithm drops as $M-1$ increases. Larger values of $M-1$ stretch the particle number and thus extend the algorithm search space, which certainly affect the optimisation methods. As shown in Table 6.32, QPSO shows better stability than the other algorithms.

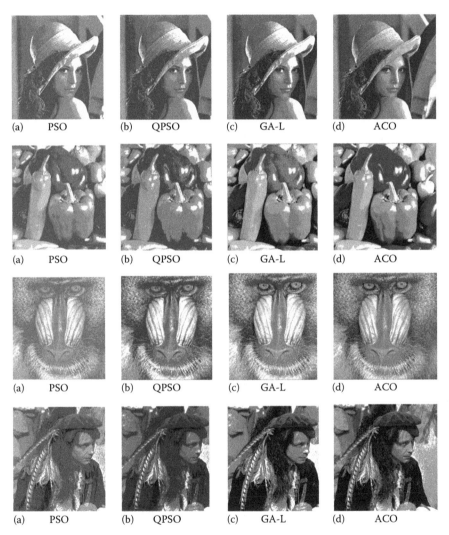

FIGURE 6.36 The thresholded images for $M - 1 = 5$ obtained by (a) PSO, (b) QPSO, (c) GA-L and (d) ACO.

6.6.4 Summary

An optimal multilevel thresholding algorithm is described in conjunction with the QPSO algorithm. The method is tested on several benchmarking images. Better performance compared to several other popular contemporary methods is also demonstrated.

6.7 IMAGE FUSION

6.7.1 Introduction

Image fusion involves integrating multiple images into a composite image that contains better information for human visual perception and computer processing tasks. The process of image information fusion can be performed at signal, feature or symbol levels. One widely used method for such information fusion is the wavelet decomposition methods [136,137]. Several transformation-based methods have also been proposed for image fusion, including PCA, ICA and HIS, etc. [138–140]. Soft computing methods were also introduced for image fusion [141] as well.

Wu et al. proposed a feature-level image fusion method based on regions. The method has the advantage that the fusion process becomes more robust and avoids some of the well-known problems in pixel-level fusion such as blurring effects and high sensitivity to noise and misregistration [139]. Since image segmentation is an important method of obtaining regions by identifying skeletons and boundaries out of an image, its effectiveness can have a significant influence on the performance of image fusion. FCM clustering algorithm has been shown to perform well in obtaining regions of images to be fused in examples extracted from [142,143]. However, FCM has the shortcoming of being prone to sticking into local minima.

In this section, a FCM method based on the QPSO algorithm (QPSO-FCM) is introduced for image segmentation [144]. The hybrid method is guaranteed theoretically to obtain good optimal solutions in the search space. The image fusion method that incorporates the QPSO-FCM algorithm is presented, and the results for two images are provided.

6.7.2 Image Segmentation Based on QPSO-FCM

In order to consider regional information, a region-based method is more suitable for the purpose of human visual perception and computer processing. Image segmentation is an essential step in image fusion at the feature level. Segmentation of an image entails the division or separation of the image into regions of similar attribute, that is, the grouping of pixels into individual clusters.

The FCM algorithm [145] is an unsupervised fuzzy clustering, which adopts the iterative algorithm to optimise the objective function based on the least squares method. Let the objective function of FCM be

$$J_{FCM}(U,V) = \sum_{k=1}^{N} \sum_{j=1}^{N_c} \mu_{jk}^m \| x_k - s^j \|^2, \tag{6.62}$$

where
$X = \{x_1, \ldots x_k, \ldots, x_N\}$ is the dataset
N is the number of data item in the dataset X
N_c is the number of clusters
m is the weighting exponent
s^j is the jth cluster centre of $S = (s^1, s^2, \ldots, s^{N_c})$
μ_{jk} denotes the membership value of the kth data in the jth cluster

The weighting exponent is usually chosen as $m \in [1, \infty)$. In the experiments described in the following, it is chosen as 2.

μ_{jk} and s^j can be obtained by a Lagrange multiplier method, respectively, as

$$\mu_{jk} = \frac{1}{\sum_{j'=1}^{N_c} \left[\frac{\| x_k - s^j \|^2}{\| x_k - s^{j'} \|^2} \right]^{1/(m-1)}}, \quad \forall_{1 \le j \le N_c} j, \forall_{1 \le k \le N} k, \tag{6.63}$$

$$s^j = \frac{\sum_{k=1}^{N} \mu_{jk}^m x_k}{\sum_{k=1}^{N} \mu_{jk}^m}, \quad \forall_{1 \le j \le N_c} j. \tag{6.64}$$

Since the FCM clustering algorithm cannot guarantee convergence to the global optimal solution, the QPSO algorithm is used to enhance its global search ability. In the QPSO-FCM clustering algorithm, a particle represents a set of N_c cluster centres. The position of the ith particle at the nth iteration is represented by $S_{i,n} = (s_{i,n}^1, \ldots, s_{i,n}^j, \ldots, s_{i,n}^{N_c})$, where $s_{i,n}^j$, the jth

Step 1: Input the sample data and produce the corresponding initial value of data;

Step 2: Confirm the number of particle swarm (pop-size) from the sample data;

Step 3: Initialize the swarm and stochastically select N_c cluster centres referring to the bound of the data for each particle.

Step 4: Initialize the personal value for each particle and the global value of the entire swarm;

Step 5: Using the fitness function (6.65) calculate the fitness value and update the current personal best position for each particle, and then update the global best position.

Step 6: Using the update equation of QPSO generate the new individual particle $S_r(t + 1)$;

Step 7: Go to Step 5 until the maximum number of iterations is reached, then the algorithm ends.

FIGURE 6.37 The procedure of the QPSO-FCM algorithm.

component of the particle, represents the jth vector of cluster centre. The fitness function of the particles is defined as

$$f(S_{i,n}) = J_m(U, S_{i,n}) = \sum_{k=1}^{N} \sum_{j=1}^{N_c} (\mu_{jk})^m d_{jk}^2 \left(x_k, s_{i,n}^j\right). \tag{6.65}$$

The detailed steps of QPSO-FCM are described as in Figure 6.37.

When an image is segmented, the output image is segmented into *num* small windows. Each window has the size of $M \times N$. A feature vector, $\vec{f}_k$ $\left(\vec{f}_k \in R^{10}, 1 \le k \le num\right)$, is then extracted from each window. Each feature vector consists of 10 features. To convolve respectively the source image with eight-channel Gabor filters, eight output images of Gabor filters in which texture features are extracted. Eight out of the 10 features are texture features, which are denoted by the window mean values of the eight output images. The other two are the x-coordinate and y-coordinate on the top left corner of the window and are used to denote the spatial location. The feature vector $\vec{f}_k$ needs to be normalised to $\hat{f}_k$, where $\hat{f}_k$ is the normalised feature vector.

Let N_c be the known number of clusters. The QPSO-FCM algorithm is used to cluster the feature vectors $\hat{f}_k(\hat{f}_k \in R^{10}, 1 \le k \le num)$. The global optimal points are the centres of the data. From the cluster centre, one can obtain $\mu_{jk} \in [0,1]$ $(1 \le j \le N_c, 1 \le k \le num)$ which denotes the membership of the feature vector $\hat{f}_k$ in the jth cluster.

After μ_{jk} is determined, the classification algorithm given in the following is used to segment the image. If $\mu_{jk} > \mu_{lk}$ ($l = 1, 2,..., N_c, j = 1, 2,..., N_c$,

$l \neq j$), then the vector $\hat{f}_k$ belongs to the jth cluster. The set of feature vectors, $F = \{\hat{f}_k \in R^{10}: 1 \leq k \leq num\}$, is partitioned into N_c groups, $\{F_1, \ldots, F_{N_c}\}$, and the image is segmented into N_c regions, $\{R_1, \ldots, R_{N_c}\}$, with R_i being the region corresponding to the feature set $F_i (1 \leq i \leq N_c)$.

6.7.3 Image Fusion Based on QPSO-FCM Segmentation

Let $\xi_1, \xi_2, \ldots, \xi_8$ be eight output images after processing using Gabor filters. Define $Mean_i^j$ $(1 \leq i \leq N_c, 1 \leq j \leq 8)$ as the mean value of the wavelet coefficients of the ith region in the jth output image as follows:

$$Mean_i^j = \frac{\displaystyle\sum_{x \in M, y \in N} \xi_i^j(x,y)}{w\_num_i^j}, \qquad (6.66)$$

where

$w\_num_i^j$ is the number of windows of the ith region in the jth output image

M and N are the size of region

Assuming the mean vectors of the eight filtered images of the source images A and B are to be fused, one obtains

$$\vec{\eta}_{i,A} = \left[Mean_{i,A}^1, Mean_{i,A}^2, \ldots, Mean_{i,A}^8 \right]^T,$$

$$\vec{\eta}_{i,B} = \left[Mean_{i,B}^1, Mean_{i,B}^2, \ldots, Mean_{i,B}^8 \right]^T. \qquad (6.67)$$

Normalise $\vec{\eta}_{i,A}$ to give $\vec{\eta}'_{i,A}$ and normalise $\vec{\eta}_{i,B}$ to give $\vec{\eta}'_{i,B}$, the similarity function is defined as

$$m_{i,AB} = 1 - \| \vec{\eta}'_{i,A} - \vec{\eta}'_{i,B} \|. \qquad (6.68)$$

The wavelet detail coefficients are used together with a decision algorithm based on an activity level and a similarity measure. Thus, the fused high- and low-frequency coefficients set in the ith region $y_{i,F}$ may be obtained by using the weighting factors

$$y_{i,F} = \omega_A y_{i,A} + \omega_B y_{i,B}, \qquad (6.69)$$

where ω_A and ω_B are the weighting factors being set such that $\omega_A = 1 - \omega_B = d_i$. Thus Equation 6.69 can be rewritten as

$$y_{i,F} = d_i y_{i,A} + (1 - d_i) y_{i,B} \qquad (6.70)$$

where the value of d_i is defined as follows. When the low-frequency coefficients are fused, $d_i = 0.5$; otherwise, d_i is defined as

$$
d_i = \begin{cases}
1 & \text{if } E_{i,A} > E_{i,B} \quad \text{and} \quad m_{i,AB} \leq T, \\
0 & \text{if } E_{i,A} \leq E_{i,B} \quad \text{and} \quad m_{i,AB} \leq T, \\
\dfrac{1}{2} + \dfrac{1}{2}\left(\dfrac{1 - m_{i,AB}}{1-T}\right) & \text{if } E_{i,A} > E_{i,B} \quad \text{and} \quad m_{i,AB} > T, \\
\dfrac{1}{2} - \dfrac{1}{2}\left(\dfrac{1 - m_{i,AB}}{1-T}\right) & \text{if } E_{i,A} \leq E_{i,B} \quad \text{and} \quad m_{i,AB} > T,
\end{cases}
\tag{6.71}
$$

$$
E_i = \sum_{h=1}^{w\_num} \bar{y}_{ih}^2,
\tag{6.72}
$$

where
  the coefficient T is the threshold of similarity measure
  w_num is the number of $M \times N$ windows in the ith region
  $\bar{y}_{ih}(1 \leq h \leq w\_num)$ is the mean value of the high-frequency coefficients
    of the hth $M \times N$ window in the ith region

6.7.4 Experimental Results

Two experiments are described here to demonstrate the algorithm. In the first experiment, the source images used in the experiment were the 'clock' multi-focus images extracted from http://www.imagefusion.org/. As shown in Figure 6.38, the input images have the left half of (a) and the right half of (b) are blurred. The second experiment was performed on several medical images as shown in Figure 6.39. The complementary features can be acquired by fusing image ci1 and image ci2. In both experiments, the size of source images is 256 × 256 pixels, the threshold of match measure T is 0.5. The window size is chosen to compromise between texture effectiveness and computation time. Smaller window size may increase the computation time. Therefore, the size of window is 16 × 16 pixels. The number of clusters is chosen by cross-validation technique and considering that the larger number of cluster will increase the computational complexity. The number of clusters was selected to be 5 in the experiments.

For the algorithmic parameter of QPSO, the population size was chosen as 30 and the maximum number of iterations was chosen as 300 for the images in Figure 6.38 and the population size was 15 and the maximum

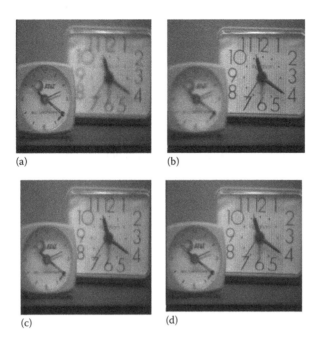

(a)   (b)

(c)   (d)

FIGURE 6.38 Fusion of out-of-focus 'clock' images. (a) Image with focus on the left. (b) Image with focus on the right. (c) The reference image. (d) The fused image with FCM.

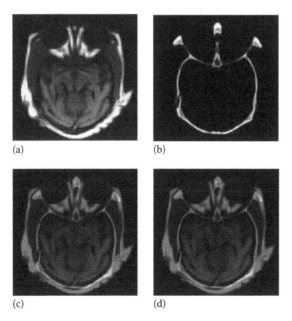

(a)   (b)

(c)   (d)

FIGURE 6.39 Fusion of medical images. (a) Input image (ci1). (b) Input image (ci2). (c) Fused image using FCM. (d) Fused image using QPSO-FCM.

number of iterations was 120 for the images in Figure 6.39. For both experiments, the contraction–expansion coefficient was chosen to decrease linearly from 1.0 to 0.5.

The best value of the object function obtained, the intra-cluster distance and the intercluster distance are the evaluation indices of the clustering quality. Figure 6.40 shows the comparison of convergence of the two methods applied to the images in Figures 6.38 and 6.39. From Figure 6.40, it is possible to conclude that FCM has better convergence speed than QPSO-FCM. However FCM could easily fall into the local optimum, and

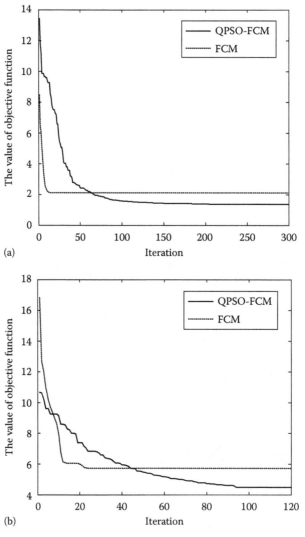

FIGURE 6.40 Convergence history for (a) the 'clock' and (b) the 'ci' images.

TABLE 6.33 Comparison of the Quality of the Clustering Results

Image	Intracluster Distance	Intercluster Distance
Figure 6.38 (FCM)	0.1533	1.3196
Figure 6.38 (QPSO-FCM)	0.1289	1.4816
Figure 6.39 (FCM)	0.1873	1.5809
Figure 6.39 (QPSO-FCM)	0.1835	1.5848

the objective function value of QPSO-FCM is better than that of FCM. Table 6.33 shows the comparison of the intracluster distance and the interclusters distance obtained by QPSO-FCM and FCM, respectively. According to the criteria of the minimum intracluster distance and the maximum intercluster distance, it is found that QPSO-FCM outperforms FCM in terms of the intracluster distance and the intercluster distance for the two sets of experiments. From the previous analysis, it is possible to conclude that the QPSO-FCM method is more efficient than FCM.

Figures 6.38 and 6.39 present the fused images of the out-of-focus images and the medical images. Tables 6.34 and 6.35 show the performance of the fusion results. For Figure 6.38, the performance of the image fusion method is evaluated by using five criteria, including root mean square error (RMSE), peek signal-to-noise ratio (PSNR), entropy (H), cross entropy (CERF) and mutual information (MI) [146]. For Figure 6.39, it is crucial to transfer important object information to the fused image from the input images. As the reference images are not available for comparison, other metric indices such as mean gradient (G), spatial frequency (SF), H, structural similarity (SSIM) and MI [147,148] are used in the numerical tests.

TABLE 6.34 Comparison of Image Fusion Using QPSO-FCM and FCM for the 'Clock'

Method	RMSE	PSNR	H	CERF	MI
QPSO-FCM	9.3462	28.7181	4.9945	0.5259	3.0126
FCM	9.5509	28.5299	5.0487	0.5694	2.9700

TABLE 6.35 Comparison of Image Fusion Results Using QPSO-FCM and FCM for the 'ci' Images

Method	G	SF	H	SSIM	MI
QPSO-FCM	6.3512	12.7481	4.3945	0.6552	4.9531
FCM	5.7170	10.5014	4.1287	0.5799	4.8834

The smaller the values of RMSE and CERF, the better the image quality of the fused images. However the opposite is true for other metrics, including the PSNR, H, G, SF, SSIM and MI. As can be seen from these results, most evaluation criteria point to the fact that the quality of Figures 6.38d and 6.39d are the best. In particular, RMSE and CERF have lowest values; PSNR and MI have highest values (see Figure 6.38d) and G, SF, SSIM and MI have highest values (see Figure 6.39d). The metric H has similar values for the fused image using QPSO-FCM and FCM as shown in Figure 6.38. In short, QPSO-FCM has better performance in clustering, which enhances the quality of fused image.

6.7.5 Summary

This section presents a fusion technique at the feature level for regional image fusion based on QPSO-FCM. It handles regional information and the relationship of pixels more than the process of a signal pixel. The algorithm is suitable for human visual perception and computer processing and can effectively enhance the robustness of fusion algorithm. Since image segmentation is an important method of obtaining image regions, it is an essential step in image fusion and has direct effects on the performance of image fusion. This section attempts to provide a discussion on the effect of image fusion using different clustering methods. Experimental results suggest that the hybrid algorithm described in this section enhances the quality of fused image and is more consistent with the human visual perception.

REFERENCES

1. O.M. Alifanov, *Inverse Heat Transfer Problems*. Springer, Berlin, Germany, 1994.
2. J.V. Beck, B. Blackwell, C.R. St-Clair Jr. *Inverse Heat Conduction: Ill-Posed Problems*. Wiley-Interscience, New York, 1985.
3. C.H. Huang, J.Y. Yan, H.T. Chen. Function estimation in predicting temperature-dependent thermal conductivity without internal measurements. *AIAA Journal of Thermophysics and Heat Transfer*, 1995, 9: 667–673.
4. C.H. Huang, M.N. Ozisik. Conjugate gradient method for determining unknown contact conductance during metal casting. *International Journal of Heat Mass Transfer*, 1992, 35: 1779–1786.
5. P. Terrola. A method to determine the thermal conductivity from measured temperature profiles. *International Journal of Heat Mass Transfer*, 1989, 32: 1425–1430.
6. S. Kim, M.C. Kim, K.Y. Kim. Non-iterative estimation of temperature-dependent thermal conductivity without internal measurements. *International Journal of Heat Mass Transfer*, 2003, 46: 1801–1810.

7. W.K. Yeung, T.T. Lam. Second-order finite difference approximation for inverse determination of thermal conductivity. *International Journal of Heat Mass Transfer*, 1996, 39: 3685–3693.
8. C.L. Chang, M. Chang. Non-iterative estimation of thermal conductivity using finite volume method. *International Communication in Heat and Mass Transfer*, 2006, 33: 1013–1020.
9. C.L. Chang, M. Chang. Inverse estimation of the thermal conductivity in a one-dimensional domain by Taylor series approach. *Heat Transfer Engineering*, 2008, 29: 830–838.
10. C.Y. Yang. Estimation of the temperature-dependent thermal conductivity in inverse heat conduction problems. *Applied Mathematical Modelling*, 1999, 23: 469–478.
11. P.C. Hansen. Analysis of discrete ill-posed problems by means of the L-curve. *SIAM Review*, 1992, 34: 561–580.
12. P.C. Hansen, D.P. O'Leary. The use of the L-curve in the regularization of discrete ill-posed problems. *SIAM Journal on Scientific Computing*, 1993, 14: 1487–1503.
13. J.D. Annan, J.C. Hargreaves. Efficient parameter estimation for a highly chaotic system. *Tellus Series A—Dynamic Meteorology and Oceanography*, 2004, 56(5): 520–526.
14. U. Parlitz. Estimating model parameters from time series by auto-synchronization. *Physical Review Letters*, 1996, 76(8): 1232–1235.
15. S.H. Chen, J. Hu, C.H. Wang, J.H. Lü. Adaptive synchronization of uncertain Rössler hyperchaotic system based on parameter identification. *Physics Letters A*, 2004, 321: 50–55.
16. L.X. Li, H.P. Peng, X.D. Wang, Y.X. Yang. Comment on two papers of chaotic synchronization. *Physics Letters A*, 2004, 333: 269–270.
17. X.P. Guan, H.P. Peng, L.X. Li, Y.Q. Wang. Parameter identification and control of Lorenz system. *ACTA Physica Sinica*, 2001, 50: 26–29.
18. J.H. Lü, S.H. Zhang. Controlling Chen's chaotic attractor using backstepping design based on parameters identification. *Physics Letters A*, 2001, 286: 148–152.
19. J. Lie, S. Chen, J. Xie. Parameter identification and control of uncertain unified chaotic system via adaptive extending equilibrium manifold approach. *Chaos, Soliton & Fractals*, 2004, 19: 533–540.
20. M. Gu, R.E. Kalaba, G.A. Taylor. Obtaining initial parameter estimates for chaotic dynamical systems using linear associative memories. *Applied Mathematics and Computation*, 1996, 76: 143–159.
21. G. Alvarez, F. Montoya, M. Romera, G. Pastor. Cryptanalysis of an ergodic chaotic cipher. *Physics Letters A*, 2003, 311: 172–181.
22. X. Wu, H. Hu, B. Zhang. Parameter estimation only from the symbolic sequences generated by chaos system. *Chaos, Soliton & Fractals*, 2004, 22: 359–366.
23. D.F. Wang. Genetic algorithm optimization based proportional–integral–derivative controller for unified chaotic system. *ACTA Physica Sinica*, 2005, 54: 1495–1499.
24. F. Gao, H.Q. Tong. Parameter estimation for chaotic system based on particle swarm optimization. *ACTA Physica Sinica*, 2006, 55: 577–582.

25. L.X. Li, Y.X. Yang, H.P. Peng, X.D. Wang. Parameters identification of chaotic systems via chaotic ant swarm. *Chaos, Soliton & Fractals*, 2006, 28: 1204–1211.
26. V. Gafiychuk, B. Datsko, V. Meleshko, D. Blackmore. Stability analysis and limit cycle in fractional system with Brusselator nonlinearities. *Physics Letters A*, 2008, 372: 4902–4904.
27. R.C. Robinson. *An Introduction to Dynamical Systems: Continuous and Discrete*. Pearson Education, Upper Saddle River, NJ, 2004.
28. E.N. Lorenz. Deterministic nonperiodic flow. *Journal of Atmospheric Science*, 1963, 20: 130–141.
29. G. Chen, T. Ueta. Yet another chaotic attractor. *International Journal of Bifurcation and Chaos*, 1999, 9: 1465–1466.
30. J. Lü, G. Chen. A new chaotic attractor coined. *International Journal of Bifurcation and Chaos*, 2002, 12: 659–661.
31. A.V. Oppenheim, R.W. Schafer, J. R. Buck. *Discrete-Time Signal Processing*. Prentice-Hall, Englewood Cliffs, NJ, 1999.
32. J.G. Proakis, D.G. Manolakis. *Digital Signal Processing*. Prentice-Hall, Englewood Cliffs, NJ, 1996.
33. D.M. Kodek. Design of optimal finite wordlength FIR digital filters using integer programming techniques. *IEEE Transactions on Acoustics, Speech and Signal Processing*, 1980, 28(3): 304–308.
34. D.M. Kodek, K. Steiglitz. Comparison of optimal and local search methods for designing finite wordlength FIR digital filters. *IEEE Transactions on Circuits and Systems*, 1981, 28(1): 28–32.
35. N. Benvenuto, M. Marchesi, A. Uncini. Applications of simulated annealing for the design of special digital filters. *IEEE Transactions on Signal Processing*, 1992, 40(2): 323–332.
36. D.J. Xu, M.L. Daley. Design of optimal digital filter using a parallel genetic algorithm. *IEEE Transactions on Circuits and Systems II: Analog and Digital Signal Processing*, 1995, 42(10): 673–675.
37. M. Öner. A genetic algorithm for optimization of linear phase FIR filter coefficients. In *Proceedings of the Asilomar Conference on Signals, Systems and Computers*, Pacific Grove, CA, 1998, pp. 1397–1400.
38. J.M.P. Langlois. Design of linear phase FIR filters using particle swarm optimization. In *Proceedings of the 22nd Biennial Symposium on Communications*, Kingston, Ontario, Canada, 2004, pp. 172–174.
39. X.P. Chen, B. Qu, G. Lu. An application of immune algorithm in FIR filter design. In *Proceedings of the 2003 IEEE International Conference on Neural Networks & Signal Processing*, Nanjing, China, 2003, pp. 473–475.
40. J. Seppänen. Audio signal processing basics. *Introductory Lectures*, Audio Research Group, Tampere University of Technology, Web Resource <HTUhttp://www.cs.tut.fi/sgn/arg/intro/basics.htmUTH>
41. R. Nambiar, P. Mars. Genetic and annealing approaches to adaptive digital filtering. In *Proceedings of the IEEE 26th Asilomar Conference on Signals, Systems & Computers*, Pacific Grove, CA, vol. 2, 1992, pp. 871–875.
42. S. Chen, R.H. Istepanian, B.L. Luk. Digital IIR filter design using adaptive simulated annealing. *Digital Signal Processing*, 2001, 11(3): 241–251.

43. J. Radecki, J. Konrad, E. Dubois. Design of multidimensional finite-wordlength FIR and IIR filters by simulated annealing. *IEEE Transactions on Circuits and Systems II: Analog and Digital Signal Processing*, 1995, 42(6): 424–431.

44. D.M. Etter, M.J. Hicks, K.H.Cho. Recursive adaptive filter design using an adaptive genetic algorithm. In *Proceedings of the IEEE International Conference Acoustics, Speech, Signal Processing*, Albuquerque, NM, 1982, vol. 7, pp. 635–638.

45. N.E. Mastorakis, I.F. Gonos, M.N.S. Swamy. Design of two-dimensional recursive filters using genetic algorithms. *IEEE Transactions on Circuits and Systems I—Fundamental Theory and Applications*, 2003, 50(5): 634–639.

46. K.S. Tang, K.F. Man, S. Kwong, Q. He. Genetic algorithms and their applications. *IEEE Signal Processing Magazine*, 1996, 13(6): 22–37.

47. S.C. Ng, S.H. Leung, C.Y. Chung, A. Luk, W.H. Lau. The genetic search approach: A new learning algorithm for adaptive IIR filtering. *IEEE Signal Processing Magazine*, 1996, 13(6): 38–46.

48. R. Thamvichai, T. Bose, R.L. Haupt. Design of 2-D multiplierless IIR filters using the genetic algorithm. *IEEE Transactions on Circuits and Systems I: Fundamental Theory and Applications*, 2002, 49(6): 878–882.

49. A. Lee, M. Ahmadi, G.A. Jullien, R.S. Lashkari, W.C. Miller. Design of 1-D FIR filters with genetic algorithm. In *Proceedings of the IEEE International Symposium on Circuits and Systems*, Orlando, FL, 1999, pp. 295–298.

50. N. Karaboga, A. Kalinli, D. Karaboga. Designing IIR filters using ant colony optimization algorithm. *Journal of Engineering Applications of Artificial Intelligence*, 2004, 17(3): 301–309.

51. A. Kalinli, N. Karaboga. New method for adaptive IIR filter design based on tabu search algorithm. *AEU International Journal of Electronics and Communications*, 2005, 59(2): 111–117.

52. N. Karaboga. Digital IIR filter design using differential evolution algorithm. *EURASIP Journal on Applied Signal Processing*, 2005, 8: 1269–1276.

53. J.-T. Tsai, J.-H. Chou. Design of optimal digital IIR filters by using an improved immune algorithm. *IEEE Transactions on Signal Processing*, 2006, 54(12): 4582–4596.

54. T. Kaczorek. *Two-Dimensional Linear System*. Springer, Berlin, Germany, 1985.

55. S.G. Tzafestas (Editor). *Multidimensional System, Techniques and Applications*. Marcel Dekker, New York, 1986.

56. W.S. Lu, A. Antoniou. *Two-Dimensional Digital Filters*. Marcel Dekker, New York, 1992.

57. G.A. Maria, M.M. Fahmy. An lp design technique for two-dimensional digital recursive filters. *IEEE Transactions on Acoustic, Speech, Signal Processing*, 1974, 22: 15–21.

58. C. Charalambous. Design of 2-dimensional circularly-symmetric digital filters. *IEE Proceedings of Electronic Circuits and Systems G*, 1982, 129: 47–54.

59. P.K. Rajan, M.N.S. Swamy. Quadrantal symmetry associated with two-dimensional digital transfer functions. *IEEE Transactions on Circuits and Systems*, 1983, 29: 340–343.

60. T. Lassko, S. Ovaska. Design and implementation of efficient IIR notch filters with quantization error feedback. *IEEE Transactions on Instrumentation and Measurement*, 1994, 43: 449–456.

61. C.-H. Hsieh, C.-M. Kuo, Y.-D. Jou, Y.-L. Han. Design of two-dimensional FIR digital filters by a two-dimensional WLS technique. *IEEE Transactions on Circuits and Systems II*, 1997, 44(5): 348–412.

62. M. Daniel, A. Willsky. Efficient implementations of 2-D noncausal IIR filters. *IEEE Transactions on Circuits and Systems II*, 1997, 44: 549–563.

63. W.-P. Zhu, M.O. Ahmad, M.N.S. Swamy. A closed-form solution to the least-square design problem of 2-d linear-phase FIR filters. *IEEE Transactions and Circuits Systems II*, 1997, 44: 1032–1039.

64. V.M. Mladenow, N.E. Mastorakis. Design of two-dimensional digital recursive filters by using neural networks. *IEEE Transactions on Neural Networks*, 2001, 12: 585–590.

65. J.-T. Tsai, J.-H. Chou, T.-K. Liu, C.-H. Chen. Design of two-dimensional recursive filters by using a novel genetic algorithm. In *Proceedings of the 2005 IEEE International Symposium on Circuits and Systems*, Kobe, Japan, 2005, vol. 3, pp. 2603–2606.

66. S. Das, A. Konar, U.K. Chakraborty. An efficient evolutionary algorithm applied to the design of two-dimensional IIR filters. In *Proceedings of the Genetic and Evolutionary Computation Conference*, Washington, DC, 2005, pp. 2157–2163.

67. I.F. Gonos, L.I. Virirakis, N.E. Mastorakis, M.N.S. Swamy. Evolutionary design of 2-dimensional recursive filters via the computer language GENETICA. *IEEE Transactions on Circuits and Systems II*, 2006, 53(4): 254–258.

68. IEEE Committee Report. Present practices in the economic operation of power systems. *IEEE Transactions on Power Apparatus and Systems*, 1971, PAS-90: 1768–1775.

69. A.J. Wood, B.F. Wollenbergy. *Power Generation, Operation, and Control*. Wiley, New York, 1984.

70. B.H. Chowdhury, S. Rahman. A review of recent advances in economic dispatch. *IEEE Transactions on Power Systems*, 1990, 5(4): 1248–1259.

71. Z.X. Lianf, J.D. Glover. A zoom feature for a dynamic programming solution to economic dispatch including transmission losses. *IEEE Transactions on Power Systems*, 1992, 7(2): 544–550.

72. S. Granville. Optimal reactive dispatch through interior point methods. *IEEE Transactions on Power Systems*, 1994, 9(1): 136–146.

73. A. Bakirtzis, V. Petridis, S. Kazarlis. Genetic algorithm solution to the economic dispatch problem. *IEE Proceedings—Generation, Transmission and Distribution*, 1994, 141(4): 377–382.

74. P.-H. Chen, H.-C. Chang. Large-scale economic dispatch by genetic algorithm. *IEEE Transactions on Power Systems*, 1995, 10(4): 1919–1926.

75. C.-L. Chiang. Improved genetic algorithm for power economic dispatch of units with valve-point effects and multiple fuels. *IEEE Transactions on Power Systems*, 2005, 20(4): 1690–1699.

76. Z.-L. Gaing. Particle swarm optimization to solving the economic dispatch considering the generator constraints. *IEEE Transactions on Power Systems*, 2003, 18(3): 1187–1195.
77. J.-B. Park, K.-S. Lee, J.-R. Shin, K.Y. Lee. A particle swarm optimization for economic dispatch with nonsmooth cost functions. *IEEE Transactions on Power Systems*, 2005, 20(1): 34–41.
78. L.S. Coelho, V.C. Mariani. Particle swarm approach based on quantum mechanics and harmonic oscillator potential well for economic load dispatch with valve-point effects. *Energy Conversion and Management*, 2008, 49(11): 3080–3085.
79. R.E. Perez-Guerrero, J.R. Cedeno-Maldonado. Economic power dispatch with non-smooth cost functions using differential evolution. In *Proceedings of the 37th Annual North American Power Symposium*, Ames, IA, 2005, pp. 183–190.
80. L.S. Coelho, V.C. Mariani. Combing of chaotic differential evolution and quadratic programming for economic dispatch optimization with valve-point effect. *IEEE Transactions on Power Systems*, 2006, 21(2): 989–996.
81. Y.M. Park, J.R. Won, J.B. Park. A new approach to economic load dispatch based on improved evolutionary programming. *Engineering. Intelligent Systems for Electrical Engineering and Communications*, 1998, 6(2): 103–110.
82. H.T. Yang, P.C. Yang, C.L. Huang. Evolutionary programming based economic dispatch for units with nonsmooth fuel cost functions. *IEEE Transactions on Power Systems*, 1996, 11(1): 112–118.
83. N. Sinha, R. Chakrabarti, P.K. Chattopadhyay. Evolutionary programming techniques for economic load dispatch. *IEEE Transactions on Evolutionary Computation*, 2003, 7(1): 83–94.
84. W.M. Lin, F.S. Cheng, M.T. Tsay. An improved tabu search for economic dispatch with multiple minima. *IEEE Transactions on Power Systems*, 2002, 17(1): 108–112.
85. J.H. park, Y.S. Kim, I.K. Eom, K.Y. Lee. Economic load dispatch for piecewise quadratic cost function using Hopfield neural network. *IEEE Transactions on Power Systems*, 1993, 8(3): 1030–1038.
86. K.Y. Lee, A. Sode-Yome, J.H. Park. Adaptive Hopfield neural network for economic load dispatch. *IEEE Transactions on Power Systems*, 1998, 13(2): 519–526.
87. T.K. Abdul Rahman, Z.M. Yasin, W.N.W. Abdullah. Artificial-immune-based for solving economic dispatch in power system. In *Proceedings of 2004 National Power and Energy Conference*, Kuala Lumpur, Malaysia, 2004, pp. 31–35.
88. K.S. Swarup, S. Yamashiro. Unit commitment solution methodology using genetic algorithm. *IEEE Transactions on Power Systems*, 2002, 17(1): 87–91.
89. H. Yoshida, K. Kawata, Y. Fukuyama, S. Takayama, Y. Nakanishi. A particle swarm optimization for reactive power and voltage control considering voltage security assessment. *IEEE Transactions on Power Systems*, 2000, 15(4): 1232–1239.

90. F.N. Lee, A.M. Breipohl. Reserve constrained economic dispatch with prohibited operating zones. *IEEE Transactions on Power Systems*, 1993, 8(1): 246–254.

91. D.W. Mount. *Bioinformatics: Sequence and Genome Analysis*. Cold Spring Harbor Laboratory Press, Cold Spring Harbor, New York, 2001.

92. L. Wang, T. Jiang. On the complexity of multiple sequence alignment. *Journal of Computational Biology*, 1994, 1: 337–348.

93. D.F. Feng, R.F. Doolittle. Progressive sequence alignment as a prerequisite to correct phylogenetic trees. *Journal of Molecular Evolution*, 1987, 25: 351–360.

94. J.D. Thompson, D.G. Higgins, T.J. Gibson. CLUSTALW: Improving the sensitivity of progressive multiple sequence alignment through sequence weighting, position-specific gap penalties and weight matrix choice. *Nucleic Acids Research*, 1994, 22: 4673–4680.

95. J. Kim, S. Pramanik, M.J. Chung. Multiple sequence alignment using simulated annealing. *Bioinformatics*, 1994, 10: 419–426.

96. A.V. Lukashin, J. Engelbrecht, S. Brunak. Multiple alignment using simulated annealing: Branch point definition in human mRNA splicing. *Nucleic Acids Research*, 1992, 20: 2511–2516.

97. K. Chellapilla, G.B. Fogel. Multiple sequence alignment using evolutionary programming. In *Proceedings of the First Congress on Evolution Composition*, Washington, DC, 1999, pp. 445–452.

98. C. Notredame, D.G. Higgins. SAGA: Sequence alignment by genetic algorithm. *Nucleic Acids Research*, 1996, 24: 1515–1524.

99. R. Thomsen. Evolving the topology of hidden Markov models using evolutionary algorithms. In *Proceedings of the Seventh International Conference on Parallel Problem Solving from Nature VII (PPSN)*, Granada, Spain, 2002, pp. 861–870.

100. P. Baldi, Y. Chauvin, T. Hunkapiller, M.A. McClure. Hidden Markov models of biological primary sequence information. *Proceedings of National Academy of Sciences USA*, 1994, 91: 1059–1063.

101. A. Krogh, M. Brown, I.S. Mian, K. Sjolander, D. Haussler. Hidden Markov models in computational biology: Applications to protein modelling. *Journal of Molecular Biology*, 1994, 235: 1501–1531.

102. K. Karplus, C. Barrett, R. Hughey. Hidden Markov models for detecting remote protein homologies. *Bioinformatics*, 1998, 14: 846–856.

103. L.E. Baum, T. Petrie, G. Soules, N. Weiss. A maximization technique occurring in the statistical analysis of probabilistic functions of Markov chains. *Annals of Mathematical Statistics*, 1970, 41: 164–171.

104. S.R. Eddy. Multiple alignment using hidden Markov models. In *Proceedings of the 2005 International Conference on Intelligent Systems for Molecular Biology*, Detroit, MI, 1995, pp. 114–120.

105. S. Kwong, C. Chau, K. Man, K. Tang. Optimisation of HMM topology and its model parameters by genetic algorithm. *Pattern Recognition*, 2001, 34: 509–522.

106. M. Slimane, G. Venturini, J.A. de Beauville, T. Brouard, A. Brandeau. Optimizing hidden Markov models with a genetic algorithm. *Lecture Notes in Computer Science*, 1996, 1063: 384–396.

107. Z.-J. Lee, S.-F. Su, C.-C. Chuang, K.-H. Liu. Genetic algorithm with ant colony optimization (GA-ACO) for multiple sequence alignment. *Applied Soft Computing*, 2008, 8: 55–78.

108. T.K. Rasmussen, T. Krink. Improved hidden Markov model training for multiple sequence alignment by a particle swarm optimization—Evolutionary algorithm hybrid. *Biosystems*, 2003, 72: 5–17.

109. L.R. Rabiner. A tutorial on hidden Markov models and selected applications in speech recognition. In *Proceedings of the IEEE*, San Diego, CA, 1989, pp. 257–285.

110. S. Henikoff, J.G. Henikoff. Amino acid substitution matrices from protein blocks. *Proceedings of the National Academy of Sciences USA*, 1992, 89: 10915–10919.

111. J. Thompson, F. Plewniak, O. Poch. A comprehensive comparison of multiple sequence alignment programs. *Nucleic Acids Research*, 1999, 27: 2682–2690.

112. J. Stoye, D. Evers, F. Meyer. Rose: Generating sequence families. *Bioinformatics*, 1998, 14: 157–163.

113. T.H. Jukes, C.R. Cantor. Evolution of protein molecules. In H.N. Munro (ed.), *Mammalian Protein Metabolism*. Academic Press, New York, vol. 3, 1969, pp. 21–132.

114. E.L. Sonnhammer, S.R. Eddy, R. Durbin. Pfam: A comprehensive database of protein families based on seed alignments. *Proteins*, 1997, 28: 405–420.

115. J. Bexdek. A convergence theorem for the fuzzy isodata clustering algorithms. *IEEE Transactions on Pattern Analysis and Machine Intelligence*, 1980, 2: 1–8.

116. D. Comaniciu, P. Meer. Mean shift analysis and application. In *Proceedings of the Seventh International Conference on Computer Vision*, Kerkyra, Greece, 1999, pp. 1197–1203.

117. P. Perona, J. Malik. Scale-space and edge detection using anisotropic diffusion. *IEEE Transactions on Pattern Analysis and Machine Intelligence*, 1990, 12(7): 629–639.

118. N.R. Pal, S.K. Pal. A review on image segmentation techniques. *Pattern Recognition*, 1993, 26(9): 1277–1294.

119. M. Sezgin, B. Sankur. Survey over image thresholding techniques and quantitative performance evaluation. *Journal of Electronic Imaging*, 2004, 13(1): 146–165.

120. L. Hertz, R.W. Schafer. Multilevel thresholding using edge matching. *Computer Vision, Graphics and Image Processing*, 1988, 44: 279–295.

121. J.C. Russ. Automatic discrimination of features in gray-scale images. *Journal of Microscopy*, 1987, 148(3): 263–277.

122. J.N. Kapur, P.K. Sahoo, A.K.C. Wong. A new method for gray-level picture thresholding using the entropy of the histogram. *Computer Vision, Graphics and Image Processing*, 1985, 29: 273–285.

123. T. Pun. Entropy thresholding: A new approach. *Computer Vision, Graphics and Image Processing*, 1981, 16: 210–239.

124. J. Kittler, J. Illingworth. Minimum error thresholding. *Pattern Recognition*, 1986, 19: 41–47.

125. Q. Ye, P. Danielsson. On minimum error thresholding and its implementations. *Pattern Recognition Letters*, 1988, 16: 653–666.

126. N. Otsu. A threshold selection method from gray-level histograms. *IEEE Transactions on Systems, Man, Cybernetics*, 1979, 9(1): 62–66.

127. P.K. Sahoo, S. Solutani, A.K.C. Wong. A survey of thresholding techniques. *Computer Vision, Graphics and Image Processing*, 1988, 41: 233–260.

128. W.B. Tao, H. Jin, L.M. Liu. Object segmentation using ant colony optimization algorithm and fuzzy entropy. *Pattern Recognition Letters*, 2008, 28(7): 788–796.

129. L. Cao, P. Bao, Z.K. Shi. The strongest schema learning GA and its application to multilevel thresholding. *Image and Vision Computing*, 2008, 26(5): 716–724.

130. J. Kennedy, R.C. Eberhart. *Swarm Intelligence*. Morgan Kaufmann Publishers, San Francisco, CA, 2001.

131. P.P. Yin. Multilevel minimum cross entropy threshold selection based on particle swarm optimization. *Applied Mathematics and Computation*, 2007, 184(2): 503–513.

132. E. Zahara, S.K.S. Fan, D.M. Tsai. Optimal multi-thresholding using a hybrid optimization approach. *Pattern Recognition Letters*, 2005, 26(8): 1082–1095.

133. A. Ratnaweera, S.K. Halgamuge, H.C. Watson. Self-organizing hierarchical particle swarm optimizer with time-varying acceleration coefficients. *IEEE Transactions on Evolutionary Computation*, 2004, 8(3): 240–255.

134. D.Y. Huang, C.H. Wang. Optimal multi-level thresholding using a two-stage Otsu optimization approach. *Pattern Recognition Letters*, 2009, 30(3): 275–284.

135. R.J. Ferrari, R.M. Rangayyan, J.E.L. Desautels, A.F. Frere. Analysis of asymmetry in mammograms via directional filtering with Gabor wavelets. *IEEE Transactions on Medical Imaging*, 2001, 20(9): 953–964.

136. S. Li, J.T. Kwok, Y. Wang. Using the discrete wavelet frame transform to merge Landsat TM and SPOT panchromatic images. *Information Fusion*, 2002, 3(1): 17–23.

137. V.S. Petrovic, C.S. Xydeas. Gradient-based multiresolution image fusion. *IEEE Transactions on Image Processing*, 2004, 13(2): 228–237.

138. J.M.P. Nascimento, J.M.B. Dias. Does independent component analysis play a role in unmixing hyperspectral data. *IEEE Transactions on Geoscience and Remote Sensing*, 2005, 43(1): 175–187.

139. T.-M. Tu, S.-C. Su, H.-C. Shyu. A new look at IHS-like image fusion methods. *Information Fusion*, 2001, 2(3): 177–186.

140. D. Kundur, D. Hatzinakos. Toward robust logo watermarking using multiresolution image fusion principles. *IEEE Transactions on Multimedia*, 2004, 6(1): 185–198.

141. S. Auephanwiriyakul, J.M. Keller, P.D. Gader. Generalized Choquet fuzzy integral fusion. *Information Fusion*, 2002, 3(1): 69–85.

142. X. Wu, D. Su, X. Luo. Multi-feature method for region based image fusion. In *Proceedings of 2008 Asiagraph*, Tokyo, Japan, 2008.

143. X. Luo, X. Wu. New metric of image fusion based on region similarity. *Optical Engineering*, 2010, 49(4): 047006-1–047006-13.

144. H. Wang, S. Yang, W. Xu, J. Sun. Scalability of hybrid fuzzy c-means algorithm based on quantum-behaved PSO. In *Proceedings of the Fourth international Conference on Fuzzy Systems and Knowledge Discovery*, Hainan, China, vol. 2, 2007, pp. 261–265.

145. R.P. Nikhil, P. Kuhu, M.K. James. A possibilistic fuzzy-C-means clustering algorithm. *IEEE Transactions on Fuzzy Systems*, 2005, 13(4): 517–530.

146. R. Hong, F. Sun. Structure similarity based objective metric for pixel-level image fusion. In *Proceedings of the Second International Symposium on Systems and Control in Aerospace and Astronautics*, Shenzhen, China, 2008, pp. 1–6.

147. W. Huang, Z.L. Jing. Evaluation of focus measures in multi-focus image fusion. *Pattern Recognition Letters*, 2007, 28(4): 493–500.

148. Z. Wang, A.C. Bovik. A universal image quality index. *IEEE Signal Processing Letters*, 2002, 9(3): 81–84.

Index

Milton Keynes UK
Ingram Content Group UK Ltd.
UKHW021829071024
449327UK00021B/1470

9 780367 381936